Gene Structure and Expression

Genetics and Biogenesis of
Mitochondria and Chloroplasts
5–7 September 1974
C. W. Birky, Jr., P. S. Perlman, and T. J. Byers

Regulatory Biology
4–6 September 1975
J. C. Copeland and G. A. Marzluf

Analysis of Ecological Systems
29 April–1 May 1976
D. J. Horn, G. R. Stairs, and R. D. Mitchell

Plant Cell and Tissue Culture:
Principles and Applications
6–9 September 1977
W. R. Sharp, P. O. Larsen, E. F. Paddock,
and V. Raghavan

Cellular Interactions in Symbiosis and Parasitism
7–9 September 1978
C. B. Cook, P. W. Pappas, and E. D. Rudolph

Gene Structure and Expression
6–8 September 1979
D. H. Dean, L. F. Johnson, P. C. Kimball,
and P. S. Perlman

EDITED BY DONALD H. DEAN,
LEE F. JOHNSON, PAUL C. KIMBALL,
AND PHILIP S. PERLMAN

Gene Structure and Expression

OHIO STATE UNIVERSITY PRESS : COLUMBUS

Copyright © 1980 by the Ohio State University Press.
All Rights Reserved.

Library of Congress Cataloguing in Publication Data

Biosciences Colloquium, 6th, Ohio State University, 1979
  Gene structure and expression.

  (Ohio State University biosciences colloquia)
  Includes Index.
  1. Gene expression—Congresses. 2. Genetic regulation—
Congresses. 3. Recombinant DNA—Congresses. I. Dean,
Donald H. II. Title. III. Series: Ohio. State University,
Columbus. Ohio State University biosciences colloquia. [DNLM:
1. Genetics, Biochemical—Congresses. 2. Genetics, Microbial—
Congresses. W3 B1563 6th 1979g / QH447 B615 1979g]
QH450.B56 1979   574.87′322   80-17606
ISBN 0-8142-0321-3

Contents

Preface

Recombinant DNA technology has had an enormous impact on research in the biological sciences. It has facilitated the development of rapid and simple methods of DNA sequence determination, thereby permitting a detailed analysis of the organization of structural genes and their regulatory sequences. It has also permitted the restructuring of genes, the purification and amplification of a single DNA fragment from a complicated mixture, and even the expression of natural and artificial genes in a variety of host organisms. The theoretical as well as practical importance of recombinant DNA research is firmly established.

The sixth Annual College of Biological Sciences Colloquium, which was held 6–8 September 1979 at the Ohio State University, explored the application of recombinant DNA technology to the analysis of gene structure and expression. Keynote presentations dealt first with recent advances in restriction enzyme analysis, cloning vehicles, and DNA sequencing strategies. The remainder of the invited presentations centered on the analysis of the structure and expression of genes in a variety of model systems ranging from viruses and bacteria to mammalian cells. Although the problems being studied varied, it was apparent that recombinant DNA technology had played an important role in each investigation. Many of the contributed presentations also involved this technology, although a variety of approaches to the over-all theme were discussed.

The Colloquium was attended by about three hundred persons from the United States, Canada, and Europe. We were pleased by the enthusiastic response. We express our gratitude to the College of Biological Sciences and to the various industrial exhibitors for their continuing financial support of these colloquia. We thank the speakers for their interesting, timely, and enthusiastic presentations. Finally, we thank the organizers, students, and others who contributed their time and efforts, and were in so many ways responsible for the success of this colloquium.

Sixth Annual Biosciences Colloquium
College of Biological Sciences
Ohio State University
6–9 September 1979

GENE STRUCTURE AND EXPRESSION

Organizers

D. H. Dean, Departments of Microbiology and Genetics, Ohio State University
L. F. Johnson, Department of Biochemistry, Ohio State University
P. C. Kimball, Departments of Microbiology and Genetics, Ohio State University
P. S. Perlman, Department of Genetics, Ohio State University

Speakers

Thomas R. Gingeras   A. M. Chakrabarty
L. W. Enquist   M. Piatak
Michael Smith   L. Chow
Lydia Villa-Komaroff   Jerry B. Lingrel
Donald H. Dean   Lee F. Johnson
Gerald R. Fink   Sarah C. R. Elgin

Gene Structure and Expression

THOMAS R. GINGERAS

The Role of Restriction Endonucleases in Genomic Analysis

1

INTRODUCTION

The list of Type II restriction endonucleases continues to grow each year. In the past year alone, more than 15 new enzymes have been found of which five have been characterized and shown to have a previously unreported recognition site. This number can be added to the current list of more than 150 Type II restriction endonucleases, which represent more than 40 different specificities (Roberts, 1979). The use of these enzymes to study the structure and function of various genomes has now expanded into virtually every area of molecular biology and is slowly finding its way into a variety of applied fields, including clinical diagnosis. The current applications depend primarily on the widely diverse recognition and cleavage properties that these enzymes possess. It will be the attempt of this paper to focus on some of these diverse properties and to review how these characteristics have been used in the past, and may be used in the future to aid in our understanding of gene structure and function.

This paper is divided into three sections. The first will summarize the recognition and cleavage properties of the Type II restriction endonucleases that have been characterized. The second section will describe a new and rapid method for the identification of restriction endonucleases with new recognition sequences. The final section will describe how these properties have provided several alternative strategies for analysis of gene structure and expression.

CLASSES OF RESTRICTION ENDONUCLEASES

The genetic information encoded in the polynucleotide chains that are the chromosomes of prokaryotic and eukaryotic cells is difficult to analyze

Cold Spring Harbor Laboratory, Cold Spring Harbor, New York 11724

TABLE I

Specific Endonucleases and Their Recognition Sequences

Tetranucleotide		Pentanucleotide		Hexanucleotide			
				Non-degenerate		Degenerate	
AluI	AG|CT	AsuI	G|GNCC	AvaIII	ATGCAT	AccI	GT|(A/C)(G/T)AC
FnuDII	CG|CG	AvaII	G|G(A/T)CC	AvrII	CCTAGG	AcyI	GPu|CGPyC
HaeIII	GG|CC	BbvI	GC(A/T)GC	BamHI	G|GATCC	AvaI	C|PyCGPuG
HpaII	C|CGG	DdeI	C|TNAG	BalI	TGG|CCA		
HhaI	GCG|C	EcoRII	|CC(A/T)GG	BclI	T|GATCA	HaeI	(A/T)GG|CC(T/A)
MnlI	CCTC	Fnu4HI	GC|NGC	BglII	A|GATCT	HaeII	PuGCGC|Py
MboI	|GATC	HinfI	G|ANTC	ClaI	AT|CGAT	HindII	GTPy|PuAC
DpnI[1]	GA*|TC	HgaI	|N10GACGCN5|	EcoRI	G|AATTC	HgiAI	G(T/A)GC(T/A)|C
RsaI	GT|AC	HphI	GGTGAN8|	HindIII	A|AGCTT	XhoII	Pu|GATCPy
TaqI	T|CGA	MboII	GAAGAN8|	HpaI	GTT|AAC		
		SfaNI	GATGC	KpnI	GGTAC|C		
				MstI	TGC|GCA		
				PstI	CTGCA|G		
				PvuI	CGAT|CG		
				PvuII	CAG|CTG		
				SacI	GAGCT|C		
				SacII	CCGC|GG		
				SalI	G|TCGAC		
				XbaI	T|CTAGA		
				XhoI	C|TCGAG		
				XmaIII	C|GGCCG		
				XmaI	C|CCGGG		

[1] DpnI recognizes the same sequence as MboI but requires 6-methyl adenosine as part of the recognition sequence.

due to the sheer size of these genomes. Restriction endonucleases have been critical tools for the dissection of these large DNA molecules into smaller and, in some cases, discrete functional units. It is the diversity of both the recognition sequences and the corresponding cleavage sites that makes restriction enzymes so valuable. The known restriction endonucleases are listed in table 1, and it can be seen that they fall into three classes, based upon the length of their recognition sites. Two restriction endonucleases not listed in table 1 are the enzymes *Eca*I from *Enterobacter cloacae*, which recognizes the sequence G↓GTNACC (H. Mayer, E. Schwarz, M. Milza, and G. Hobom, unpublished observations) and *Bgl*I from *Bacillus globigii* (Wilson and Young, 1976), which recognizes the sequence GCCN₄↓N₁GGC (T. Bickle, personal communication). These enzymes, which apparently interact with a sequence longer than six nucleotides, should probably be classified with the hexanucleotide set since the specific part of the recognition site is, in each case, six nucleotides. Similarly, *Hinf*I (and related enzymes; see table 7) could be considered a member of the tetranucleotide set, by virtue of the four specific nucleotides present in its recognition sequence. Whether these enzymes are exceptions or represent the prototypes of yet uncharacterized sets is not clear. Because of its unusual recognition site, *Bgl*I offers certain possibilities for genetic engineering, and this will be discussed in a later section.

Some of the enzymes listed in table 1 are worthy of special note. For a long time it has been known that the activity of restriction endonucleases can be inhibited by the presence of either 6-methyl adenosine or 5-methyl cytosine at certain sites within the recognition sequence. *Dpn*I from *Diplococcus pneumoniae* is unusual in this regard in that it is the only restriction endonuclease that absolutely requires a methylated base (6-methyl adenosine) to be present within its recognition sequence (GA↓TC) for cleavage to occur (Lacks and Greenberg, 1975; Geier and Modrich, 1979). This is interesting because the enzyme *Dpn*II, produced by another *D. pneumoniae*, recognizes the same sequence but is inhibited by this same methylated base. The role of modified nucleotides in DNA has recently been examined. It is still unclear what relationship exists between modification and gene expression, but studies have been initiated by Waalwijk and Flavell (1978), Bird and colleagues (1979), and others to investigate this area. If methylation proves to be of importance for the control of gene expression, then appropriate restriction endonucleases could prove most valuable to probe the genome for the presence of modified nucleotides. Such a screening method has been described by Bird and Southern (1978). Several restriction endonucleases are known that would be useful for this purpose. The restriction endonuclease *Mbo*I recognizes the sequence GATC (Gelinas et al., 1977), and is inhibited by the presence of 6-methyladenosine within this sequence. However, the restriction enzyme *Sau*3A can cut this same sequence when the A is modified (Sussenbach et al., 1976). Similarly, the restriction endonuclease *Hpa*II recognizes the sequence CCGG and is inhibited from cleavage by modification at the internal cytosine (Mann and Smith, 1977; Garfin and Goodman, 1974). However, the modified sequence can be cut by the enzyme *Msp*I (Waalwijk and Flavell, 1978). A similar pair of enzymes exist for the sequence CC(A/T)GG. The enzyme *Eco*RII is inhibited by modification at the internal cytosine of this sequence, whereas the enzyme *Bst*NI is not (J. Brooks, personal communication). Each of these pairs of restriction endonucleases could be used to investigate the modification pattern present within a genome.

An important addition to the tetranucleotide group of enzymes in table 1 was made with the discovery of the *Rsa*I enzyme (J. Gardner and S. Kaplan, unpublished observations). Eight of the sixteen possible tetranucleotide palindromic combinations have now been discovered. Notably missing from this group are those with recognition sequences containing only A and T residues. The significance of this is unknown; however, it is becoming less and less likely that it is merely a statistical aberration.

In the class of endonucleases that recognize pentanucleotide sequences, the enzyme *Hga*I has properties that seem of unusual interest for genetic engineering. *Hga*I recognizes a pentanucleotide sequence, but cleaves 5 nucleotides away from this sequence on one strand and 10 nucleotides away on the other strand (Brown and Smith, 1977). This results in a fragment with a 5′ terminal pentanucleotide extension that is unique for each fragment generated by this enzyme because it is unrelated to the recognition sequence. This means that an attempt to religate a collection of DNA fragments produced by *Hga*I digestion would lead only to the joining of fragments that lay adjacent to one another in the original genome. Among the religated population one might expect to find a reconstruction of the original genome. Manipulations of individual *Hga*I fragments, such as site-specific mutagenesis, followed by genomic reconstruction, offers interesting synthetic opportunities (Brown and Smith, 1977). *Bgl*I, which leaves a unique 3′ trinucleotide extension, could also be utilized in this same way.

The enzymes *Mbo*II (Brown et al., 1979) and *Hph*I (Kleid et al., 1976) are restriction endonucleases that share an important property with the enzyme *Hga*I. *Mbo*II and *Hph*I are enzymes that recognize a pentanucleotide sequence and cleave distally from their recognition site. Thus, fragments obtained by digestion with either *Hga*I, *Mbo*II, or *Hph*I could be inserted into a cloning vector and excised readily because the ligation would not affect the recognition sequence. Often the excision of new inserts is difficult, unless the recognition site is maintained during ligation. Although in principle the use of *Hph*I, *Mbo*II, and *Hga*I fragments avoids any inherent difficulties of this sort, it must be noted that the cleavage by *Hph*I and related enzymes is a directed process and so the recognition sequence will not necessarily lie within the fragment cloned.

The final group in table 1 are those enzymes that recognize hexanucleotide sequences. Because all these enzymes recognize palindromic sequences, they may contain within the central four nucleotides of their recognition sites a sequence identical with one that has been described for a tetranucleotide-recognizing enzyme. Of particular interest are companion tetranucleotide and hexanucleotide enzymes that cleave at the same position relative to the common tetranucleotide sequence (e.g., GATC group) (table 2). The usefulness of this feature can be best illustrated by experiments involving *in vitro* selection of recombinant DNA molecules (Roberts, 1977).

The results of religating fragments derived from a *Bam*HI and *Bgl*II double digest would include a new set of sequences not previously observed. Either the sequence GGATCT or AGATCC could result from such a ligation of mixed fragments. Such sequences are not recognized by either the *Bam*HI or the *Bgl*II enzymes. Thus, recombinants could be selected from a mixture by redigestion with *Bam*HI and *Bgl*II. The formation of such recombinants could be tested by digestion with *Xho*II

TABLE 2

RESTRICTION ENDONUCLEASES PRODUCING
IDENTICAL TERMINAL EXTENSIONS

Extension	Enzyme	Recognition Site
5′ ¡GATC	*Mbo*l	¡GATC
	*Bam*Hl	G¡GATCC
	*Bgl*ll	A¡GATCT
	*Xho*II	Pu¡GATCPy
5′ ¡CG	*Hpa*ll	C¡CGG
	*Taq*l	T¡CGA
5′ ¡TCGA	*Sal*l	G¡TCGAC
	*Ava*l	C¡PyCGPuG
	*Xho*l	C¡TCGAG
5′ ¡CCGG	*Xma*l	C¡CCGG
	*Ava*l	C¡PyCGPuG

(PuGATCPy), an enzyme that will cleave the novel recombinant sequences. Thus, it is possible to select *in vitro* for recombinant molecules prior to cloning.

COMPUTER-ASSISTED METHODS FOR THE DISCOVERY
OF NEW RESTRICTION ENDONUCLEASE RECOGNITION SITES

The search for new and useful endonucleases continues by many different laboratories, including our own. The characterization of these new enzymes has led to the development of a computer-assisted strategy that leads to a unique and small number of predictions for the recognition sequence of almost any Type II restriction endonuclease. This is achieved by cleaving a DNA molecule of known sequence (e.g., ϕX174 RF) with the new restriction endonuclease and comparing the fragment pattern obtained with a computer-generated set of patterns. This approach has provided the recognition sequence of several enzymes, including *Bbv*I, *Sfa*NI, *Pvu*I, *Mst*I (Gingeras et al., 1978), *Xho*II, and *Dde*I (unpublished observations).

The method is illustrated for the enzyme *Bbv*I from *Bacillus brevis*. Figure 1 shows a digest of ϕX174 DNA by *Bbv*I. The fragment lengths were estimated from their mobilities in this gel and were fed into the computer as illustrated by figure 2. The input consists of a fragment length and a percent error estimate. The program now takes the fragment length and the percent error estimate and searches for a match in a preformed master table. This master table contains a listing of all fragment sizes that can be produced from the ϕX174 sequence by putative enzymes which recognize all possible combinations of tetra-, penta-, or hexanucleotide sequences. Next to each fragment length in this table is a listing of the possible tetranucleotide, pentanucleotide, or hexanucleotide sequences that are capable of generating such a fragment length. When an experimental fragment length is entered, the program retrieves from the master table a collection of tetra-, penta-, and hexanucleotide sequences that would be capable of generating that group of fragment lengths equivalent to the fragment length and its associated error. This set of sequences that could theoretically produce that fragment length now constitutes the data base against which the remaining fragment lengths will be compared. The input of the second fragment length repeats this process, and the computer then finds which of the sequences in the new set could also produce a fragment of the new length. Further fragment lengths serve to reduce the set of possibilities until only one or a few possible sequences remain. Table 3 indicates that after the submission of six of the twelve fragments produced by *Bbv*I the computer

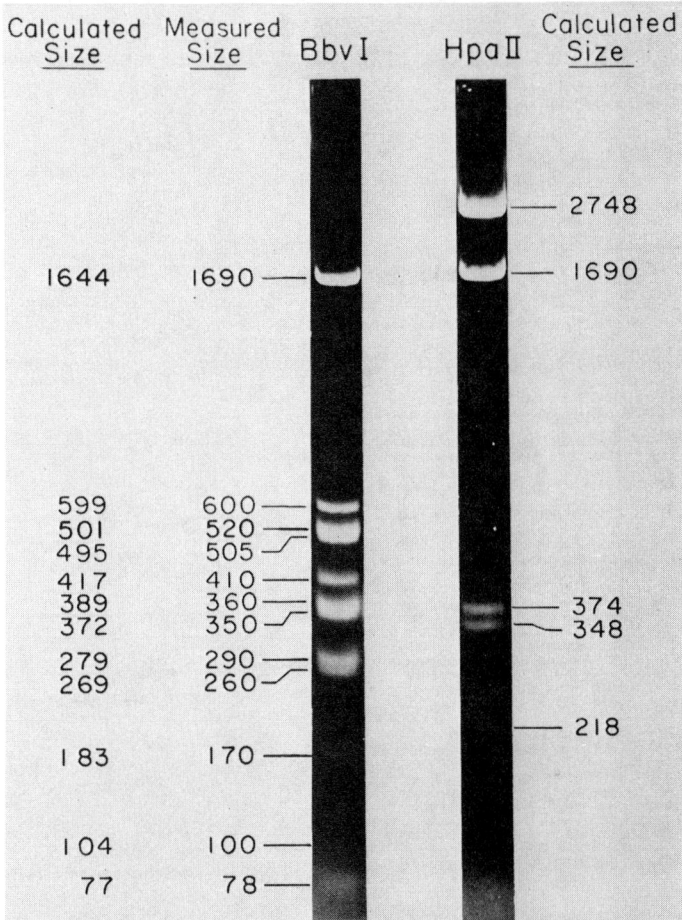

Fig. 1. *Bbv*I on ϕX174 Rf DNA fractionated on a 2% agarose gel. ϕX174 RF DNA was cut *Hpa*II and used as size markers. The experimentally determined fragment lengths are listed alongside the *Bbv*I digest as well as the theoretical lengths as measured by computer scans of the ϕX174 sequence looking for the sequence GC(A/T)GC (Gingeras et al., 1978).

arrived at a unique prediction for its recognition sequence. This prediction was upheld, and hence checked, even after additional fragment lengths from the digest were supplied to the program.

The recognition sequence predicted in this way can be easily checked by two independent means. The first of these is to use a computer program to search through the ϕX174 sequence (Sanger et al., 1977) for the occurrence of the unique sequence obtained. In the case of *Bbv*I, this is the putative

```
PLEASE KEY IN LENGTH AND PERCENT SEPARATED BY A BLANK           Step #1
>2748 0.1
SET TO BE SEARCHED RANGES FROM 2473 TO 3023
IF YOU WISH TO OMIT THIS SET OF LIMITS TYPE IN OMIT                 1st fragment length as input
>NO
THE NUMBER OF OCCURRENCES IS   281
DO YOU WISH TO SEE THE BASES CHOSEN..ANS YES OR NO
>NO
PLEASE KEY IN LENGTH AND PERCENT SEPARATED BY A BLANK           Step #2
>1690 0.05
SET TO BE SEARCHED RANGES FROM 1605 TO 1774
IF YOU WISH TO OMIT THIS SET OF LIMITS TYPE IN OMIT                 2nd fragment length as input
>NO
THE NUMBER OF OCCURRENCES IS    234
DO YOU WISH TO SEE THE BASES CHOSEN..ANS YES OR NO
>NO
THE NUMBER OF BASES IN THE INTERSECTION IS    31
DO YOU WISH TO SEE THE BASES CHOSEN..ANS YES OR NO
>NO
DO YOU WISH TO STOP THIS ITERATION..ANS YES OR NO
>NO
PLEASE KEY IN LENGTH AND PERCENT SEPARATED BY A BLANK           Step #3
>374 0.05
SET TO BE SEARCHED RANGES FROM  355 TO  393
IF YOU WISH TO OMIT THIS SET OF LIMITS TYPE IN OMIT                 3rd fragment length as input
>NO
THE NUMBER OF OCCURRENCES IS    361
DO YOU WISH TO SEE THE BASES CHOSEN..ANS YES OR NO
>NO
THE NUMBER OF BASES IN THE INTERSECTION IS     5
DO YOU WISH TO SEE THE BASES CHOSEN..ANS YES OR NO
>NO
DO YOU WISH TO STOP THIS ITERATION..ANS YES OR NO
>NO
PLEASE KEY IN LENGTH AND PERCENT SEPARATED BY A BLANK           Step #4
>348 0.05
SET TO BE SEARCHED RANGES FROM  330 TO  365
IF YOU WISH TO OMIT THIS SET OF LIMITS TYPE IN OMIT                 4th fragment length as input
>NO
THE NUMBER OF BASES IN THE INTERSECTION IS     1
DO YOU WISH TO SEE THE BASES CHOSEN..ANS YES OR NO
>NO
DO YOU WISH TO STOP THIS ITERATION..ANS YES OR NO
>NO
PLEASE KEY IN LENGTH AND PERCENT SEPARATED BY A BLANK           Step #5
>218 0.05
SET TO BE SEARCHED RANGES FROM  207 TO  229
IF YOU WISH TO OMIT THIS SET OF LIMITS TYPE IN OMIT                 5th fragment length as input
>NO
THE NUMBER OF OCCURRENCES IS    298
DO YOU WISH TO SEE THE BASES CHOSEN..ANS YES OR NO
>NO
THE NUMBER OF BASES IN THE INTERSECTION IS     1
DO YOU WISH TO SEE THE BASES CHOSEN..ANS YES OR NO
>NO
DO YOU WISH TO STOP THIS ITERATION..ANS YES OR NO
>YES
  CCGG
DO YOU WISH TO RESTART ENTIRE PROCEDURE YES OR NO              PREDICTED RECOGNITION SITE
>NO
```

Fig. 2.   This is an example of the interaction between the user and the computer program to predict the recognition sites of restriction enzymes. The theoretical lengths of the *Hpa*II fragments from a φ174 RF digest were used as input. Errors of 10% were assumed for the largest fragment and 5% for the remaining fragments (Gingeras et al., 1978).

recognition site GC(A/T)GC. The total number and sizes of fragments found during this search should exactly match those found experimentally. The second method of checking the prediction involves the scanning of the DNA sequences of other genomes or plasmids (e.g., SV40 [Reddy et al., 1978; Fiers et al., 1978], fd [Schaller et al., 1978], pBR322 [Sutcliffe, 1978], and G4 [Godson et al., 1978]). The numbers and sizes of the fragments should match those predicted by the computer.

TABLE 3

USE OF EXECJCL PROGRAM TO DETERMINE THE
RECOGNITION SITE OF *Bbv*1.

Experimental Lengths	Number of Sequence Possibilities		
	5% error	10%	15%
1,690	234	449	640
600	27	204	207
520	4	37	99
505	4	14	34
410	2	8	19
360	1*	4	19
350	0	1*	4
290		1*	3
260		1*	2
170		1*	2
100		1*	2
78		1*	1*

* The sequence 5' GCAGC 3'

Although many of the restriction enzymes available at the moment give patterns from which the total number of fragments and their lengths can be determined with some accuracy, this is not always the case. Contaminating nonspecific nucleases can sometimes lead to degradation of fragments and cause the loss of bands from the digest, and low enzyme concentrations can lead to partial digestion. Thus, for many enzymes, it is difficult to obtain a complete digest from which unambiguous assignment of fragment number and length can be determined. Many of the uncharacterized enzymes have remained so for precisely these reasons. It was of some interest to ask whether a putative recognition sequence could be predicted from an incomplete digest by the use of these programs. The first set of experiments that were conducted using these computer programs varied the order with which the fragment lengths were provided to the program. The results are shown in table 4.

Using a 5% error limit, it can be seen that no matter what order the fragments are entered into the program, a unique solution can always be generated with four of the five fragments and that in one case (2), a unique solution is generated after only three fragments are entered. It is also clear from this example that the largest fragment length produced by the digest has the most dramatic effect on reducing the number of possibilities. Unfortunately, a fragment of this length is also subject to the greatest possibility of error in its length determination. As expected, the order with which the fragments are provided to the program has no effect on the total

TABLE 4

EFFECT OF VARYING THE ORDER OF FRAGMENT LENGTHS
PROVIDED TO EXECJCL

(5% error limit on all length values[1])

Case 1		Case 2		Case 3	
Fragment Length	Possibilities	Fragment Length	Possibilities	Fragment Length	Possibilities
2,748	131	218	410	378	436
1,690	16	2,748	8	1,690	39
378	2	378	1*	2,748	2
348	1*	1,690	1*	348	1*
218	1*	348	1*	218	1*

(10% error limit on all length values)

2,748	326	278	779	218	804
1,690	54	218	229	2,748	30
378	12	1,690	37	378	4
348	7	348	27	1,690	2
218	2*	2,748	2*	348	2*

[1] These fragment lengths are from a *Hpa*ll digest of ϕX174 RF DNA.
* This number represents or contains within its members the correct recognition sequence.

number of possibilities finally predicted. However, the rate with which those possibilities are reduced can be affected significantly.

A second experiment consisted of systematically omitting one or more fragments from the input data and led to the results shown in table 5. The omission of a single fragment from an *Hpa*II digest of ϕX174 DNA gives a unique answer in all cases except that in which the largest fragment is omitted. In that case examination of the four possibilities remaining indicated that only the sequence CCGG could have been the recognition site because in the three other cases more fragments were predicted than occurred in the digest. But, in particular, a large fragment of a length 2,700 nucleotides was missing. This experiment showed the relative value of a large fragment in reducing the number of possibilities. By omitting more fragments from the digest, it was still possible to obtain a unique and correct answer in certain cases, although the most useful information to emerge from this experiment was that a small number of possibilities can still be generated when rather sparse information is available and from which the correct recognition sequence might be deduced by further experimentation.

One final experiment was carried out to determine the accuracy with which fragment lengths must be known in order that correct recognition

TABLE 5

EFFECT OF OMITTING ONE OR MORE FRAGMENTS
FROM DIGEST PATTERNS

Fragment Length	Omission of One Fragment					
	0	1	2	3	4	5
2,748	131	...	131	131	131	131
1,690	16	274	...	16	16	16
378	2	39	10	...	2	2
348	1	12	5	1	...	1
218	1	4*	1†	1	1	...

Fragment Length	Omission of Two Fragments			
	1	2	3	4
2,748	...	131	131	131
1,690	...	...	...	...
378	350	...	10	10
348	124	7	...	5
218	46	1	1	...

Fragment Length	Omission of Three Fragments			
	1	2	3	4
2,748	131	131	131	131
1,690	...	16	...	...
378	...	...	10	...
348	...	...	...	7
218	8	...	...	...

For this experiment the known fragment lengths generated by *Hpa*II were used as input and a uniform error of 5% applied. The numbers above the columns indicate the fragment(s) omitted.

* Sites selected were CCGG, ACTCA, ATGTC, and AATGTC. Of these, only CCGG could be the recognition site based on total number of fragments in digest.

† CCGG is the predicted site.

sites can be deduced. Using computer-determined fragment lengths, the error limits were systematically increased until the point was reached that a unique solution was still generated by the computer but that, by increasing the error, more than one possibility remained. The results with several enzymes of known sequence that cleave ϕX174 DNA are shown in table 6. It can be seen that with the exception of the *Eco*RII endonuclease, errors in the range from 15% to 34% still lead to the correct deduction of the recognition sequence. The rather low error needed for the *Eco*RII fragment reflects the fact that this enzyme only cleaves the ϕX174 DNA in two positions.

As can be seen in table 7, the majority of known restriction enzyme recognition sites consists of a linear array of 4, 5, or 6 nucleotides.

TABLE 6

TOLERABLE ERRORS IN FRAGMENT LENGTH DETERMINATION

Enzyme	Site	Number of Fragments	Maximum % Error Allowing a Unique Prediction
1. *Alu*l	AGCT	24	23
2. *Bbv*l	GC(A/T)GC	14	21
3. *Eco*Rll	CC(A/T)GG	2	0.2
4. *Hae*lll	GGCC	11	16
5. *Hga*l	GACGC	14	15
6. *Hpa*ll	CCGG	5	17
7. *Hha*l	GCGC	18	>25
8. *Hin*1056l	CGCG	14	18
9. *Hph*l	GGTGA	9	18
10. *Mbo*ll	GAAGA	11	19
11. *Mnl*l	CCTC	35	>30
12. *Taq*l	TCGA	10	34

TABLE 7

SEQUENCE PATTERNS RECOGNIZED BY RESTRICTION ENDONUCLEASES

Pattern	Example	Sequence	Total No. of Unique Examples*
NNNN	*Hpa*ll	CCGG	9
NNNNN	*Hph*l	GGTGA	4
NNNNNN	*Eco*Rl	GAATTC	21
NNXNN	*Hin*fl	GANTC	5
NNPyPuNN	*Hind*ll	GTPyPuAC	1
NPyNNPuNN	*Ava*l	CPyCGPuG	2
PuNNNNPy	*Hae*ll	PuGCGCPy	2
NN(A/C)(G/T)NN	*Acc*l	GT(A/C)(G/T)AC	1
(A/T)NNNN(A/T)	*Hae*l	(A/T)GGCC(A/T)	1
N(A/T)NN(A/T)N	*Hgi*l	G(A/T)GC(A/T)C	1

In this table the following abbreviations are used: N = any one of the four deoxyribonucleotides, but with a specific value assigned for any given restriction enzyme; X = any one of the four deoxyribonucleotides and no specific value is necessary; Pu = A or G can be present at this point in the sequence; Py = C or T can be present at this point in the sequence; A/C = A or C can be present at this point in the sequence; G/T = G or T can be present at this point in the sequence; A/T = A or T can be present at this point in the sequence.

* R. J. Roberts (1979).

Consequently, we have been able to develop computer programs that are now capable of determining the recognition sequence of restriction endonucleases that are members of each of the classes indicated in table 7. In addition, these computer programs will be singularly helpful in reducing the laborious process that was needed to derive the recognition sequence of enzymes such as *Hga*I, *Hph*I, and *Mbo*II, which cut at sequences distal from the recognition site.

Because of the numerous sequences that are available from several viral genomes and plasmids and those that will be available soon (e.g., the adenovirus-2 DNA sequence), the utility of these computer programs should be greatly increased. At present the data base is still too small to guarantee success for a hexanucleotide recognition sequence.

TWO SPECIFIC APPLICATIONS OF RESTRICTION ENDONUCLEASES
FOR GENETIC ANALYSIS

Two particular applications of restriction endonucleases have been of particular interest in our laboratory over the past year. The first of these involves the use of restriction endonucleases for the purpose of assessing the accuracy of the primary sequence of the Ad2 genome. The second application involves the use of restriction enyzmes to clone the bacterial genes coding for these endonucleases and their associated methylases.

Our strategy for sequencing the Ad2 genome (total length approximately 35,000 base pairs) has involved the use of the Sanger dideoxy chain termination procedure using, as template, whole Ad2 DNA that has been resected by exonuclease III($3'{\rightarrow}5'$) or T7 exonuclease ($5'{\rightarrow}3'$). In this way sequence can be obtained from both strands of the genome, and the complementary data usually allows ambiguities, which arise on one strand, to be resolved. Also, for most of the genome, the second-strand sequence serves to check the accuracy of the sequence derived from the first strand. However, because ambiguities and problem spots occur with some finite frequency, the final sequence derived still contains some sections where the sequence is only derived from one strand. Consequently, we have found it extremely useful to check the sequence by showing that the restriction enzyme sites, predicted from the sequence, do in fact occur at the predicted positions. This may be looked upon as a random spot check of the fidelity of the sequence, in much the same way as the finished products from an assembly line are sampled for quality.

The specificity of each restriction enzyme is used to detect the presence and location of its recognition sequence. Multiple digestions with a variety of enzymes permit a sizable sampling of the sequences present in the Ad2 genome. This strategy is exemplified in table 8 (Gingeras et al., 1979). The

TABLE 8

LIST OF PREDICTED AND OBSERVED RESTRICTION
ENDONUCLEASE RECOGNITION SITES PRESENT
BETWEEN MAP POSITIONS 97.3 TO
100%[a] IN THE AD2 GENOME

Enzyme	Occurrences Predicted	Occurrences Observed[b]	Recognition Sequence
*Alu*l	2	2[c]	AGCT
*Asu*l	1	nt	GGNCC
*Ava*l	1	nt	CpyCGpuG
*Bbv*l	3	nt	GC(A/T)GC
*Eca*l	1	nt	GGTNACC
*Eco*RI'	2	nt	PuPuATPyPy
*Fnu*DII	3	3	CGCG
*Hae*ll	1	1	PuGCGCPy
*Hae*lll	2	2	GGCC
*Hha*l	3	3	GCGC
*Hin*dll	1	1	GTPyPuAC
*Hin*dlll	1	1[d]	AAGCTT
*Hin*fl	2	1[e]	GANTC
*Hpa*l	1	1	GTTAAC
*Hpa*ll	4	4	CCGG
*Hph*l	2	nt	GGTGA
*Mbo*ll	1	1	GAAGA
*Mnl*l	6	nt	CCTC
*Taq*l	1	1	TCGA
*Xma*l	1	1	CCCGGG

[a] This region of Ad2 contains 1,009 nucleotides.

[b] The observed occurrences shown in this column not only agree in number but also in the size of fragments predicted by a computer analysis of the sequences in Region N. (nt = not tested.)

[c] One *Alu*l site is from within the *Hin*dlll site and will not generate a new fragment.

[d] This *Hin*dlll site is the site used to generate the fragment from 97.3–100%.

[e] One *Hin*fl fragment would not be expected to be observed because it is predicted to be only 14 base pairs in length.

*Hin*dIII K fragment (map positions 97.3 to 100%) was sequenced and determined to be 1,009 nucleotides long. The sequence of this fragment was scanned by a computer program in order to determine the number and sizes of fragments predicted to arise from each of several restriction enzyme digestions of this Ad2 fragment. The number of fragments predicted to arise for each enzyme is recorded in table 8. These predictions were checked by digesting the *Hin*dIII K fragment with 13 different restriction enzymes.

The numbers and sizes of the fragments resulting for each digestion are shown in table 8 and compared with the values predicted by the computer after scanning the 1,009 nucleotides of this fragment. By summing the number of nucleotides checked by each digest, it can be seen that more than 10% of the original sequence of 1,009 nucleotides is directly confirmed by this simple analysis. If a total of 30–40 restriction endonuclease digestions were performed on the complete Ad2 virus genome, 20% of the total sequence could be directly checked in this way.

Although several reviews have appeared that cover the specific role played by restriction endonucleases in genetic engineering, even the most recent (Zabeau and Roberts, 1979) is already out of date. New and ingenious strategies have appeared with amazing frequency, and one of particular interest has recently been reported by Bahl and colleagues (1978). The concept of oligonucleotide "adaptors" to convert one restriction site to another has been around for some time (Heyneker et al., 1976) and has been used extensively to prepare genomic shotguns (Maniatis et al., 1978). The old method required that a partial digest of the genome be prepared and the fragments methylated with, for example, the *Eco*RI methylase. A synthetic duplex containing the *Eco*RI site was then ligated to these fragments, and upon cleavage with *Eco*RI the partial digest could then be cloned into an *Eco*RI vector. In the method described by Bahl and colleagues (1978), illustrated in figure 3, the same principle is used but the adaptor already contains preformed, staggered ends, thus eliminating the need for methylation and the subsequent recutting. One further refinement of the adaptor is that it contains not only the *Bam*HI sequence at which joining takes place but also an *Sma*I (or *Xma*I) sequence internally. Thus, either enzyme could be used to release the inserted DNA.

The cloning of the genes for bacterial restriction endonucleases and their associated methylases is a project being carried out by Dr. J. Brooks in our laboratory at present. Although direct selection for a restriction enzyme gene has been successful in one case (*Hha*II from *Haemophilus haemolyticus*) (Mann et al., 1978), other attempts have been fruitless (J. Brooks, K. Murray, H. O. Smith, personal communications). This is probably because, in general, the transfer of a restriction enzyme gene to a bacterial cell, which does not already contain the appropriate methylase, would prove lethal to that cell. Consequently, the strategy being used by Dr. Brooks is to first isolate clones containing just the methylase gene. This may be directly selected by first preparing a shotgun of the bacterial DNA in a phage or plasmid vector and then isolating the mixed population of recombinant DNA. This DNA is then subjected to restriction *in vitro*, and the surviving DNA is used to retransfect *E. coli*. Only that DNA which contains an expressing methylase gene could survive such a selection.

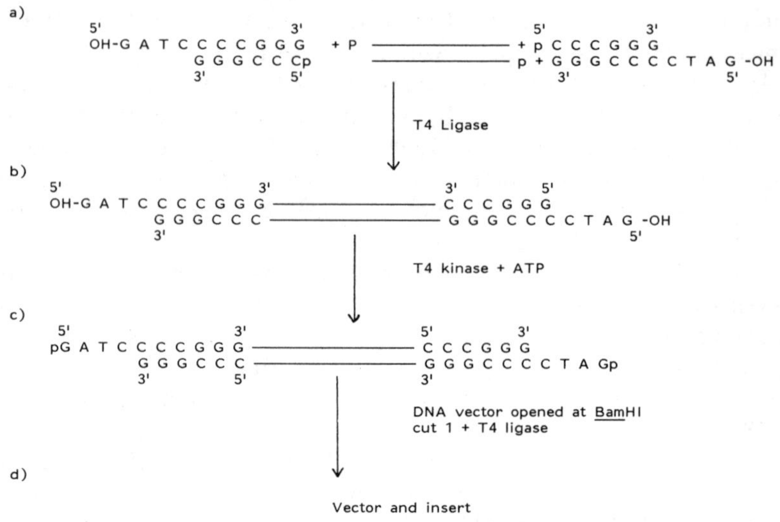

Fig. 3. A scheme as outlined by Bahl et al. (1978) for the insertion of blunt-ended DNA partial digest fragments into a *Bam*HI opened cloning vector using a preformed (ready-made) adaptor. Note that the *Mbo*I/*Xma*I adaptor in steps a and b contain 5′ hydroxyl groups at the *Mbo*I end of the adaptors. This condition will favor ligation of the insert to adaptor in only one orientation, thus leaving the *Mbo*I sequence available for ligation to the vector DNA.

CONCLUDING REMARKS

The utilization of restriction endonucleases depends upon individual properties inherent in each of these enzymes. These enzymes have been used to map the physical composition of a DNA genome, to discover the functions encoded in the sequences mapped, and to isolate and thus manipulate functional segments of the genome by means of the rapidly advancing recombinant DNA technology. Despite the fact that so much has been learned about the structure and function of genomes using these enzymes as biochemical tools, little is known about chemical properties of restriction endonucleases as enzymes. It is tempting to predict that work in this little explored area of enzymology might well add to the versatility and value of these enzymes.

ACKNOWLEDGMENTS

I am grateful to R. J. Roberts for his helpful comments and useful discussions, to J. Brooks who has allowed me to quote some of her unpublished observations, and to M. Moschitta for her assistance in the preparation of this manuscript.

This work was supported by grants to TRG and R. Gelinas from the Whitehall Foundation, to R. J. Roberts from the National Science Foundation (PCM76-82448) and to TRG as a National Institutes of Health Postdoctoral Fellow.

REFERENCES

Bahl, C. P., R. Wu, R. Brousseau, A. K. Sood, H. M. Hsiung, and S. A. Narang. 1978. Chemical synthesis of versatile adaptors for molecular cloning. Biochem. Biophys. Res. Commun. 81: 695–703.

Bird, A. P., and E. M. Southern. 1978. Use of restriction enzymes of study eukaryotic DNA methylation. I. The methylation pattern in ribosomal DNA from *Xenopus laevis*. J. Mol. Biol. 118: 27–47.

Bird, A. P., M. H. Taggart, and B. A. Smith. 1979. Methylated and unmethylated DNA compartments in the sea urchin genome. Cell 17: 889–901.

Brown, N. L., and M. Smith. 1977. Cleavage specificity of the restriction endonuclease isolated from *Haemophilus gallinarium* (*Hga*I). Proc. Nat. Acad. Sci. USA 74:3213–16.

Brown, N. L., and C. A. Hutchison, III, and M. Smith. 1979. J. Mol. Biol. (in press).

Fiers, W., R. Cantreras, G. Haegeman, R. Rogiers, A. Van de Voorde, H. Van Heuverswyn, J. Van Herreweghe, G. Volckaert, and M. Ysaebaert. 1978. Complete nucleotide sequence of SV40 DNA. Nature 273:113–20.

Garfin, D. E., and H. M. Goodman. 1974. Nucleotide sequences at the cleavage sites of two restriction endonucleases from *Haemophilus parainfluenzae*. Biochem. Biophys. Res. Commun. 59:108–16.

Geier, G. E., and P. Modrich. 1979. Recognition sequence of the *dam* methylase of *Escherichia coli* K12 and mode of cleavage of *Dpn*I endonuclease. J. Biol. Chem. 254:1408–13.

Gelinas, R. E., P. A. Myers, and R. J. Roberts. 1977. Two sequence-specific endonucleases from *Moraxella bovis*. J. Mol. Biol. 114:169–79.

Gingeras, T. R., J. P. Milazzo, and R. J. Roberts. 1978. A computer assisted method for the determination of restriction enzyme recognition sites. Nucl. Acids Res. 11:4105–28.

Gingeras, T. R., J. P. Milazzo, D. Sciaky, and R. J. Roberts. 1979. Computer programs for the assembly of DNA sequences. Nucl. Acids Res. 7:529–45.

Godson, G. N., B. G. Barrell, R. Staden, and J. C. Fiddes. 1978. Nucleotide sequence of bacteriophage G4 DNA. Nature 276:236–47.

Heyneker, H. L., J. Shine, H. M. Goodman, H. Boyer, J. Rosenberg, R. E. Dickerson, K. Narang, K. Itakura, S. Lin, and A. D. Riggs. 1976. Synthetic *lac* operator DNA is functional *in vivo*. Nature 263:748–52.

Kleid, D., Z. Humayan, A. Jeffrey, and M. Ptashne. 1976. Novel properties of a restriction endonuclease isolated from *Haemophilus parahaemolyticus*. Proc. Nat. Acad. Sci. USA 73:293–96.

Lacks, S., and B. Greenberg. 1975. A deoxyribonuclease of *Diplococcus pneumoniae* specific for methylated DNA. J. Biol. Chem. 250:4060–66.

Mann, M. B., and H. O. Smith. 1977. Specificity of *Hpa*II and *Hae*III DNA methylases. Nucl. Acids Res. 4:4211–21.

Mann, M. B., R. Nagaraja, and H. O. Smith. 1978. Cloning of restriction and modification genes in *E. coli*: the *Hha*II system from *Haemophilus haemolyticus*. Gene 3:97–112.

Maniatis, T., R. C. Hardison, E. Lacy, J. Lauer, C. O'Connell, and D. Quon. 1978. The isolation of structural genes from libraries of eukaryotic DNA. Cell 15:687–701.

Reddy, V. B., R. Thimmappaya, K. N. Dhar, B. Subramanian, S. Zain, J. Pan, P. K. Ghosh, M. L. Celma, and S. M. Weissman. 1978. The genome of simian virus 40. Science 200:494–502.

Roberts, R. J. 1977. The role of restriction endonucleases in genetic engineering. *In* R. F. Beers and E. G. Bassett (eds.), Recombinant molecules: impact on science and society, pp. 21–32. Raven Press, New York.

Roberts, R. J. 1979. Directory of restriction endonucleases. *In* L. Grossman and K. Moldave (eds.), Methods in enzymology, 65:1–15. Academic Press, New York.

Sanger, F., G. M. Air, B. G. Barrell, N. L. Brown, A. R. Coulson, J. C. Fiddes, C. A. Hutchison III, P. M. Slocombe, and M. Smith. 1977. Nucleotide sequence of bacteriophage ϕX174 DNA. Nature 265:687–95.

Schaller, H., E. Beck, and M. Takanami. 1978. The sequence and regulatory signals of the filamentous phage genome. *In* D. T. Denhardt, D. Dressler, and D. S. Ray (eds.), The single-stranded DNA phages, pp. 139–63. Cold Spring Harbor Press, Cold Spring Harbor, New York.

Sussenbach, J. S., C. H. Monfoort, R. Schephof, and E. E. Stobberingh. 1976. A restriction endonuclease from *Staphylococcus aureus*. Nucl. Acids Res. 3:3193–3202.

Sutcliffe, G. (1978). Nucleotide sequence of the ampicillin resistance gene of *Escherichia coli* plasmid pBR322. Proc. Nat. Acad. Sci. USA 75:3737–41.

Waalwijk, C., and R. Flavell. 1978. DNA methylation at a CCGG sequence in the large intron of the rabbit β-globin gene: tissue specific variations. Nucl. Acids Res. 5:4631–41.

Wilson, G. A., and F. E. Young. 1976. Restriction and modification in the *Bacillus subtilis* genospecies. *In* D. Schlessinger (ed.), Microbiology—1976, pp. 350–57. American Society of Microbiology, Washington, D.C.

Zabeau, M., and R. J. Roberts. 1979. The role of restriction endonucleases in molecular genetics. *In* J. H. Taylor (ed.), Molecular genetics, pt. 3, pp. 1–63. Academic Press, New York.

L. W. ENQUIST

Special Derivatives of Phage λ and Their use for Cloning Fragments of Complex Genomes

2

INTRODUCTION

Coliphage λ has been intensively studied for more than twenty years (see Hershey, 1971). One of the attractive properties of λ used by the molecular biologist is its ability to pick up or transduce bits of the *Escherichia coli* chromosome. Many such transducing derivatives of λ have been used to produce *E. coli* genes in large quantity for genetic and biochemical analysis. Recent advances now allow us to introduce into phage λ, DNA fragments from almost any source. The knowledge and techniques that previously were available only for study of *E. coli* DNA fragments are now applicable, in theory, to virtually any DNA fragment. The proof of this statement can be seen readily in the almost unbelievable rate of recent advancement in our understanding of complex genomes.

One reason phage λ has played a major role in such recombinant DNA activities is that special mutants (vectors) can be constructed to carry foreign DNA fragments. One class of vectors is able to grow only if it carries a foreign DNA fragment. Such a positive selection enabled scientists to screen large numbers of phages knowing that almost every one carried a bit of new DNA. Another reason was that it was possible to add to these special λ vectors, certain mutations that greatly lowered the chances of survival outside the laboratory. This twofold capacity, positive selection for phages with foreign DNA and biological safety, made phage λ the choice for the initial experiments that cloned a single specific DNA

Laboratory of Molecular Virology, National Cancer Institute, National Institutes of Health, Bethesda, Maryland 20205 fragment from millions of fragments of several complex eukaryotic genomes.

I will discuss, in general, the λgtWES and λ *D*am vector systems that have proved useful for my colleagues and me in the analysis of several interesting genomes. Certainly, other excellent vector systems exist that have been quite successful, but I cannot do justice to them here. The most notable of these systems are the λ Charon phages (Blattner et al., 1977) and the many plasmid vectors. Specifically, the topics that will be discussed are: basic features of phage λ biology; the construction of two biologically contained λ vectors; the concepts and methods involved in using *in vitro* packaging of recombinant DNA; the use of the λ*D*am mutation in recombinant DNA technology; and finally, a summary of how many of these concepts were used to clone specific DNA fragments from mouse, RNA tumor virus-transformed mink cells, and herpes simplex virus.

BASIC FEATURES OF λ BIOLOGY

Of the many phages that grow in *E. coli*, phage λ has attracted much attention because of the ease of genetic and biochemical manipulation. The phage requires its host, not only for energy and raw materials, but also for many functions utilized during the phage life cycle. This sharing of components means that λ can be propagated only in a limited number of hosts. In the laboratory the K-12 subline of *E. coli* is used regularly.

Phage λ is a temperate phage with two distinct modes of growth (fig. 1). One mode (lytic) results in the duplication of the λ chromosome several hundred times with most of this DNA packaged into mature virus particles. The other mode (lysogenic) ensues when the λ repressor protein turns off all lytic functions and the circular λ chromosome is inserted into the *E. coli* chromosome. The inserted phage DNA (called prophage) is kept quiescent by continuous synthesis and action of the repressor. This cell with integrated, repressed prophage is called a lysogen. The prophage can be replicated passively as part of the host DNA almost indefinitely. If, however, the repressor is inactivated (induction), the phage enters the lytic mode, excision occurs, and the normal sequence of development follows, resulting in cell death and release of many mature progeny phage. Spontaneous induction is infrequent, but agents that damage DNA (thymine starvation, mitomycin, ultraviolet irradiation, for example) act as extremely efficient inducing agents. It is possible to grow λ by infection or by induction of a lysogen.

Phage particles can be purified easily from the lysed cultures by a variety of methods. Usually the lysates are cleared of debris by low-speed

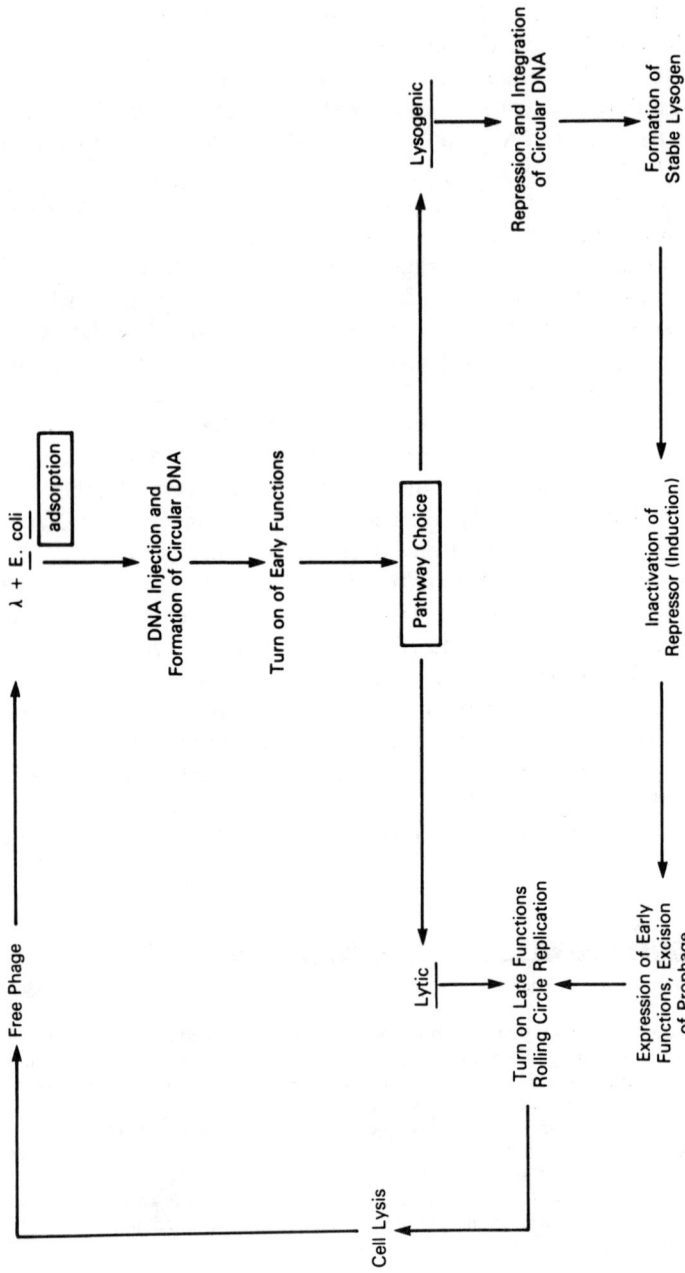

Fig. 1. The life cycle of coliphage lambda. This flow diagram depicts several important stages during infection of a λ-sensitive *E. coli* strain. It demonstrates the two possible life styles of λ at the *Pathway choice*. The *lytic* response yields rapid phage production and cell lysis. The *lysogenic* response gives rise to integrated, repressed prophage that can be *induced* to enter the *lytic* cycle. Phage released after lytic growth can begin the cycle again if λ-sensitive bacteria are present.

centrifugation, the phages are concentrated by high-speed ultracentrifuga-
tion or polyethylene glycol (PEG) precipitation, and the concentrated
particles are banded by isopycnic centrifugation in CsCl. It is common to
obtain yields of 10^{10}–10^{11} phage/ml of lysate under normal conditions, and
certain lysis-inhibited phage can give 10- to 100-fold more. Since one
expects about 1 μg of λ DNA from 2×10^{10} particles, yields of more than a
milligram of DNA per liter are possible.

λ phage particles consist of about equal amounts of protein and DNA.
The DNA inside a phage particle is resistant to pancreatic DNase, which
enables one to digest most contaminating *E. coli* DNA with ease. DNA
inside the λ capsid can be released simply by treatment with EDTA
followed by subsequent deproteinization with phenol. The DNA so
obtained from λ particles is double-stranded and linear, and contains
about 49,500 base pairs (49.5 kbp). At each 5' end are 12 unpaired bases
that are complementary. These "sticky ends" can cohere readily to form
linear multimers or circular molecules, depending on the DNA concentration. Circular molecules formed in this way are often called "Hershey circles" after their discoverer, A. D. Hershey. One useful facet of λ DNA is
that the strands can be separated and purified in quantity.

The λ chromosome carries information for about 50 genes and a variety
of control elements. A striking feature of λ chromosome organization is
that related functions are clustered (fig. 2). For example, capsid protein
genes are all adjacent and next to the tail gene cluster. The insertion-
excision system, the repressor-operator control complex, the replication
functions, and the lysis genes are found in similar clusters.

During the past two decades, many mutations, deletions, rearrangements, and substitutions have been isolated and characterized in almost
every known gene and region (see Hershey, 1971). This wealth of genetic

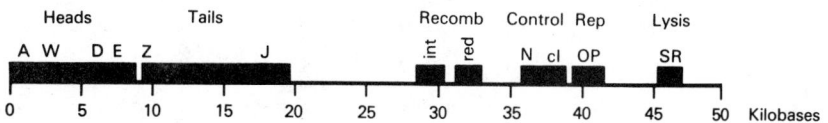

Fig. 2. Organization of the lambda chromosome. A depiction of the linear lambda DNA
molecule showing the clustering of functions and the approximate size in kilobases. The dark
boxes are particular functions and the letters immediately above the boxes are gene names.
Only those genes referred to in the text are indicated. For example, A, W, D, E are some of the
genes involved in head formation. Z and J are two genes involved in tail production. Two gene
clusters are found in the recombination region; the integration-excision system (the *int* gene
is noted) and the general recombination system (the *red* genes are noted). Two important
genes in the repressor-operator control region are the positive regulator N and the repressor
cI. The replication functions O and P and the replication origin are located in a cluster. The
lysis functions S and R are near the right end of the lambda chromosome.

and biochemical information has proven invaluable for constructing and utilizing λ as a vector for recombinant DNA experiments.

THE λGTWES SYSTEM

The λ vector system, λgtWES, was developed in Philip Leder's laboratory. It was based on elegant work by Davis and coworkers, who constructed the basic vector designated λgt (Thomas et al., 1974). The letters "gt" stand for generalized transducer, and were perhaps more prophetic than Davis and coworkers realized. λgt, like several other vectors designed during this period, was primarily able to accept DNA fragments produced by *Eco* R1. This was because *Eco* R1 was one of the first restriction enzymes available. In addition, because the enzyme could be produced in a λ sensitive *E. coli* strain, it was a simple matter to isolate λ mutants lacking *Eco* R1 restriction sites.

Davis and colleagues realized an important fact: when λ packages its DNA into capsids, the DNA substrate is a long concatemer (many λ monomers joined together). The capsid is not simply filled to capacity with DNA; rather, DNA between *two* cohesive end sites (*cos*) is packaged. There is a minimum amount (about 36 kb) and a maximum amount (about 53kb) of DNA λ can package and still form a viable phage. Conveniently there are certain regions of the λ chromosome not required for phage growth, and these can be deleted with little difficulty. It was reasoned that if one could remove enough nonessential DNA, one could construct a λ variant too small to be packaged. This would be an ideal vector because the small vector genome would not form a plaque unless its size was increased. Simple addition of any DNA fragment would enable the phage to grow and form a plaque. Finding a λ variant so small it could not be packaged, keeping it alive and yet be a vector for *Eco* R1 fragments, was not as difficult for Davis's group as it would seem. Normal λ and the phage they constructed are shown in figure 3. They removed all but two *Eco* R1 sites by mutations and deletions, added a small deletion called *nin*5 in the *Eco* R1 E-fragment, and removed the entire nonessential Eco R1 B-fragment. This phage was the source for λgt, a DNA molecule too small to be packaged. Its precursor is the phage on line 2 or line 3 (fig. 3) that has 3 Eco R1 fragments. The fragments are called the left arm, the central fragment, and the right arm. The left and right arms have all the known λ genes essential for growth. This means that if the central fragment is removed and the left and right arms joined together by their *Eco* R1 cohesive termini, the resulting DNA should be able to replicate and give rise to plaque-forming phage. In fact, the joint left and right arms will not form plaques because they are too

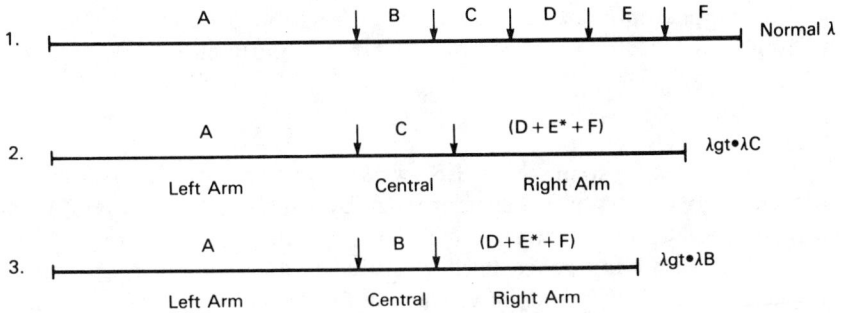

Fig. 3.   The λgt vectors. This figure shows wide-type or normal λ DNA (line 1), λgt·λC DNA (line 2), and λgt·λB DNA (line 3). The arrows indicate *Eco* R1 restriction sites, and the letters describe the particular Eco R1 fragment. The fragments are not drawn to scale. In λgt·λC (line 2) the B fragment is deleted and the restriction sites on both sides of fragment E have been mutated. Consequently the D, E, and F fragments are fused. In addition, the *nin*5 deletion has been added, and it shortens the E fragment by 2.8 kb. The E* notation is to indicate the presence of this *nin*5 deletion. In λgt·λB (line 3) the C fragment is removed and the left and right arms are identical to λgt·λC. The B and C fragments are not essential to λ growth. They are included in the λgt vectors only to provide sufficient length for phage packaging.

small to be packaged. They will make plaques *only* if another Eco R1 fragment is inserted between them. Thus, λgt DNA is produced *in vitro* from either of the phages shown in lines 2 and 3 by removing the central fragment.

Thomas et al. (1974) proposed a useful nomenclature for λgt vectors and hybrids. Eco R1 fragments carried by λgt are written after "gt" with a dot (·) between λgt and the fragment. Thus λgt·λC indicates the λ *Eco* R1 "C" fragment is inserted between the two arms of λgt. Another useful vector of the same series was also developed by Davis and colleagues. It was called λgt·λB, and had the λ *Eco* R1 "B" fragment inserted between the arms of λgt (line 3, fig. 3).

These two vectors were used by Leder's group as a basis for construction of so-called safe vectors (EK2 or HV2). Such safer vectors were, and still are, required for certain recombinant DNA procedures. As detailed elsewhere, wild-type lambda itself has a very low probability of surviving in nature (Enquist and Szybalski, 1978). We assumed it would be a straightforward matter to increase further the biocontainment properties by a variety of steps including: (a) abolishing its capacity to form prophage or plasmids; (b) introducing conditional lethal mutations so that the phage can only be propagated under laboratory conditions; and (c) adding mutations that increased phage yields so that smaller volumes of culture fluid were necessary. What was not so simple was to test the hypothesis that

λ vectors with these additional mutations were indeed biologically contained. In 1975, when our first "safe" vector was constructed, the requirements for safe vectors were not well established, and no one had good experience in testing biological containment. The inevitable result was long delays and frustration. The problems we and others experienced for certification of the new vectors hopefully are a thing of the past. More efficient procedures are now available for testing and approval of new vectors.

The modifications of λgt that we introduced to make the λgtWES vectors were straightforward and are described in detail in several publications (Enquist et al., 1976; Tiemeier et al., 1976; Leder et al., 1977a; Leder et al., 1977b). Briefly, we constructed the first vector from λgt · λC. This phage already carried two mutations that increased biological containment. These were a temperature-sensitive repressor mutation, cIts857, and the $nin5$ deletion. The cIts857 mutation affects the ability of the phage to make stable lysogens. At temperatures below 34° C the phage makes repressor at rates and levels approaching normal. Increasing the temperature results in denaturation of the repressor protein so that at 37° little active repressor remains and the phage cannot form stable lysogens. It is therefore committed to lytic development. The $nin5$ deletion removes about 2.8 kb of the right arm of λgt between genes P and Q. The DNA removed encodes a major transcription stop signal so that phages with this deletion tend to make late functions (lysis proteins, capsid components, and tail proteins) independent of normal λ control mechanisms. This greatly reduces the chance of plasmid formation (stable, nonintegrated phage DNA molecules that replicate without killing the cell). For example, plasmids can be formed from phage λ by a single mutation in the regulatory N gene that results in poor expression of lytic functions, yet allows low-level replication. Such N^- lambda plasmids cannot be maintained if the phage carries a $nin5$ deletion (nin stands for N independent).

We sought to improve the containment of λgt · λC in a way that would not result in a phage that grew poorly. Our choice was to use nonsense (*amber*) mutations in essential genes. These mutations are conditional in that phages with such *amber* mutations will grow only in certain *E. coli* strains carrying *amber* suppressor t-RNAs (sup^+). The phages will not grow in any *E. coli* host lacking the *amber* suppressor (sup^-). We introduced three *amber* mutations in λgt · λC, two in the left arm and one in the right. In the left arm, the mutations were in the W and E genes (*W*am403 and *E*am1100). The E gene product is one of two major capsid proteins and also is required for formation of the λ cohesive ends. The W gene product is the protein required to join capsids to tails. The right arm mutation was in gene *S* (*S*am 100). The *S* gene product cooperates with phage endolysin to produce lysis and also plays a role in DNA synthesis. We introduced the *S*am100 mutation into λgt·C by mutagenesis and then crossed in the *W* and *E* mutations. This first vector was called λgt *W*am403 *E*am1100 *S*am100· λC or λgtWES· λC for convenience (fig. 4). This phage grew well on *E. coli* strains with the amber suppressor, *sup*F; it did not grow at all in *sup⁻* strains. What happens in *sup⁻* cells is thought to be the following: λgtWES· λC infects *sup⁻* cells and begins replicating normally. At temperatures above 37°, a true lytic infection starts (see fig. 1). Because no *E*, *W*, or *S* proteins can be made because of the amber mutations, several striking changes occur during the lytic cycle. No phage heads could form because of the absence of the *E* protein. The completed tails could not be joined to phage heads, even if there were any, because the head-tail joining protein, *W*, also would be missing. Because there were no capsids, the replicating λ DNA would never be packaged or cleaved at *cos*, and there would be an accumulation of immature, high molecular weight λ DNA. Finally, *S* protein would not be present so no cell lysis would occur, and there would be continued rounds of λ replication.

The *S*am100 mutation was useful in another way. Cells infected with λ*S* mutants will not lyse but will continue to accumulate intracellular phage particles—as many as 1,000 per cell. These phages can be released by adding a few drops of chloroform; lysis then occurs almost immediately. Contrary to this, chloroform kills but does not lyse normal uninfected *E. coli* cells. The convenience of λ*S* mutants is that infected cells can be collected and concentrated prior to chloroform-induced lysis. It was possible to use this lysis inhibition in the λgtWES vectors as follows. The amber mutations in genes *W* and *E* could be suppressed by *sup*F and *sup*E

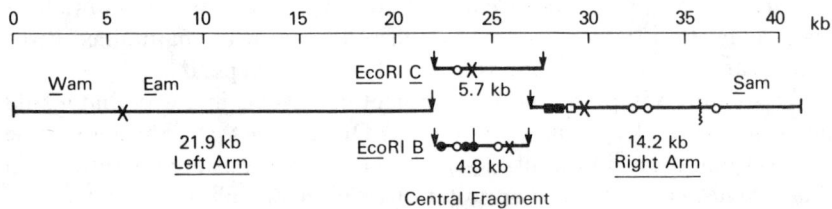

Fig. 4.   The λgtWES vectors. This diagram gives the approximate size and characteristic features of two λgtWES vectors, λgtWES·λb (central fragment in *Eco* RI B) and λgtWES·λC (central fragment is *Eco*RI C). The vectors were derived from the λgt vectors (see fig. 3). The left arm containing the W and E amber mutations and the right arm containing the S amber mutation are indicated. The vertical wavy line in the right arm indicates the *nin*5 deletion. Restriction endonuclease sites are indicated as follows: ↓, *Eco*RI; x, *Bam*HI; o, *Hind*III; •, *Sac*I or *Sst*I; ↓, *Xba*I; □, *Hho*I and ■, *Sal*I. Note that in λgtWES·λB the central fragment is as shown and is inverted from the normal orientation of λ *Eco*RI B.

suppressors, but *S*am 100 could only be suppressed efficiently by *sup*F and not by *sup*E. Thus when λgtWES ·λC infects a *sup*E cell, both *W* and *E* products are made but the suppressed *S* product is poorly active. This results in accumulation of completed phage particles that can be released by addition of chloroform. (Chloroform is also a very good agent for killing *E. coli*, making this system doubly attractive from the biosafety point of view.) The method worked best, however, by induction of λgtWES · λC lysogens. Recall that the entire λ integration-excision system is located within the λC fragment. Thus λgtWES· λC can integrate in *E. coli* and form stable lysogens. Because the vector also carries the *c*Its857 temperature-sensitive repressor mutation, such lysogens are stable only at 30–32°C. Raising the temperature causes induction and production of λgtWES· λC. If the lysogen is a *sup*E derivative of *E. coli*, then, after induction, the *S*am100 mutation is poorly suppressed, lysis is delayed, and chloroform is needed to complete lysis. It was possible to make λgtWES · λC in large quantity by this method.

An additional feature of λgtWES · λC involved the λC fragment and the λ integration-excision system. Weisberg and I developed a sensitive plaque-color test that gave red plaques if a phage carried the λ integration-excision functions and white plaques if these functions were inactive or absent. Details of the assay are described in Enquist and Weisberg (1976, 1977). Thus in this test λgtWES · λC made red plaques, and recombinants that lost the λC fragment made white plaques. We have used this method not only for screening recombinants but also to assay the purity of λgtWES after the λC fragment had been removed. The method could detect one parental λgtWES·λC in 10^6 λgtWES recombinants.

λgtWES ˙λC was certified for use as an EK2 vector in January 1976. However, we constructed a second EK2 vector to overcome a problem inherent in λgtWES˙λC, namely, the C fragment itself. Recall that to form λgt from λgt · λC (or λgtWES · λC), the λC fragment must be removed physically. The λC fragment contains most of the λ general recombination system (the *red* genes, for *re*combination *d*efective) as well as the integration-excision system. It is well established that λ *red⁻* phages grow poorly when compared with normal λ *red⁺* phages because replication is reduced and DNA packaging is less efficient (Enquist and Skalka, 1973). The phage yield consequently drops by one-half to as much as one-fifth compared with λ *red⁺*. The result is that, in a mixed population, λ *red⁺* phages outgrow λ *red⁻* quite readily (Cameron and Davis, 1977). Thus any parental forms of λgtWES · λC would take over a hybrid population rapidly. This is of concern because biochemical removal of the λC fragment is never absolute—some λC fragment always remains. It seemed logical to begin with a *red⁻* vector and our choice was λgt · λB (fig. 3). λgt · λB is missing the λC fragment and is therefore *red⁻*; there can be no problem of parental *red⁺* phages overgrowing the *red⁻* hybrids.

We also became aware of another useful facet of λgt·λB. The λB fragment contains two *Sst* I restriction endonuclease sites about 1 kb apart (the enzyme *Sac* I also cleaves the same sites), so that if the vector is digested with *Sst* I (or *Sac* I) as well as with *Eco* R1, the λB fragment is effectively shattered. Addition of *Eco* R1 fragments to this double-digest followed by ligation and phage formation will result in many authentic hybrid phage and a reduced number of parental λgt·λB phage. The double-digest method generated λgt with no biochemical purification steps.

These facts induced us to make λgtWES · λB by the procedures described in Tiemeier et al. (1976). λgtWES · λB was tested and approved as an EK-2 vector in February 1977 and supersedes λgtWES·λC. It should be noted that the λgt portion of λgtWES·λB and λgtWES·λC are identical; they differ only in the central fragment. We have found λgtWES·λB to be a most versatile vector (fig. 4). From experience we now know that it can carry DNA fragments from about 2 kb to 17 kb (see Enquist et al., 1979) rather than from 1 kb to 15 kb as thought earlier. We have used the vector to clone purified fragments with one end formed by *Hind* III and the other by *Eco* R1 (Enquist et al., 1979). Similarly, I have cloned *Bam* H1-*Eco* R1 fragments following the same scheme (unpublished). The method described by Maniatis and colleagues (1978) to form "libraries" of complex genomes using λ hybrids is also applicable to the λgtWES vectors. The vector used by Maniatis was Charon 4A because it has a larger insert capacity (about 19.3 kb). Libraries with the λgtWES vectors should be made with slightly smaller fragments—about 15 kb is optimal.

The λgtWES vectors accept primarily *Eco* R1 fragments. This limitation can be overcome using synthetic *Eco* R1 linkers added to blunt-ended fragments produced by a variety of restriction enzymes. These can then be joined to the *Eco* R1 arms of the λgtWES vectors (see Maniatis et al., 1978). Alternatively, it may be possible to add synthetic sites to the vector adapting it to other restriction enzymes. *Sst* I (*Sac* I), *Xba* I, *Xho* I, and *Sal* I fragments can be inserted in λgtWES·λB because single sites or two adjacent sites for these enzymes exist in nonessential regions. Thus the vector is not only a replacement vector (foreign DNA replaces λB fragment) but also an insertion vector (no vector DNA lost, foreign DNA inserted). A derivative of λgtWES·B containing only one *Sst* 1 site in the central λB fragment has also been approved as an EK2 vector.

IN VITRO PACKAGING

In general, once a suitable vector is constructed, the next few steps are straightforward. The restricted foreign DNA is mixed in a proper ratio with the restricted vector DNA and this mixture joined together with DNA ligase. This rather complex, ligated mixture of DNA molecules contains, among other things, the desired hybrid. Initially, this DNA was introduced back into *E. coli* by a process called transfection. Two methods were in vogue. In one the cell wall and outer mucopolysaccharides of *E. coli* were partially removed yielding fragile spheroplasts (Henner et al., 1973). These spheroplasts took up λ DNA after osmotic shock and produced infectious phage particles. The technique worked well enough but was inefficient (about 10^5 plaques/μg added λ; theoretical value should be 10^{10} plaques/μg added λ). Another transfection system was also available. Here *E. coli* was treated with $CaCl_2$, λ DNA was added, and phage particles were produced (Mandel and Higa, 1970). The method was as inefficient as spheroplasts, but it proved more reproducible and was less tedious. Both methods were rather strain-specific in that some *E. coli* stocks were much worse than others. It was clear the introduction of DNA back into *E. coli* to form phage particles was a limiting step in the analysis of complex genomes by recombinant DNA methods.

For example, in transfection with intact λgtWES · λB DNA, we usually obtained ~ 10^5 plaques/μg. When λgtWES arms were mixed with foreign DNA, ligated, and transfected back into *E. coli*, the yield always dropped about 10- to 100-fold yielding ~10^3–~10^4 plaques/μg, at best! If one considers that a complex genome like mouse has more than a million *Eco* R1 fragments, this combined inefficiency of ligation and transfection is serious. At least 10^6–10^7 plaques would be required to approach finding a single unique fragment. Blattner et al. (1978) calculated that to produce 10^6–10^7 individual phages from a ligation mixture by $CaCl_2$ transfection, one would need 240 mg of vector DNA, 24 mg of foreign DNA, and 400 liters of $CaCl_2$-treated *E. coli*! The amounts were clearly prohibitive. The initial solution to this problem was development of a powerful procedure to fractionate and purify specific DNA fragments. The use of RPC-5 chromatography and preparative gel electrophoresis gave 1,000-fold purification of the mouse globin genes (Leder et al., 1977a; Tilghman et al., 1977) and made possible the first cloning of a mammalian gene even under these inefficient conditions.

Sternberg and I realized that A. Becker and M. Gold (1975) had another solution to the problem of inefficient return of DNA to *E. coli*. They were studying particle assembly of λ *in vitro* and, much to our interest, were getting as many as 10^6–10^7 plaques from a μg of added λ DNA. We received the basic protocol from them and began adapting it for packaging recombinant DNA *in vitro* into infectious phage particles (Sternberg et al., 1977). Other variations of the Becker and Gold procedure now exist for *in vitro* packaging and are described in Hohn and Murray (1977) and Blattner et al. (1978).

In vitro packaging is really quite an amazing process. Briefly, recombinant DNA is made as before with foreign DNA and a λ vector, and this is added to crude extracts containing partially assembled λ virion proteins. The added DNA is packaged *in vitro* into the phage head, and tails are joined to the filled heads yielding an infectious particle. We were able to obtain 10^7–10^8 plaques/μg of λ DNA with this system—as much as 3 orders of magnitude better than any transfection system! Another positive feature was that the phages formed *in vitro* never had replicated or been exposed to selective pressures. They were stable and could be stored as any conventional phage lysate. This is in contrast to transfection methods in which the DNA had to replicate before it was encapsidated in phage particles and the resulting plaques had to be harvested prior to storage. Phages that grew less well were selected against by transfection methods.

The packaging extracts are crude lysates made from two induced λ lysogens. The prophages in both lysogens carry several mutations and are defective. One important mutation is the *c*Its857 temperature-sensitive repressor: the lysogens can be induced by a simple temperature shift. An *S* amber mutation is also present so that the induced lysogens do not lyse, resulting in many rounds of λ replication along with accumulation of phage proteins. An important mutation introduced by us was the *b*2 deletion. This mutation blocked prophage λ excision from the *E. coli* chromosome. The induced λ*b*2 prophage is effectively trapped in the *E. coli* chromosome even though many rounds of prophage replication occur. It was established that trapped prophage DNA could not be packaged into plaque-forming virions. Therefore, induced λ*b*2 extracts contain no source of packagable DNA; the only DNA that can be packaged is added exogenous DNA. Two more prophage mutations were needed to complete the basic packaging extract lysogens. One lysogen carries an amber mutation in prophage capsid gene *A*; the other carries an amber mutation in prophage capsid gene *E*. Finally, to ensure that no exchange of information occurred by recombination of endogenous DNA with exogenous recombinant DNA, we added the *E. coli recA* and λ*red*3 mutations. These mutations effectively inactivate the major general recombination systems present in the packaging extracts.

The two packaging lysogens, while complicated in construction, are straightforward to use (fig. 5). Upon raising the temperature to induce each lysogen, the trapped prophage replicates *in situ*, and fills the cell with capsid and tail components ready to assemble upon addition of DNA. The cells are concentrated by centrifugation (they do not lyse because of the *S* amber mutation) and two separate extracts are prepared. In one an equal mixture of *both* induced lysogens is mixed and sonicated. This sonic extract (SE) is primarily a source of partially assembled heads ready to take up DNA. The other extract (FTL) is made by freezing and thawing the *E*am lysogen with lysozyme to produce a viscous mixture containing primarily the *A* gene product. Both extracts contain assembled tail components. These crude lysates are immediately frozen in liquid nitrogen, where they are stable for many months. Actual *in vitro* packaging is done by thawing out an aliquot of the sonic extract (SE), mixing it with a tris buffer containing EDTA, MgCl$_2$, ATP, spermidine, mercaptoethanol, and the recombinant λ DNA. After a 15 min incubation on the lab bench, an

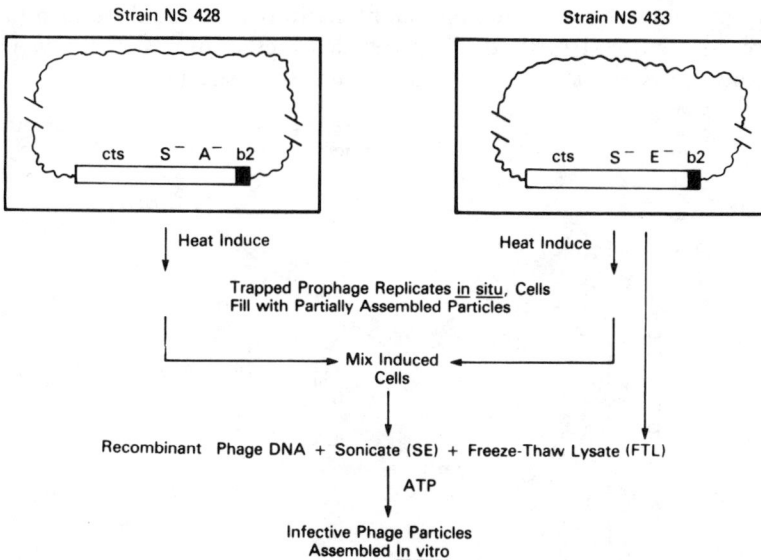

Fig. 5.   A schematic depiction of the steps involved in *in vitro* packaging. The two *E. coli* lysogens, NS428 and NS433 are indicated as rectangles. Inside each rectangular cell, the defective prophage specific for each lysogen is shown as part of the *E. coli* chromosome. The temperature sensitive repressor is indicated by *cts*, the lysis defective mutation is noted by *S*⁻ and the specific *A* protein and capsid protein defects are given by *A*⁻ in NS428 and *E*⁻ in NS433, respectively. The b2 deletion that prevents prophage excision is indicated by the black shading. The subsequent steps in extract formation and *in vitro* packaging are outlined.

aliquot of freshly thawed FTL is added and incubated at 37°C for 1 hr. Finally, DNase is added in a tris-MgSO₄ buffer to digest any unpackaged recombinant DNA as well as *E. coli* DNA. This mixture now can be treated as a conventional phage lysate. Aliquots can be plated anytime for assay as the mixture is stable for months at 4°C.

During our characterization of the *in vitro* packaging system, we found that if λ DNA molecules of different lengths were packaged, the *in vitro* packaging efficiency decreased markedly as the size of the DNA decreased (Sternberg et al., 1977; fig. 6). This size selection could be reduced if putrescine as well as spermidine was used in the reaction. We have found that the size selection is particularly useful when a small vector phage is used, because the packaging system then provides a strong selection for those larger phage with inserted fragments. This is convenient for vectors with only one site of insertion (for example, λgtWES · λB when used for *Xho* I, *Xba* I, *Sst* I, or *Sal* I fragments). As discussed previously, a positive selection of this sort was available only for replacement vectors like the λgt series. We have also found that the size selection aids in isolation of vectors carrying multiple fragments and combinations that could be missed by more conventional methods. The detailed methodology of our *in vitro* packaging system is found in Enquist and Sternberg (1979).

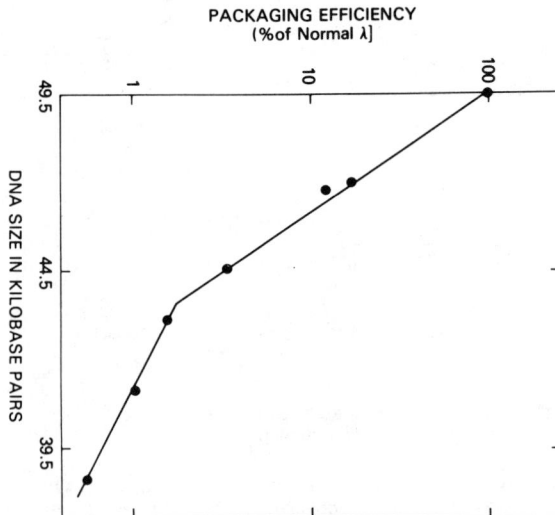

Fig. 6. The size of λDNA affects the efficiency of in vitro packaging. This diagram is derived from data of Sternberg et al., 1977. Lambda DNA molecules of known sizes were packaged *in vitro*. The packaging efficiency was expressed as percent of the value obtained for normal λDNA. The packaging efficiency falls markedly as the size of the DNA decreases.

USE OF THE λ DAM MUTATION

There are two major proteins in the λ capsid: the products of genes *E* and *D*. Sternberg and Weisberg made the surprising discovery that when the λ chromosome size is less than 41 kb (about 82% of normal λ), the *D* protein is no longer needed. For example, λ with an amber mutation in gene D (λ *D*am) normally cannot grow unless suppressed. However, λ *D*am will grow on *sup⁻* hosts when it contains deletions of 8.5 kb or more (table 1). The *D*-

TABLE 1

GROWTH OF λDAM PHAGE ON
SUPPRESSOR PLUS AND MINUS
E. COLI

Size of λ Dam Phage DNA	Plating on	
(kb)	*sup⁻*	*sup⁺*
49.5 (wild type)	–	+
41	+ / –	+
Less than 41	+	+

deficient particles are extremely sensitive to EDTA, but in other respects behave as normal λ phage. We have introduced *D*am mutations into two vectors and have demonstrated that this mutation facilitates identification and selection of certain fragments based on size and also provides a simple way to isolate deletions of the cloned fragment (Sternberg et al., 1979; Enquist and Sternberg, 1979).

We first added a *D*am mutation to a λ insertion vector for *Eco* R1, *Sal* I, and *Xho* I fragments (λ Dam sr1λ3; Sternberg et al., 1977). This vector could accept fragments up to about 15 kb in size. Because the vector carried two deletions removing about 11.2 kb, it could grow in the absence of D protein (it formed plaques on *sup⁻ E. coli*). We found that the vector could carry fragments up to about 2.5 kb and still grow on *sup⁻ E. coli*. Thus, by plating a pool of λ recombinant phages on *sup⁻ E. coli*, we were able to select the hybrids carrying fragments 2.5 kb or smaller.

We also introduced the *D*am15 mutation into λgt·λC (see fig. 3). λgt*D*am hybrids carrying fragments of 5 kb or less formed plaques on *sup⁻ E. coli*, whereas hybrids with larger fragments grew only on *sup⁺* hosts.

It was apparent that this same approach could be used to select deletions in λ *D*am phages. In any given λ stock, spontaneous deletions can be found at a frequency of 1 in 10^4 or 10^5 phages. The *D*am mutation greatly facilitates detection of these deletions *if* the starting DNA size is 41 kb or larger. Any deletion that lowers the DNA content below 41 kb will enable the λ *D*am phage to grow on *sup⁻ E. coli*. True revertants of the *D*am mutation can readily be identified because they make larger plaques than *D*am-containing phages. In addition such revertants are rare, occurring less than 1 in 10^6 λ *D*am phages. Prior to the *D*am selection, the only technique available for selection of λ deletions was resistance to heat or chelating agents (EDTA or pyrophosphate). Normal λ is sensitive to these agents, but deleted phage are not. A limitation was that multiple cycles of selection were required because about 1 in 1,000 particles were phenotypically resistant. Multiple cycles are not needed using the *D*am mutations. Deletions are selected simply by direct plating on *sup⁻ E. coli*. Independent deletions can be obtained by making a series of 1 ml minilysates in *sup⁺* bacteria, each lysate derived from a single plaque. By plating an aliquot of each lysate on *sup⁻ E. coli* and choosing one plaque from each, the independence of each isolate is assured. The use of the *D*am method for analysis of cloned fragments is described in Enquist and Sternberg (1979) and Sternberg et al. (1979).

USE OF THE λGTWES SYSTEM: CLONING SPECIFIC FRAGMENTS FROM
MOUSE, HERPES SIMPLEX VIRUS AND RNA TUMOR VIRUSES

My own experience in using this versatile vector system has come from two laboratories: initially with Philip Leder and colleagues in cloning globin genes from the mouse and now with George Vande Woude and colleagues in cloning DNA fragments of herpes simplex viruses and RNA tumor proviruses.

Even though the work of construction and obtaining approval for use of λgtWES was encumbered by the new bureaucracy surrounding recombinant DNA, it was clearly worthwhile when the system was used by Leder and colleagues to clone the first mammalian DNA fragment carrying the mouse β-globin gene (Leder et al., 1977; Tilghman et al., 1977). We barely had celebrated this event when it was discovered that the gene contained *intervening sequences*. This surprise, coupled with similar observations in the adeno and SV40 viruses opened a new door for molecular biology of higher organisms. In a few months Leder's group isolated λgtWES hybrids carrying another mouse β-globin gene as well as the α-globin gene, both carrying their own intervening sequences (see Tiemeier et al., 1978). It soon was shown that the intervening sequences were transcribed and subsequently spliced out yielding the mature mRNA. This "RNA splicing" is now an active area of research with recombinant DNA procedures playing a major role.

More recently, in George Vande Woude's group, we are applying the λgtWES system to DNA and RNA tumor viruses. The interesting features of the initial experiments were that we had to work under P4 maximum containment conditions, first at Fort Detrick and then in Building 41 at the N.I.H. The λgtWES system was the first λ vector to be used under such stringent conditions, and it proved itself well. While we were operating under P4 containment, the N.I.H. guidelines for recombinant DNA research were changed. The level of containment for the DNA and RNA tumor virus work was reduced drastically, and many more kinds of experiments became possible. What follows is a brief accounting of our recent experiences in cloning these viruses.

Herpes viruses present a variety of problems to the clinician and to the molecular biologist. For example, the viruses are common infectious agents of many vertebrates including man. Persistence in a latent form for many years only to give rise to typical overt disease is a hallmark of herpes. A role for these viruses in certain malignancies also has been suggested. At the molecular level herpes viruses have large genomes with an unusual sequence arrangement (see below). How this large genome is replicated, regulated, and rearranged is only now being examined.

We have begun a molecular analysis of one group of herpes, herpes simplex virus I (HSV-I). The genome of HSV-I consists of a double-stranded linear DNA molecule about 160 kb in size. The molecule can be divided into two portions, a large 133 kb segment (L) and a small 26 kb segment (S) (fig. 7). L and S are terminally redundant and are capable of inverting with respect to the other by mechanisms not well understood. Because of this L-S inversion, the HSV-1 genome exists in four approximately equal permutations in a given population of DNA molecules. Another sequence rearrangement occurs during the production of defective HSV-1 particles (dHSV) that arise after repeated passage at high multiplicity. One class of dHSV we are studying consists of a tandem repeat of only one end (the S region) of the parental HSV-1 (fig. 7).

We have cloned in λ, *Eco*R1 fragments of HSV-1 and S-region defective HSV-1 (Enquist et al., 1979). The parental DNA used for our studies is shown schematically (not to scale) in figure 7. The *Eco* R1 sites are denoted by arrows. For the normal HSV-1 genome, we have clones representing most of the DNA molecule. The HSV thymidine kinase gene (located near *Eco* R1 fragment N) was of some interest because of its use as a selectable marker for DNA transfer in certain thymidine kinase–defective mammalian cell lines. We cloned the active gene on a *Bam* HI fragment using the plasmid pBR322 in collaboration with Summers and colleagues.

Defective HSV-1 has a much-simplified genome (fig. 7). dHSV-1 consists

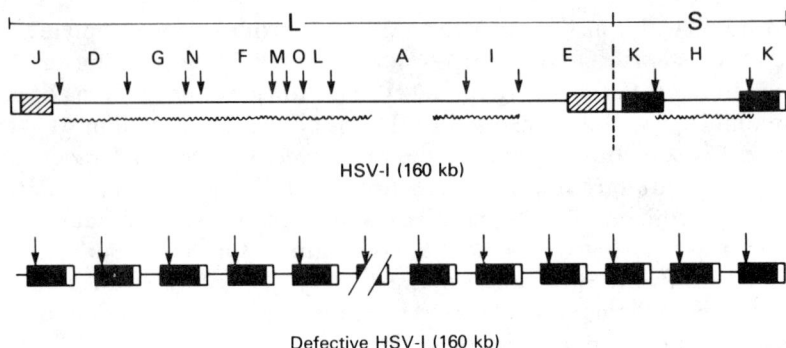

Fig. 7.   Simplified map of HSV-1 and defective HSV-1 DNA. Normal HSV-1 DNA (Patton strain) is given on top, and defective HSV-1 DNA is depicted below. The figure is not drawn to scale. The arrows indicate *Eco* R1 restriction sites. The letters denote specific fragments. The L and S inverting regions are noted for normal HSV-1. Only one permutation is shown for simplicity. The boxed regions indicate the repeated regions. Similar shading patterns designate homologous regions among the repeats. The wavy lines below the HSV-1 map give the regions we have cloned in λ or plasmids. The defective HSV-1 genome (below) is much simpler and consists of about 17 repeats of the extreme right end of the normal HSV-1 molecule.

of about 17 iterations of a portion of the extreme end of S. The repeat unit carries a single *Eco* R1 site; thus, virtually all the genome can be cloned. We discovered two general classes of S region–defective *Eco* R1 fragments based on abundance and size of the iterated DNA. These were denoted major and minor dHSV fragments, respectively (fig. 8). The major class had a repeat unit of 8–9 kb, and both minor classes were about 2 kb smaller. The minor dHSV fragments fell into two distinct groups. The minor B1 deletion class seems to have a specific deletion within the terminal repeat of the S region. The minor B3 deletion class contains more variable deletions in a neighboring region outside the terminal repeat. We are using these dHSV clones as tools for locating and sequencing the regions involved in dHSV genesis as well as reagents for studying HSV transcription and translation.

The RNA tumor viruses present another set of biological problems. For example, the single-stranded viral RNA genome is converted to a double-stranded DNA form that integrates into the host genome. Another example is the formation of an unusual defective variant in which specific host sequences are picked up by the RNA tumor virus. This defective variant can now transform cells. We have chosen the murine RNA tumor viruses as tools for studying these two particular phenomena. Specifically, we use Moloney-murine leukemia virus (M-MuLV) and its defective

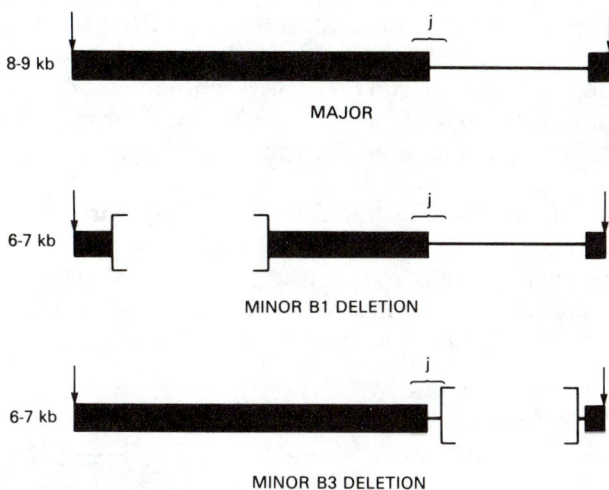

Fig. 8.   Three classes of dHSV *Eco* R1 fragments. When defective HSV-1 (dHSV) DNA is cleaved with *Eco* R1, two size classes of fragments are seen. The most abundant class is the 8-9 kb *major* class and the least abundant class the 6-7 kb *minor* class in size. The vertical arrows are *Eco* R1 sites, and subsequent information was obtained by cloning the major and minor fragments in λgtWES·λB (K. D.-Thompson, L. Enquist, and G. Vande Woude, in preparation). The shaded regions of each fragment represent the S-region terminal repeat (see fig. 7). The heterogeneity in major class fragments is located at the region denoted j. This heterogeneity at the junction of repeating units is probably present in minor class defectives as well. The minor class fragments contain two deletion types, both in equal abundance. The regions deleted are shown by brackets.

variant, Moloney murine sarcoma virus (MSV). We have cloned in phage λ, the integrated DNA proviral forms of two MSV isolates (Vande Woude et al., 1979a,b). We used mink cells transformed by these viruses as a source of integrated viral DNA to reduce problems arising from the many related endogenous viruses in the normal mouse cells. One such cloned fragment is diagrammed in figure 9, line 1. The entire MSV genome is present including flanking mink sequences. One notable feature of integrated MSV is the presence at both ends of a 600 base pair direct repeat of MSV information. This repeat is created by the reverse transcription process where RNA is copied into DNA. In itself it is a novel structure, for it contains about 150 base pairs of information from the 5′ end of the virion RNA molecule and 450 base pairs from the 3′ end of the virion RNA genome. The other region of interest is the block of mouse information (~1 kb) picked up during formation of MSV from M-MuLV in BALB/c mice (abbreviated "src"). The entire cloned fragment in line 1 is biologically active; that is, it efficiently causes morphological transformation of mink and mouse cells in culture. The proper MSV virus can subsequently be obtained from these transformed cells. We are cloning smaller and smaller fragments in plasmids to localize the region that causes morphological transformation. Our results to date indicate that the "src" region alone transforms at a low frequency, but the addition of a 600 bp repeat region increases this more than three orders of magnitude.

A useful variant of the integrated MSV clones appeared upon analysis of the hybrid phage lysates. A fraction of the phage carrying the integrated MSV fragment apparently recombined during growth in *E. coli* at the 600 base pair repeats, resulting in a deletion of the MSV sequences between them (fig. 9, line 2). The new fragment has identical left and right mink

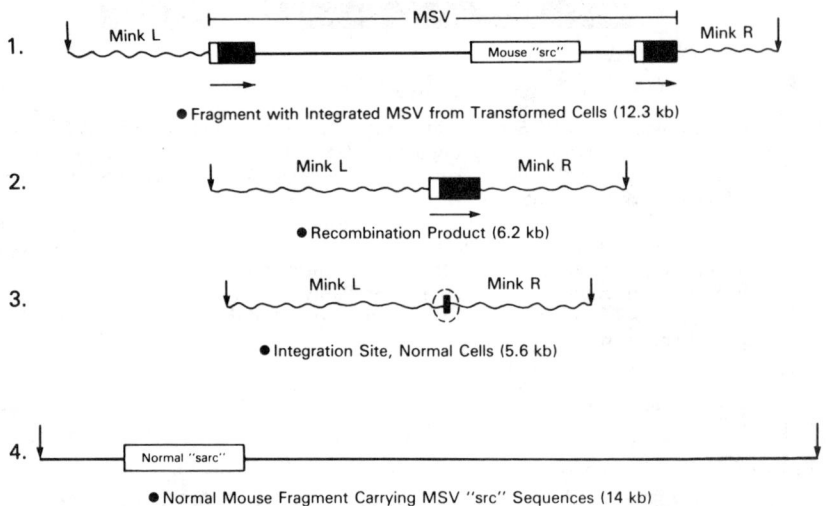

Fig. 9.   λgtWES clones of Moloney-Murine sarcoma virus. These four *Eco* R1 fragments have been cloned in λgtWES · λB by Oskarsson, McClements, Enquist, and George Vande Woude. Analysis using electron microscopy was done by Sullivan and Maizel. The 12.3 fragment in line 1 was isolated from an MSV transformed mink cell line. The entire MSV DNA genome is present, integrated between two mink regions designated *mink*L (left) and *mink*R (right). The boxed regions bracketing the integrated MSV are the 600 base pair repeats of MSV (see text). The arrows below each repeat indicate that they are direct, not inverted. The internal boxed region labeled mouse "scr" represents those specific *mouse* sequences picked up and carried by the MSV virus. The 6.2 kb fragment in line 2 was derived from that in line 1 after growth in *E. coli*. Presumably recombination occurred between the 600 base pair repeats deleting the entire MSV sequence. The 5.6 kb fragment in line 3 was obtained from normal mink cells using the line 1 fragment as a specific probe. This fragment presumably represents a site of integration for MSV. The 14 kb fragment in line 4 was obtained from normal mouse DNA using a bit of the MSV mouse "scr" DNA from the line 1 fragment as probe. Only the boxed region (labeled normal "sarc" as opposed to "src") was homologous to line 1 fragment and then only in the "src" region.

sequences to those of the parental fragment, but a single copy of the repeat is retained between them.

We used the mink sequences bracketing the integrated MSV as probes to locate this site of MSV integration in normal cells. The fragment was cloned and is shown in line 3, figure 9. It has identical left and right mink sequences and no evidence for anything between them. The dotted circle in the figure should contain the site of MSV integration. These clones (lines 1, 2, 3) will identify, among other things, the critical DNA sequences involved in the integration process.

The fragment from normal Balb/c mice harboring the information carried by MSV (the "src" sequence) was cloned using as probe a restriction fragment derived solely from the MSV "src" region (fig. 9, line 1). A single 14 kb fragment hybridized with MSV "src." This large cloned fragment contains only ~1 kb of DNA homologous to MSV (called "sarc"), and it hybridizes to the MSV "src" region alone and to no other part of the virus. What the remaining 13 kb encodes is unknown. The 14 kb fragment is not biologically active; i.e., it cannot transform. With the clones diagrammed in figure 9, we can begin to attack several important questions including how MSV integrates, how "sarc" was acquired by the M-MuLV genome, and how normal cell sequences acquire malignant potential.

Certainly these λgtWES hybrids now provide tools for further detailed analysis of large DNA viruses like HSV and small RNA viruses like the murine tumor viruses. It is safe to say that the coming year will be full of surprises, and λ vector systems like λgtWES will play an important role.

ACKNOWLEDGMENTS

I thank my colleagues who were instrumental in construction and testing of the λgtWES system: Dr. Philip Leder, Dr. David Tiemeier, Dr. Robert Weisberg, and Dr. Nat Sternberg. I gratefully acknowledge the patient advice and assistance of Dr. David Tiemeier, Dr. Shirley Tilghman, and Dr. John Seidman, who took time to teach me recombinant DNA methods. It has been a pleasure to share in the collaborative efforts of Dr. George Vande Woude's laboratory in using the λgtWES system for MSV and HSV experiments. I thank M. J. Madden, Drs. W. McClements, and K. D. Thompson for their critical review of this manuscript.

REFERENCES

Becker, A., and M. Gold. 1975. Isolation of bacteriophage lambda A-gene protein. Proc. Nat. Acad. Sci. USA 72:581–85.

Blattner, F. R., B. G. Williams, A. E. Blechl, K. Denniston-Thompson, H. E. Faber, L.-A. Furlong, D. I. Greenwald, D. O. Keifer, D. D. Moore, J. W. Schumm, E. L. Sheldon, and O. Smithies. 1977. Charon phages: safer derivatives of bacteriophage lambda for DNA cloning. Science 196:161–69.

Blattner, F. R., A. E. Blechl, K. Denniston-Thompson, H. E. Faber, J. E. Richards, J. L. Slightom, P. W. Tucker, and O. Smithies. 1978. Cloning human fetal ϒ-globin and mouse α-type globin DNA: preparation and screening of shotgun collections. Science 202:1279–83.

Cameron, J. R., and R. W. Davis. 1977. The effects of Escherichia coli and yeast DNA insertions on the growth of lambda bacteriophage. Science 196:212–15.

Enquist, L., and A. Skalka. 1973. Replication of bacteriophage λ DNA dependent on the function of host and viral genes. I. Interaction of *red*, *gam*, and *rec*. J. Mol. Biol. 75:185–212.

Enquist, L., and R. A. Weisberg. 1976. The red plaque test: a rapid method for identification of excision defective variants of bacteriophage lambda. Virology 12:147–55.

Enquist, L., D. Tiemeier, P. Leder, R. Weisberg, and N. Sternberg. 1976. Safer derivatives of bacteriophage λgt.λC for use in cloning of recombinant DNA molecules. Nature 259:596–98.

Enquist, L., and R. Weisberg. 1977. A genetic analysis of the *att-int-xis* region of bacteriophage λ. J. Mol. Biol. 111:97–120.

Enquist, L., and A. Skalka. 1978. Replication of bacteriophage lambda DNA. Trends in Biochem. Sciences. 3:279–83.

Enquist, L. W. and W. Szybalski. 1978. Coliphage λ as a safe vector for recombinant DNA experiments. *In* E. Kurstak and K. Maramorosch (eds.), Viruses and environment, pp. 625–52. Academic Press, New York.

Enquist, L. W., M. J. Madden, P. Schiop-Stansly, and G. F. Vande Woude. 1979. Cloning of Herpes simplex type 1 DNA in a bacteriophage lambda vector. Science 203:541–44.

Enquist, L., and N. Sternberg. 1979. *In vitro* packaging of a λ Dam vector and its use in recombinant DNA experiments. *In* R. Wu (ed.), Methods in enzymology. Academic Press, New York. (In press.)

Henner, W. D., I. Kleber, and R. Benzinger. 1973. Transfection of *Escherichia coli* spheroplasts. III. Facilitation of transfection and stabilization of spheroplasts by different basic polymers. J. Virol. 12:741–47.

Hershey, A. D. 1971. The bacteriophage lambda. Cold Spring Harbor Press, Cold Spring Harbor, New York.

Hohn, B., and K. Murray. 1977. Packaging recombinant DNA molecules into bacteriophage particles *in vitro*. Proc. Nat. Acad. Sci. USA 74:3259–63.

Leder, P., D. Tiemeier, and L. Enquist. 1977a. EK-2 derivatives of bacteriophage lambda useful in cloning of DNA from higher organisms: the λgtWES system. Science 196:175–77.

Leder, P., D. Tiemeier, S. Tilghman, and L. Enquist, 1977b. Use of an EK-2 vector for the cloning of DNA from higher organisms. *In* Molecular cloning of recombinant DNA, p. 205–17. Academic Press, New York.

Leder, P., S. M. Tilghman, D. C. Tiemeier, F. I. Polsky, J. G. Seidman, M. H. Edgell, L. W. Enquist, A. Leder, and B. Norman. 1978. The cloning of mouse globin and surrounding gene sequences in bacteriophage λ. Cold Spring Harbor Symp. on Quant. Biol. 42:915–20.

Mandel, M., and A. Higa. 1970. Calcium-dependent bacteriophage DNA infection. J. Mol. Biol. 53:159–62.

Maniatis, T., R. C. Hardison, E. Lacy, J. Laver, C. O'Connell, D. Quon, G. K. Sim, and A. Efstratiadis. 1978. The isolation of structural genes from libraries of eukaryotic DNA. Cell 15:687–701.

Murialdo, H., and H. Echols. 1978. Genetic map of bacteriophage lambda. Bacteriol. Rev. 42:577–91.

Sternberg, N., D. Tiemeier, and L. Enquist. 1977. *In vitro* packaging of a λ *D*am vector containing Eco R1 fragments of *E. coli* and phage P1. Gene 1:255–80.

Sternberg, N., D. Hamilton, L. Enquist, and R. Weisberg. 1979. A simple technique for isolation of deletion mutants of phage lambda. Gene, in press.

Thomas, M., J. R. Cameron, and R. W. Davis. 1974. Viable molecular hybrids of bacteriophage lambda and eukaryotic DNA. Proc. Nat. Acad. Sci. USA 71:4579–83.

Tiemeier, D., L. Enquist, and P. Leder. 1976. Improved derivative of a phage λ EK2 vector for cloning of recombinant DNA. Nature 263:526–27.

Tiemeier, D., S. Tilghman, F. Polsky, J. Seidman, A. Leder, M. Edgell, and P. Leder. 1978. A comparison of two cloned mouse β-globin genes and their surrounding and intervening sequences. Cell 14:237–45.

Tilghman, S. M., D. C. Tiemeier, F. Polsky, M. H. Edgell, J. G. Seidman, A. Leder, L. W. Enquist, B. Norman, and P. Leder. 1977. Cloning specific segments of the mammalian genome: bacteriophage λ containing mouse globin and surrounding gene sequences. Proc. Nat. Acad. Sci. USA 74:4406–10.

Vande Woude, G. F., M. Oskarsson, W. L. McClements, L. W. Enquist, D. Blair, P. Fischinger, J. V. Maizel, and M. Sullivan. 1979a. Characterization of integrated Moloney sarcoma proviruses and flanking host sequences cloned in bacteriophage λ. Cold Spring Harbor Symp. Quant. Biol., in press.

Vande Woude, G. F., M. Oskarsson, L. W. Enquist, S. Nomura, M. Sullivan, and P. Fischinger. 1979b. Cloning of integrated Moloney sarcoma proviral DNA sequences in bacteriophage λ. Proc. Nat. Acad. Sci. USA 76:4464–68.

MICHAEL SMITH

New Strategies for DNA Sequence Determination

3

INTRODUCTION

The first section of this article is a brief review of the development of the strategies for biological macromolecule sequence determination. Emphasis is on those developments that have been pivotal in advancing the range, sensitivity, and precision of experimental methods and that have in turn greatly expanded our understanding of the relationships of macromolecule structure and biological function. Although this review is selective, with a personal perspective, it does provide a background for a description of the remarkable present-day (ladder) methods for DNA sequence determination. It also provides a background for the strategies used in DNA sequence determinations carried out in the author's laboratory; these studies are the subject of the later sections of this paper.

PROTEIN AND RNA SEQUENCE DETERMINATION

The lineage of the development of methods for DNA sequence determination descends from protein to RNA to DNA sequences. This order is the reverse of the direction of the flow of genetic information as expressed in the conventional dogma of molecular biology, and reflects increasing technical difficulties in macromolecule isolation and purification as one passes from protein to RNA to DNA. The basic strategy of protein sequence determination was developed by Sanger (1959) in his attack on the sequence of beef insulin. Although there have been continual refinements in the methodology of protein sequence determination (Hirs

Department of Biochemistry, University of British Columbia, 2075 Westbrook Place, Vancouver, British Columbia V6T 1W5 and Timasheff, 1977), the same strategy is evident in the determination of the largest complete sequence of a protein that is presently known, the 1,021 amino acid sequence of β-galactosidase from *E. coli* (Zabin and Fowler, 1978). That strategy is to fragment the protein in two or more separate experiments using enzymatic or chemical methods that cleave at specific but different peptide bonds (specific fragmentation). The fragments obtained in each of these experiments are separated, purified, and their amino acid sequences determined by stepwise degradation. Unique overlaps between the sequences of different fragments are next established, leading to the complete sequence of the protein.

This same basic strategy was employed for determination of small RNA molecules, principally t-RNAs (Holley, 1968). A significant departure from the methodology of protein sequence determination was made possible by the presence of phosphorus in each nucleotide of an RNA. Radioactive labeling, using ^{32}P introduced *in vivo* or *in vitro* led to an enormous increase in the sensitivity of the methodology, and also permitted the development of an important ion-exchange chromatographic method, homochromatography, where separation of ^{32}P-labeled RNA fragments is induced by an excess of unlabeled RNA fragments (Brownlee and Sanger, 1967, 1969). The direct extension of the basic strategy of protein sequence determination to t-RNAs was possible because of their small size (70–90 nucleotides per molecule) and the availability of endoribonucleases that cleave in a base-specific way adjacent to pyrimidine or guanine nucleotides. The occurence of modified nucleosides in t-RNAs (Hall, 1971) also was significant because the modified nucleosides are essential to the identification of the specific location of different fragments within the over-all structure of the t-RNA. Larger RNAs contain, in the main, only the four nucleosides, adenosine, cytidine, guanosine, and uridine. This has made the determination of longer sequences much more difficult for two interrelated reasons: (1) a given sequence (oligonucleotide) will occur more than once in an RNA, and (2) it is difficult to establish unique overlaps between two sets of fragments. Simply stated, it is more difficult to establish by specific fragmentation the sequence of an RNA built of permutations of four monomers than it is to establish that of a protein containing permutations of 20 monomers. Despite this, some very long RNA sequences have been determined, principally for bacteriophage and *E. coli* ribosomal RNA, but also for mRNA. (Weissman et al., 1973; Fiers et al., 1976; Fellner, 1974; Platt, 1978). The problems inherent in such structural determinations are illustrated by recent revisions in *E. coli* ribosomal RNA sequences (Woese et al., 1975; Ehresmann et al., 1977; Brosins et al., 1978).

EARLY STRATEGIES FOR DNA SEQUENCE DETERMINATION

Initial studies on DNA sequences used the strategy of specific fragmentation, the two principle methods being degradation with diphenylamine-formic acid to produce pyrimidine oligonucleotides (Burton and Petersen, 1960) and degradation using the nucleotide-specific T4 endonuclease IV (Sadowski and Bakyta, 1972; Ling, 1971). The most important technical development in this phase of DNA sequence determination was a new method for direct determination of the sequence of an oligodeoxyribonucleotide from the positions of its partial degradation products after two-dimensional electrophoresis-homo-chromatography (Ling, 1972). This is possible because there are only four nucleotides, each with a different pK in the acid range. Hence, movement in the electrophoretic dimension is a reflection of the relative base composition of an oligonucleotide. Mobility in the homo-chromatography dimension is primarily a reflection of the length of the oligonucleotide. The advantages of this direct reading (wandering spot) method are speed and sensitivity. The method has been a very useful technique, subject to some ambiguities, for the determination of the sequences of fragments of both RNA and DNA. There is one unrelated but very important point to be made about these early determinations of DNA sequence: they provided rigorously established DNA sequences that were essential controls for the development of the more recent one-dimensional gel-electrophoretic (ladder) methods for nucleic acid sequence determination. It should be noted at this point that a general technical principle is common to the wandering spot and ladder sequencing strategies. Both methods employ ^{32}P-labeled substrates; if the label is distributed throughout the fragment under investigation, then the homologous series of shorter fragments required by the sequencing method, which all must have one end (5' or 3') in common, cannot be contaminated with fragments (other than mononucleotides) derived from internal regions. This means that the set of fragments has to be generated by kinetically controlled partial exonucleolytic degradation from the distant end or synthetically by starting synthesis at the fixed point (5' end). When a fragment whose sequence is being determined is ^{32}P-labeled at one end, then the detectable set of shorter fragments also can be generated by chemical or enzymatic random single-hit endonucleolytic cleavage. Mention in the preceeding paragraph of the possibility of generating specific ^{32}P-labeled fragments by *in vitro* synthesis brings us to an important strategy available to RNA and DNA sequence determination. It is consequent on the ability of a single strand of a polynucleotide to act as a template for the enzymatic synthesis of a ^{32}P-labeled complementary strand. An early application of this approach was in the sequence determination of bacteriophage Qβ RNA using its own replicase as the synthesizing enzyme (Weissmann et al., 1973). RNA sequence can also be derived by synthesis of complementary DNA using either a DNA polymerase or a reverse transcriptase to generate the complementary sequence (Proudfoot et al., 1976; Marotta et al., 1976). Extensive use of *E. coli* RNA polymerase to produce specific RNA fragments, *in vitro*, using DNA templates has resulted in important DNA sequence information. (Reznikoff and Abelson, 1978). Particularly important, historically and because of applications in the ladder methodologies, were the development of synthetic sequencing methods using *E. coli* DNA polymerase I and T4 DNA polymerase with DNA primer-templates (Wu and Kaiser, 1968; Englund, 1972). All these *in vitro* synthetic methods have the common feature of generating a relatively short ^{32}P-labeled nucleic acid fragment whose sequence is determined by nearest-neighbor or wandering-spot techniques. These methods are still very useful, indeed essential, in the determination of the sequences at the termini of nucleic acids. However, it is clear from the above discussion that methods for reliable determination of extended sequences in one step are required for an all-out assault on the major problems of DNA sequence.

RAPID DNA SEQUENCE DETERMINATION: THE LADDER METHODS

During experiments where a chemically synthesized oligodeoxy-ribonucleotide was used as a primer for DNA polymerase with ϕX174 viral DNA as template where one deoxyribonucleotide 5'-triphosphate was present at low concentration, it was observed that certain fragments accumulated because the next nucleotide that ought to be added, as determined by the template DNA, was derived from the limiting deoxyribonucleoside 5'-triphosphate (F. Sanger, personal communication). This observation was used to develop the "minus" method of DNA sequence determination (Sanger and Coulson, 1975). This method involves a two-step set of enzymatic reactions. First, a primer, in complex with the complementary template DNA whose sequence is being determined, is extended by *E. coli* DNA polymerase I in the presence of all four deoxyribonucleoside 5'-triphosphates (one of which is ^{32}P-labeled). By kinetic control a complete mixture of products representing the primer extended by one nucleotide up to a few hundred nucleotides is produced. The unreacted deoxyribonucleoside 5'-triphosphates are then removed. The ^{32}P-labeled mixture of fragments (all with the same 5'-end, the 5'-terminus of the primer), still complexed with the DNA template, is treated with DNA polymerase in the presence of only three deoxyribonucleoside 5'-triphosphates. Any given fragment is extended until the missing nucleotide should be added, at which point synthesis stops. Thus, the total mixture of fragments is converted into a mixture whose 3'-ends define the positions where the missing nucleotide should be added. This procedure is repeated in three further reactions with a different nucleotide missing in turn. The four sets of products are analyzed in parallel by acrylamide gel electrophoresis, which can resolve a wide range of fragment sizes, their mobilities being inversely related to fragment length. Hence, the position and identity of the missing nucleotides and, therefore, the sequence of the newly synthesized nucleic acid can be determined by examining the four parallel patterns of DNA fragments that are revealed by autoradiography.

The "minus" method does not always provide all the sequence of a given DNA tract for two technical reasons: (1) because of the uneven kinetics of DNA synthesis, not all fragments are produced in equivalent amount, and (2) the method only defines the first nucleotide in a tract containing two or more residues of the same mononucleotide. To overcome this problem, the "plus" method was developed (Sanger and Coulson, 1975). In this procedure the initial set of ^{32}P-labeled fragments is again subdivided into four experiments that contain the template DNA and only one of the four possible deoxyribonucleoside 5'-triphosphates together with T4 DNA polymerase. This enzyme contains a powerful 3' → 5' exonuclease activity (Englund, 1972). Consequently, the combination of polymerase and exonuclease in the presence of one deoxyribonucleoside 5'-triphosphate results in a set of fragments, with a common 5'-end (the 5'-end of the primer), 3'-terminated with the nucleotide corresponding to the triphosphate. The "plus" experiments complement the "minus" experiments in identifying nucleotides and also define the last nucleotide in a tract of identical residues. The exact length of such a tract is defined by using a graticule provided by parallel gel electrophoretic separation of the original mixture of ^{32}P-labeled fragments alongside the four "minus" and the four "plus" experiments (Brown and Smith, 1977).

The "plus and minus" method is described in some detail because of its intrinsic ingenuity and because of its germinal role. Its simplicity and rapidity allow the determination of the majority of the sequence of the 5,386 nucleotides of ϕX174 DNA (Sanger et al., 1978). This was the first complete sequence of a DNA genome and resulted in the discovery of previously unsuspected overlapping genes.

An entirely different approach to rapid DNA sequence determination is provided by base-specific chemical cleavage of an end-labeled DNA strand (Maxam and Gilbert, 1977, 1979). Again, simultaneous fractionation of the four sets of products by gel electrophoresis allows the unambiguous identification of each of the nucleotides in a DNA sequence. One particularly useful feature is that the method provides specific identification of each nucleotide in a tract of identical residues.

A second enzymatic method has been developed more recently (Sanger et al., 1977). This also allows positive identification of all members of a tract of identical residues. The method uses a primer template combination with *E. coli* DNA polymerase I. Sets of fragments whose 3'-ends correspond to a specific nucleotide are generated in four separate reactions by using chain terminating analogs of deoxynucleoside 5'-triphosphates. The 2', 3'-dideoxynucleoside 5'-triphosphates are used more commonly, although arabinoside 5'-triphosphates also are effective (Sanger et al., 1977). The enzymatic synthesis is carried out in the presence of all four of the normal deoxyribonucleoside 5'-triphosphates. In each of the four parallel reactions, only one analog is present at a concentration where it chain terminates at an efficiency of about 1%. Apart from its technical simplicity, the method is also effective because it does not involve kinetic control and hence produces fairly equivalent amounts of fragments.

In many cases the choice between using the chemical cleavage or the enzymatic terminator method is a matter of personal preference. Under equivalent conditions of gel-electrophoresis to generate ladder patterns, the two methods produce equivalent amounts of sequence with similar accuracy. The chemical method is easy for the neophyte to set up. The enzymatic method is simpler technically; however, care must be exercised in the use of enzymes that are variable in activity and purity. Strategies for use of the terminator method have been developed that avoid the need for preparation of a single-stranded template and that allow the use of duplex DNA fragments as substrates for sequence determinations (Maat and Smith, 1978; Smith, 1979 a, b).

The most significant development in DNA sequencing methodology beyond the procedures described above has been the use of very thin acrylamide gels to generate the ladder patterns (Sanger and Coulson, 1978). Their use increases the resolution of the autoradiographic detection of fragments at least twofold. It is also important to carry out gel-electrophoresis under conditions where secondary structures, due to hairpin structures formed by inverted repeats (palindrones), are denatured. Although 7M urea is used in the buffer, it is also important to carry out the electrophoresis at high temperature (Brown and Smith, 1977).

It is of interest that analogs of the methods for DNA sequence determination have now been developed for RNA sequences. These include a chemical cleavage method (Peattie, 1979) and an enzymatic terminator method (Zimmern and Kaesberg, 1978; McGeoch, 1979). A novel modification of the enzymatic method, which has been applied to RNA sequence determination, is to substitute a hypoxanthine triphosphate for the guanine triphosphate. This is significant because the I-C base pair is less stable than the G-C base pair; consequently, I-containing inverted repeats form less stable hairpin duplex structures that can easily be denatured during gel electrophoresis (Mills and Kramer, 1979). Presumably the same strategy will be beneficial in DNA sequence determinations using the enzymatic method.

The use of a synthetic oligodeoxyribonucleotide primer in DNA sequence determination was described above, and such synthetic oligodeoxyribonucleotides have been used extensively in DNA sequence determinations (Wu et al., 1978). However, the most important source of fragments of DNA used in sequence determination is restriction endonuclease fragmentation. The properties and uses of restriction endonucleases have been reviewed recently (Smith, 1979), and are also discussed in detail elsewhere in this volume (Gingeras, Enquist). At this point it is appropriate to list the uses of the enzymes important to DNA sequence determinations:

1. In making recombinant DNA, particularly plasmid clones, which are the most convenient sources of DNA for sequence determination for genomes more complex than the more simple bacteriophage and viruses (the self-cloning of mitochondrial DNA in the petite mutants of yeast is an interesting exception to the need for man-made recombinants; see Perlman et al., this volume). The recombinant DNA is essential for the sequencing of segments of complex genomes because it both reduces the background and also provides pmole amounts of DNA required by the methodologies.

2. To provide small fragments suitable for direct sequence determination or for use as specific primers of DNA polymerase.

3. To specifically cleave long primer fragments prior to the analysis of the products of enzymatic termination sequence detemrinations.

4. As independent checks on the accuracy of DNA sequences.

An additional use of restriction endonucleases is to define the location of some methylated bases, which occur in DNAs at a low frequency at specific sites (Hall, 1971; Waalwijk and Flavell, 1978; Lui et al., 1979; Rae and Steele, 1979).

A completely different and very promising technique for characterization of all types of modified bases in DNA involves the use of mass-analyzed ion kinetic energy spectrophotometry (Schoen et al., 1979). 5-

methylcytodine can be detected by its failure to cleave in the chemical method of sequence determination (Ohmori et al., 1978).

One further technique that is an essential component of DNA sequence determination is the use of a computer for data processing. A variety of programs have been written to facilitate editing, searches, and comparisons of DNA sequences (McCallum and Smith, 1977; Staden, 1977, 1978, 1979; Korn et al., 1979; Gingeras, this volume).

The discussion to this point has sought to present a summary of the development of the modern rapid methods of DNA sequence determination. The next sections will deal with some specific problems of DNA sequence determination that illustrate applications of various strategies and that provide some insights into the relationship between DNA sequence and biological function.

AUTONOMOUS (NON-DEFECTIVE) MAMMALIAN PARVOVIRUS DNA SEQUENCES

The DNA of this class of viruses, which is linear and single-stranded, contains about 4,500 nucleotides (Tattersall and Ward, 1978). The viral DNA is complementary (i.e., the minus strand) to the RNA transcript and hence provides a convenient probe for viral gene expression. transcript and hence provides a convenient probe for viral gene expression. The initial transcript corresponds to about 95% of the DNA, and subsequent processing results in excision of about 30% of the transcript from an internal region near the 5'-end (Tal et al., 1979). The initial translation product is a protein (later modified by proteolysis) that corresponds in size to the coding capacity of the mature transcript (Tattersall, 1978; Tal et al., 1979). Each end of the viral DNA is a duplex hairpin structure containing 100 to 200 nucleotides (Astell et al., 1979a). Viral replication, which involves linear polymeric duplexes, is dependent on the S phase of the host cell cycle (Berns and Hanswirth, 1978). All these biological properties make the autonomous mammalian parvoviruses attractive targets for studies on the relationship between DNA sequence and eukaryote DNA replication and gene expression. The availability of several viruses that are related but serolgically and biologically distinct (Tattersall and Ward, 1978) allows the possibility of assigning biological function by comparison of DNA sequences. Because the viral DNA is the minus strand, sequence determination starting at the 3'-end not only provides information on the sequence involved in the initiation of double-stranded replicative DNA synthesis, but also defines the sequences involved in the initiation of transcription and translation. Our first studies, therefore, have been directed at the sequence of this region of the viral

DNAs of minute virus of mice (MVM), Kilham rat virus (KRV), and two hamster viruses (H-1 and H-3). The technical problems in carrying out the sequence determinations lie in the following areas: (1) identification of the nature of the 3′-end of the DNAs, (2) unique labeling of the 3′-end at high specific activity, (3) unambiguous determination of the sequence within the stable hairpin duplex, and (4) further extending the sequences in the absence of sites for specific end-labeling and with only small amounts of DNA being available.

The logical start to sequencing a DNA at its 3′-terminus is specific 3′-labeling. In the case of a DNA with a 3′-terminal hairpin, this can be achieved using a single ^{32}P-deoxyribonucleoside 5′-triphosphate and a primer-template dependent DNA polymerase. In the case of the parvoviruses, labeling was much less than stoichiometric, and attempts to improve yields by manipulating reaction conditions resulted in nonspecific labeling consequent to endo- and exo-nucleolytic degradation of viral DNA (Astell et al., 1979a). Comparative studies, using T4 DNA polymerase and an exonuclease-free reverse transcriptase together with different ^{32}P-deoxyribonucleoside 5′-triphosphates revealed that the 3′-end of the DNA of parvoviruses is heterogeneous (Astell et al., 1979b). Presumably, this is due to exonuclease action subsequent to viral DNA synthesis.

Resolution of this problem allowed determination of the first few nucleotides at the 3′-end of the four viruses by the wandering-spot method (fig. 1), followed by further sequences using the chemical ladder method. This was technically difficult because of the great stability of the hairpin duplex. It was only possibly to resolve the GC-rich sequence at the apex of the hairpin and also the sequence at the distal and of the hairpin by carrying out gel electrophoreses at temperatures close to 100° (fig. 2).

The DNA sequence immediately adjacent to the 3′-hairpin was determined using a synthetic oligodeoxyribonucleotide, d(pTTC-TAAAAA), as a primer for sequence determination by the enzymatic terminater method (fig. 3). The intrinsic primer-template of the viral DNA was inactivated by the double-strand specific exonuclease III of *E. coli* (Astell et al., 1979b). The oligodeoxyribonucleotide was synthesized enzymatically using *E. coli* polynucleotide phosphorylase in the presence of $MnCl_2$. This method is particularly convenient for obtaining relatively simple, but unique, oligodeoxyribonucleotide sequences that occur quite frequently in DNA. The sequence determination revealed a *Hin*fI restriction endonuclease site in the vicinity of nucleotide 230 in all the viral DNAs except that of H-I. This allowed sequence determination by the enzymatic terminator method using viral DNA as a primer-template

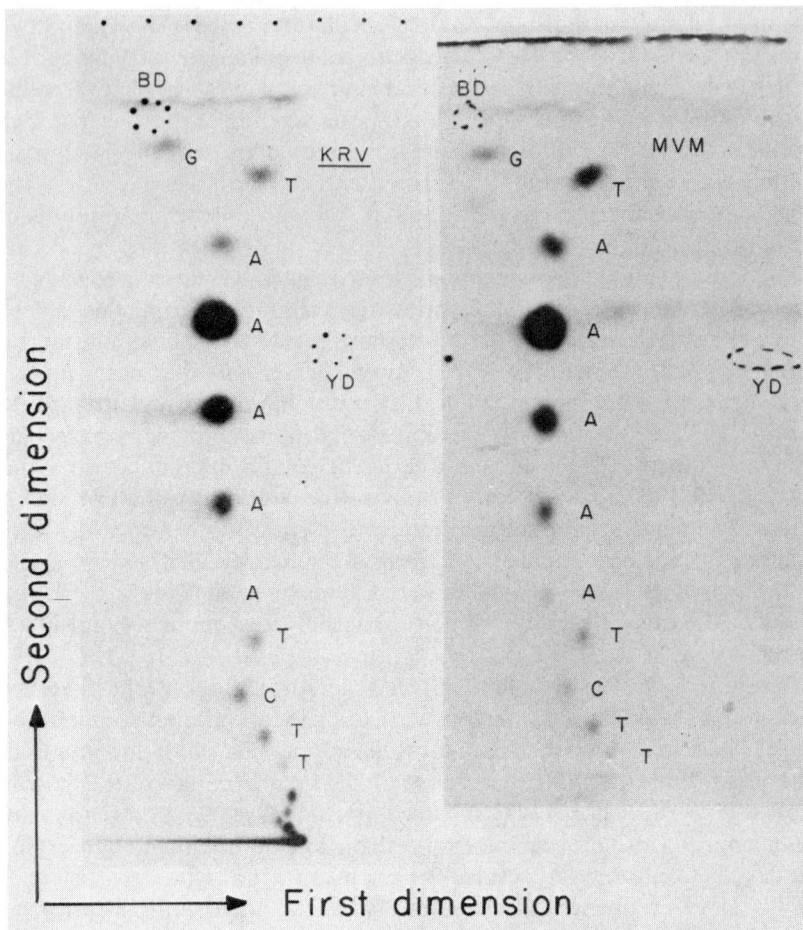

Fig. 1.   Wandering-spot sequencing of 3'-end-labeled KRV and MVM DNAs (Astell et al., 1979a). The DNAs were end-labeled using ^{32}P-dGTP and T4 DNA polymerase and the products partially digested with endonuclease P1 (Astell et al., 1979a). The sequences were analyzed using the two-dimensional electrophoresis-homochromatography method (Brownlee and Sanger, 1969). The markers are blue dye (BD) and yellow dye (YD).

followed by *Hin*fI cleavage (fig. 4). The corresponding sequence in H-I was determined using the synthetic primer d(pTTTCATTT). An *Eco*RII site in all four DNAs, near nucleotide 340, allowed extension of the sequence to near nucleotide 450 (Astell et al., 1979c). The sequences of the four DNAs are shown in figure 5.

Comparison of the four sequences provides considerable insight into the biological function of this region of the viral DNAs. The region from nucleotides 250 to 430 is typified by differences in sequence corresponding to nucleotide substitutions, and these differences are regularly spaced at three nucleotide intervals. The most obvious interpretation is that the region codes for a polypeptide whose amino acid sequence is conserved with silent nucleotide changes at the wobble position of the codons. Examination of the RNA complementary to the DNA predicts an initiation codon at nucleotide 260. Assuming that the transcript start is in the region of nucleotides 150–250 (Tal et al., 1979), the proposed translation initiation is in accord with the model described by Kozak (1978).

Precise assignment of the position of the start of transcription awaits characterization of the initial transcript (Tal et al., 1979). The sequence in the region 175–220 is similar to putative eukaryote promoter sequences (Smith et al., 1979; Astell et al., 1979b), and this would predict a transcript start in the region of nucleotide 220. It is of interest that this region of the viral DNAs not only has differences equivalent to nucleotide substitutions but also differences resulting from insertions or deletions (fig. 5).

The DNA sequences allow quite specific predictions about the detailed structure of the 3'-terminal hairpins, assuming that there is maximal Watson-Crick pairing (fig. 6). These predicted structures are supported by the behavior of the DNA fragments in the sequencing experiments (fig. 2), and also by the products of partial digestion by a single-stranded specific nuclease (fig. 7). Although the function of the hairpin is not yet understood, the strong conservation of sequence and structure suggests that there is some major biological significance, presumably connected either with the initiation of synthesis of the complementary DNA strand or with the production of mature viral DNA.

The above results provide a strong incentive to complete the sequence determination of the genome of at least one viral DNA, and so define the nature of the intervening sequence, of the coding region, and of the 5'-hairpin structure and their relationship to viral and host cell biology.

SEQUENCE OF YEAST DNA AT THE CYC1 AND CYC7 LOCI

An indirect yet completely reliable method, the comparison of peptide sequences from frameshift mutants, was used to determine the sequence of 42 nucleotides at the CYC1 locus of the yeast genome that codes for iso-1-cytochrome c (Stewart and Sherman, 1974). This made it possible to synthesize a specific oligodeoxyribonucleotide probe (Gillam et al., 1977) that was used to monitor the isolation, as recombinant DNA, of the fragment of yeast DNA containing the CYC1 locus (Montgomery et al., 1978). Two new strategies were developed, using the enzymatic terminator

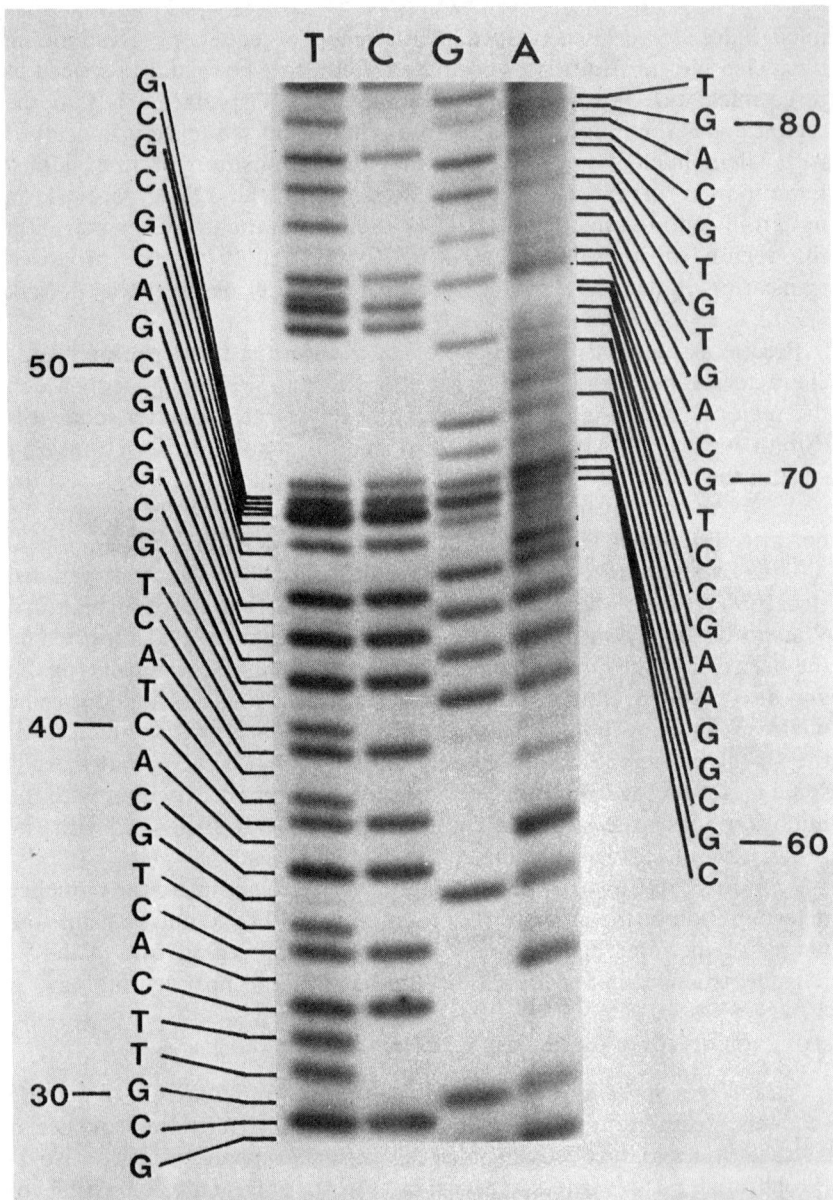

Fig. 2. Ladder sequencing by chemical cleavage (Maxam and Gilbert, 1977) of part of the 3'-end of KRV DNA (Astell et al., 1979a) from nucleotides 28 to 81; (a) products analyzed on a 12% polyacrylamide gel (0.5 mm thick; Sanger and Coulson, 1978) at 30 watts for 4 hr, (b [*opposite*]) an identical experiment with electrophoresis at 50 watts for 2 hr. The higher power

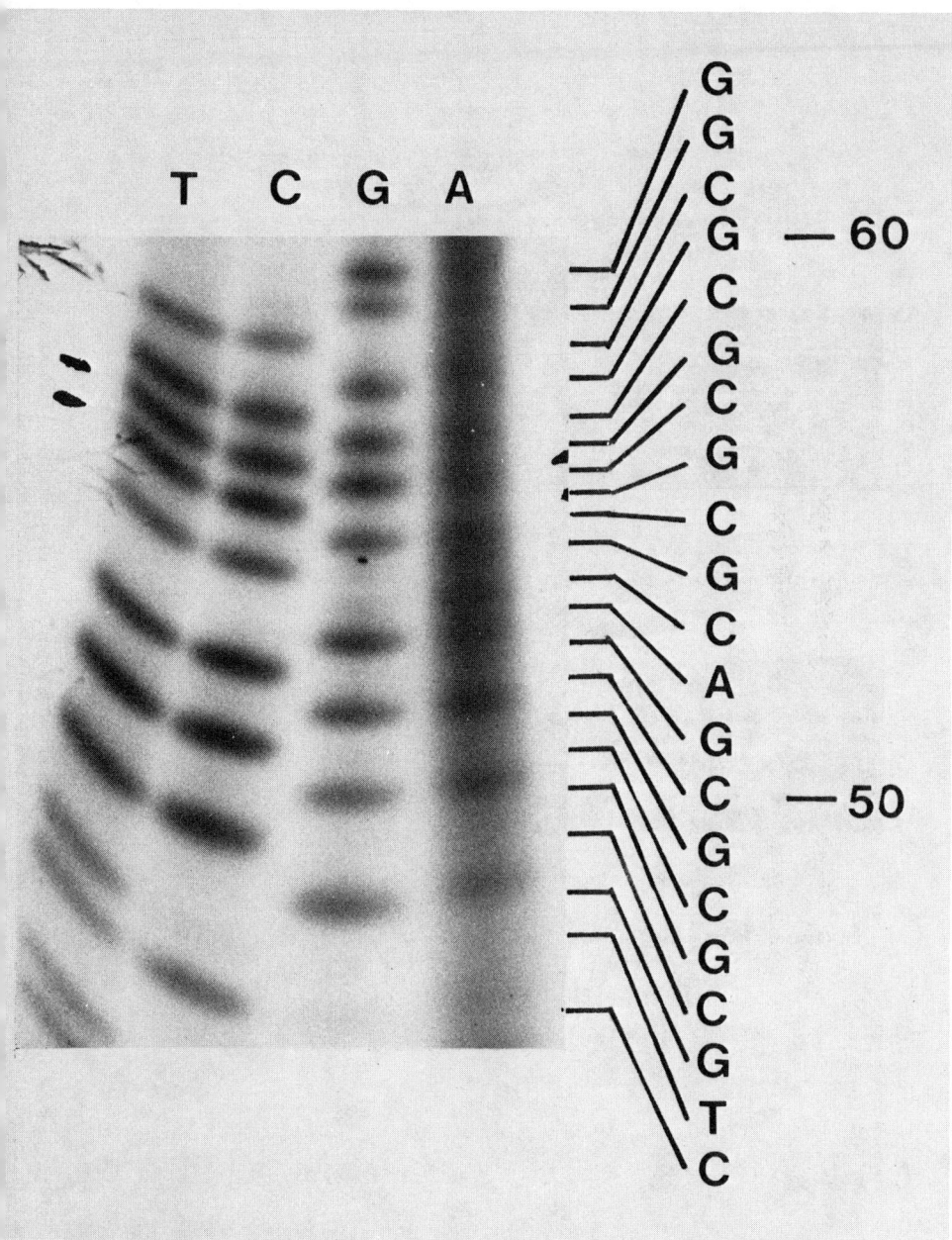

produced a temperature of >90° in the gel, resulting in complete resolution of nucleotides 53 to 61.

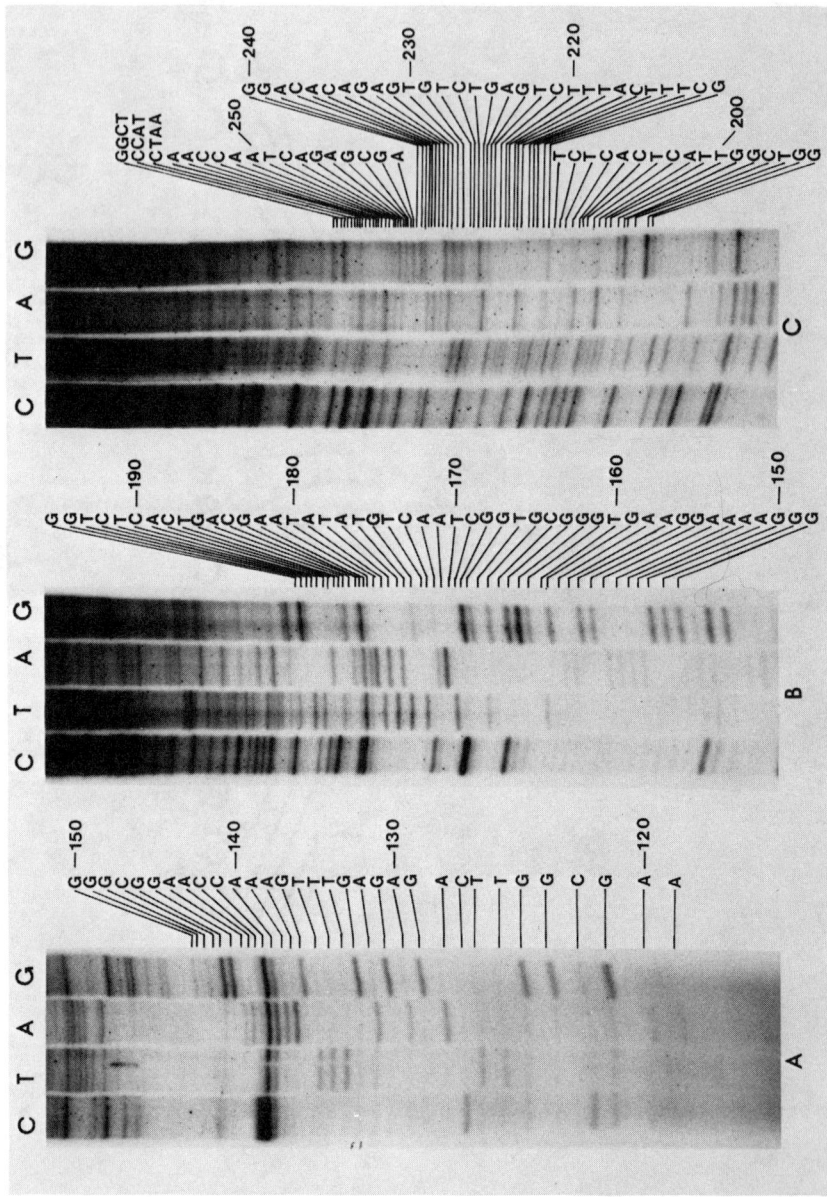

Fig. 3. Ladder sequencing by the enzymatic terminator method (Sanger et al., 1977) of nucleotides 119–267 of KRV DNA. The viral DNA was treated with E. coli exonuclease III to prevent endogenous priming, and the synthetic oligodeoxyribonucleotide d (pTTCT-AAAAA) was used as the primer to

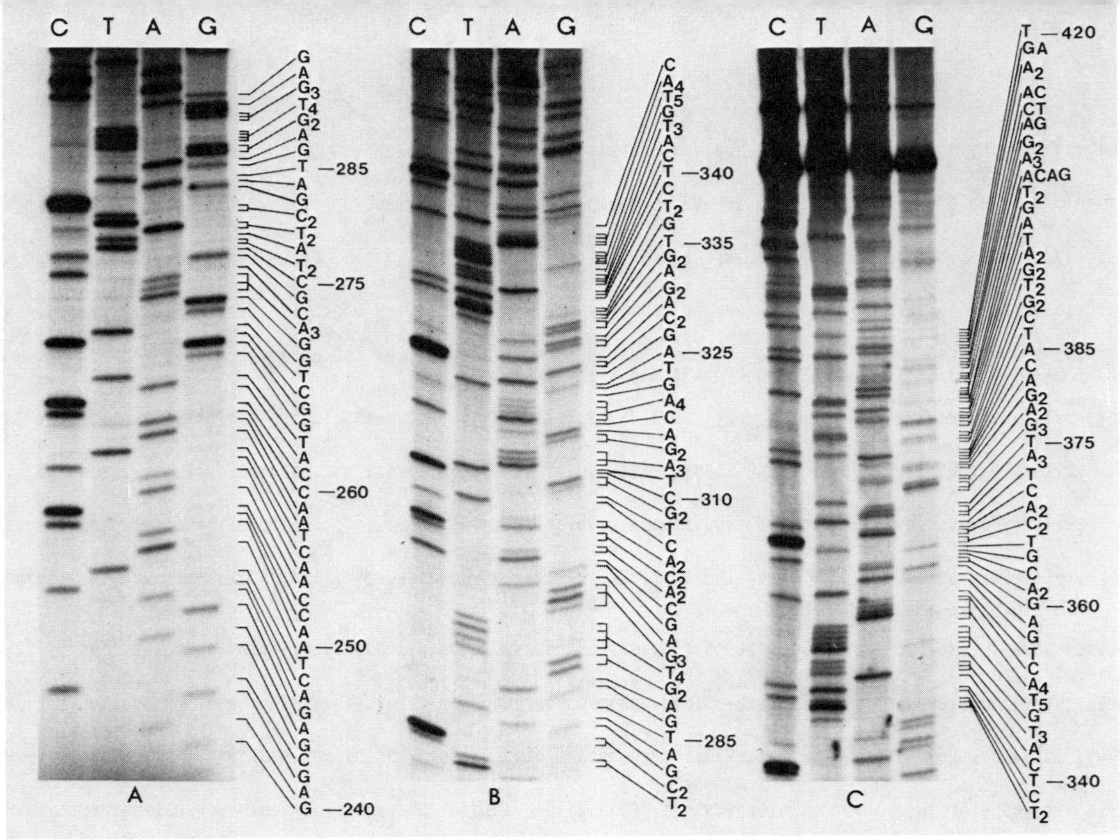

Fig. 4. Ladder sequencing by the enzymatic terminator method of nucleotides 240–420 of KRV DNA, using the 3' end of the hairpin as primer; fragments were released by digestion with *Hin*FI: (A) 12% polyacrylamide gel, (B) 8% gel, and (C) 6% gel.

KRV 3'-TAAAAATCTTGACTGGTTGGTACAAGTGCGTTCACTGCGCACTACTGCGCCGGCGCGGAAGCCTGCAGTGTGCAGTGAA CGCAAGTGTACCAACCAGTCAAGATTTTTACTAT
                               FnuDII,HhaI FnuDII,HhaI

H-1 3'-TAAAAATCTTGACTGGTTGGTACAATGCCGTTCACTGCGCACTACTGCGCCGGCGCGGAAGCCTGCAGTGTGCAGTGAATCGCAAAGTGTACCAACCAGTCAAGATTTTTACTAT
                               FnuDII,HhaI FnuDII,HhaI

H-3 3'-TAAAAATCTTGACTGGTTGGTACAAGTGCGTTCACTGCACTACTGCGCCGGCGCGGAAGCCTGCAGTGTGCAGTGA TCGCAAAGTGTACCAACCAGTCAAGATTTTTACTAT
                               FnuDII,HhaI FnuDII,HhaI

MVM 3'-TAAAAATCTTGACTGGTTGGTACAAGTGCATTCACTGCACTACTGCGCCGGCGCGGAAGCCTGCAGTGTGCAGTGAAT GCAAAGTGTACCAACCAGTCAAGATTTTTACTAT
                               FnuDII,HhaI FnuDII,HhaI

KRV TCGCCAAGTCTCTCAAACTTTGGTTCCGCCCTTTTCCTTCACCC GCACC GATTGACATATATTCGTCAGTGAGACC AGCCAATGAGT GAGACGAAAGTAAAG ACTCAGACACTC
                                                                                                                HinfI

H-1 TCGCCAAGTCTCTCAAACTTTGGTTCCGCCCTTTGCCTTCACCCCGCACC GATTGACATATATCCGTCAGTGAGACC AGCCAATGAGT GAGACGAAAGTAAAG ACTCAACACTC

H-3 TCGCCAAGTCCCCTCAAACTTTGGTTCCGCCCTTTTCCTTCACCC GCACC GATTGACATATATTCGTCAGTGAGT AGCCAATGAGT GAGATGAAAGTAAAG ACTCAGACACTC
                                                                                                               HinfI

MVM TCGCCAAGTCCCCTCAAA TTTGGTTCCCGCGCTTTTCCTTCACCC GCACCAAATTT CATATATTCGTTGATGACTTC AGTCAATGAATAGAAATGAATAGAAAGTAA GACACTCAG CTC
                  HhaI,FnuDII   HinfI TaqI

            (Met)Ala Gly Asn Ala Tyr Ser Asp Val Val Leu Gly Ala Thr Asn Trp Lys Asp Lys Ser Gln
    5'....UAACCAACUAACCAUG GCU GGA AAC GCU UAU UCC GAU GUG GUU UUG GGA GCA ACC AAC UGG CUA AAG GAC AAA AGU AGC CAG

KRV T GTGTCCTCGCTCTGATTGGTAC CGA CCT TTG CGA ATA AGG CTA CTC CAA AAC CCT CGT TGG TTG ACC GAT TTC CTG TTT TCA TCG GTC
                                                                                                                                   EcoRII

H-1 T GTGTCCTCGCTCTGATTGGTAC CGA CCT TTG CGA ATG AGG CTA CTC CAA AAC CCT CGT TGG TTG ACC CAT TTC CTG TTT TCA TCG GTC
                                                                                                                                   EcoRII

H-3 T GTGTCCTCGCTCTGATTGGTAC CGA CCT TTA CGA ATG AGG CTA CTC CAA AAC CCT CGT TGG TTG ACC GAC TTC CTG TTT TCA TCG GTC
                                                                                                                                   EcoRII

MVM TGCGTGTCTTT CTCTCATTGGTTGATTGGTAC CGA CCT TTG CGA ATG AGA CTT AGA CTA CGA ATG AGA CTT CAA AAC CCT CGT TGG TTG ACC AAT TTC CTT TTT TCA TTG GTC
    HgaI   EcoRII

```
                                                                      360
        Glu  Val  Phe  Ser  Phe  Val  Lys  Thr  Asn  Val  Gln  Leu  Asn  Gly  Lys  Asp  Ile  Gly  Trp  Asn  Ser  Tyr  Arg  Lys
        GAG  GUG  UUC  UCA  UUU  GUU  AAA  ACU  AAC  GUC  CAA  CUA  AAU  GGG  AAG  GAC  AUC  GGU  UGG  AAU  AGU  UAC  AGA  AAG
KRV     CTC  CAC  AAG  AGT  AAA  CAA  TTT  TGA  CTC  GTT  CAG  GTT  TTA  CCC  TTC  CTG  TAG  CCA  ACC  TTA  TCA  ATG  TCT  TTC
        MnlI
H-1     CTC  CAC  AAG  AGT  AAA  CAA  TTT  TTA  CTT  TTA  CTC  GAT  GAT  CCC  TTC  CTG  TAG  CCA  ACC  TTA  TCA  ATG  TCT  TTC
        MnlI
H-3     CTC  CAC  AAG  AGT  AAA  CAA  TTT  TTA  CTC  TTA  CTC  GAT  GAT  CCC  TTC  CTG  TAG  CCA  ACC  TTA  TCA  ATG  TCT  TTC
        MnlI
MVM     CTT  CAC  AAG  AGT  AAA  CAA  TTT  TTA  CTT  CTA  CAA  GTT  GAC  CCT  TTA  CTA  TAG  CCT  TTA  TCA  ATG  TTT  TTT
        MnlI

                          420
        Glu  Leu  Gln  Asp
        GAG  CUA  CAA  GAU  G
KRV     CTC  GAT  GTT  CTA  C
        AluI
H-1     CTC  GAT  GTT  CTA  C
        AluI
H-3     CTC  GAT  GTT  CTA  C
        AluI
MVM     CTC  GAC  GTC  CTC  C
        AluI PstI MnlI
```

Fig. 5. Sequences of the 3' ends of the DNAs of KRV, H-1, H-3, and MVM (Astell et al., 1979 a, b, c). Overlined nucleotides represent differences compared with KRV, and gaps indicate nucleotides not present in that DNA. The putative mRNA and polypeptide sequences for KRV are shown.

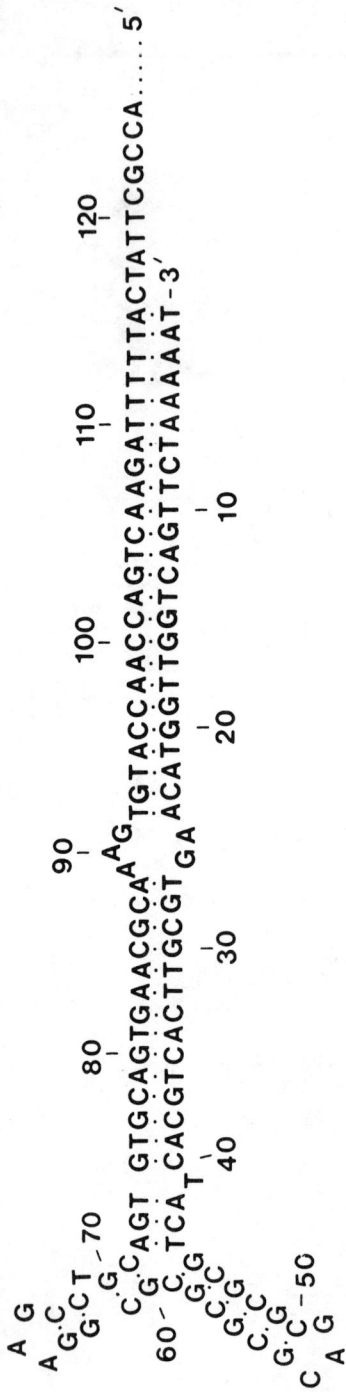

Fig. 6. The hairpin structure deduced for the 3' end of KRV DNA (Astell et al., 1979a, b) together with the differences in H-1, H-3, and MVM DNAs.

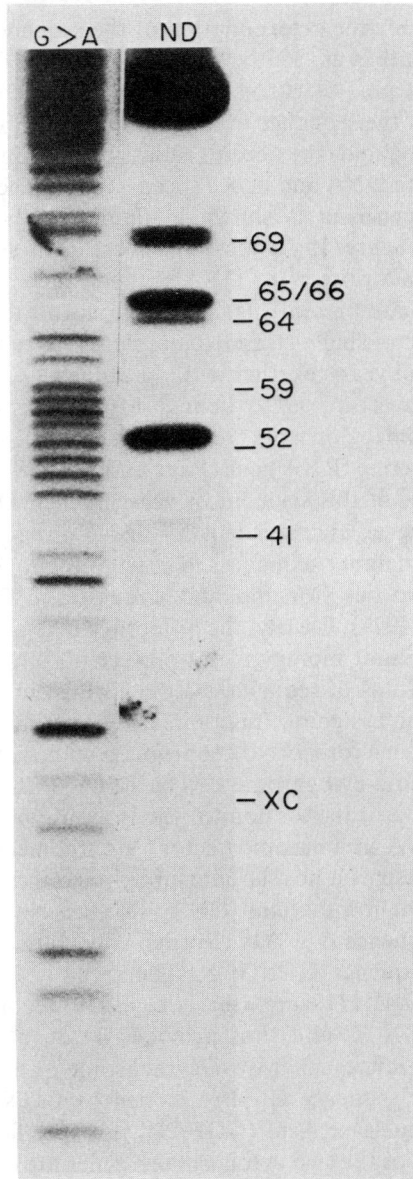

Fig. 7. Alignment of fragments produced by partial digestion of 3'-end-labeled MVM DNA by the single-stranded specific mung bean endonuclease alongside a ladder sequencing gel (for the G > A chemical cleavage reaction). The numbers indicate the size of the 3'-end-labeled fragments, and XC denotes the Xylene Cyanol marker dye (Astell et al., 1979a).

method, to simplify the determination of the sequence of the DNA at the CYC1 locus (Smith et al., 1979). The first strategy is diagrammed in figure 8, and makes it possible to use plasmid vector DNA as a primer for determination of the sequence of a recombinant DNA. The novelty of the strategy is that it avoids the need to isolate restriction fragments or single-stranded template DNA and hence speeds up sequence determination; the result of an experiment is shown in figure 9. The second strategy is diagrammed in figure 10; it allows the use of a synthetic oligodeoxy-ribonucleotide as a primer for DNA sequence determination, when both strands of DNA are present (fig. 11). Here, again the method avoids the need for isolation of single-stranded template DNA. The total sequence of the CYC1 locus of yeast together with the adjacent 5′-sequence is shown in figure 12 (Smith et al., 1979; Leung, Montgomery et al., 1979). This sequence spans the region from the 3′ side of the iso-1-cytochrome c gene to the far side of a serine tRNA gene (Page and Hall, 1979). When the general sequence features of this segment of yeast nuclear DNA, which includes two structural genes and adjacent 5′- and 3′-intercistronic regions, are compared with analgous sequences in yeast mitochondrial DNA, there are impressive similarities (Macino and Tzagoloff, 1979; Bos et al., 1979; Henogens et al., 1979). Clearly, the homology between mitochondrial and prokaryote molecular biology is not perfect.

Since comparisons of sequence between different DNA sequences is a powerful tool for assigning function, it was of interest to compare the sequence of the gene for iso-1-cytochrome c with that of the second yeast cytochrome c, iso-2-cytochrome c. The DNA of the iso-1-cytochrome c gene was used as a probe to monitor the isolation of the second gene as a recombinant DNA in a plasmid vector (Montgomery et al., 1979). Iso-2-cytochrome c presents a novel opportunity for recognizing the location of its gene within the recombinant DNA. The protein contains, near its C-terminus, the sequence Ala-Ala (Borden and Margoliash, 1976), which is coded by the sequence GCNGCN. The recently discovered restriction endonuclease *Fnu*4HI (Leung, Lui, et al., 1979) recognizes and cleaves at the sequence GCNGC and thus provides a convenient probe for gene location. It is of interest that iso-1-cytochrome c and iso-2-cytochrome c both contain the sequence Gly-Pro, coded by GGNCCN for which the restriction endonuclease AsuI (GGNCC) is a specific probe.

The sequences of the two cytochrome c genes are compared in figure 13 (Montgomery et al., 1979). There are many interesting comparisons that can be made, and only some of these will be discussed here. Neither coding sequence contains an intervening sequence. It will be interesting to see if

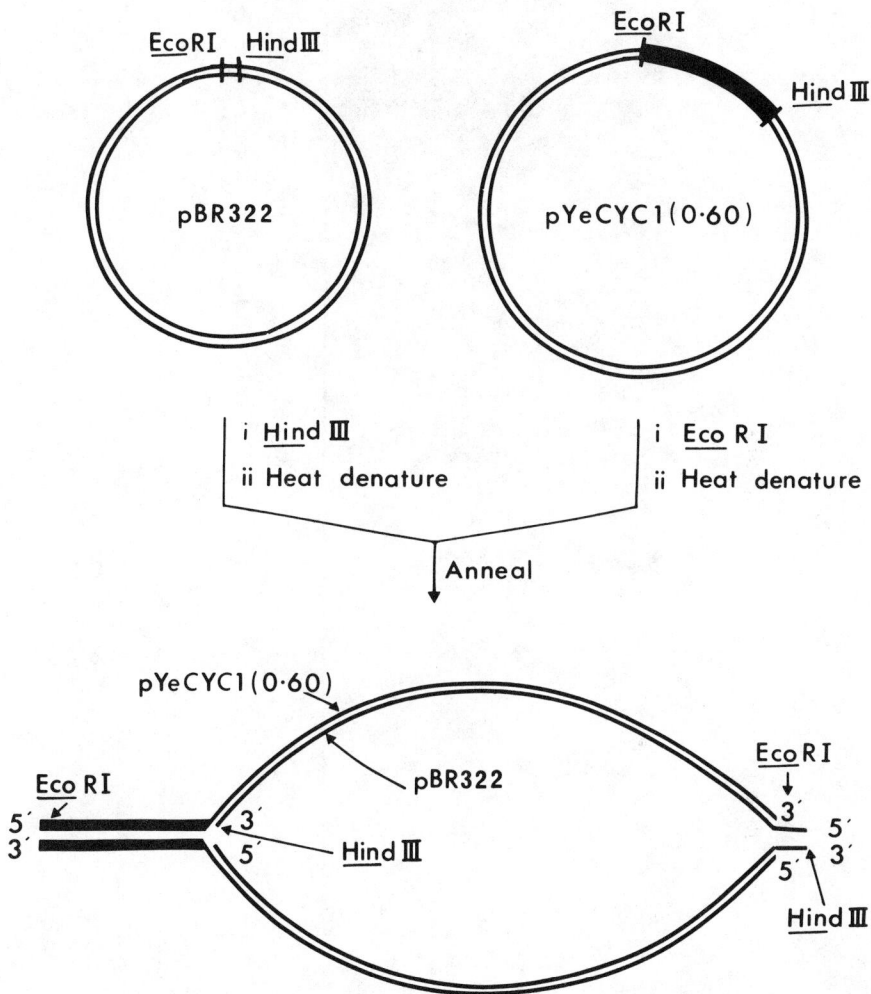

Fig. 8. Strategy for using *Hind III*-cleaved pBR322 as a primer for enzymatic terminator sequence determination using *Eco* RI- cleaved recombinant DNA [pYeCYC1(0.60)] as template. The heteroduplex between these two DNAs has two potential sites for DNA synthesis: the two 3′ termini in the upper pair of strands in the heteroduplex. After DNA synthesis using the terminator method, cleavage with *Hind*III releases the fragments corresponding to yeast DNA (heavy lines) while the other labeled fragments remain attached to pBR322 DNA and hence do not migrate during polyacrylamide gel electrophoresis (Smith et al., 1979).

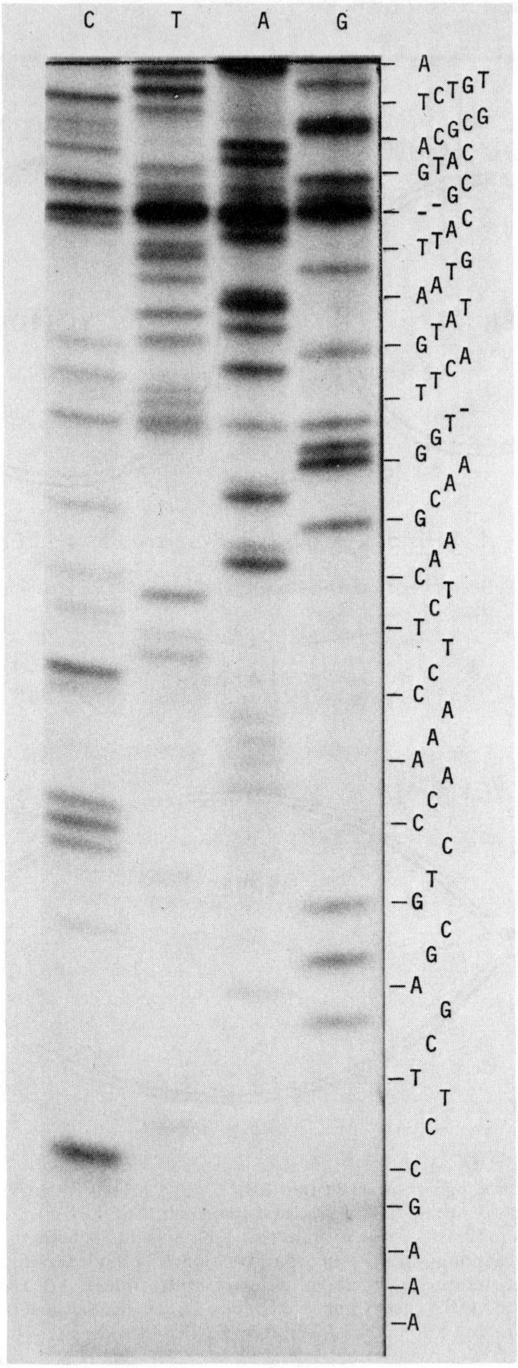

Fig. 9. Gel electrophoresis (12% polyacrylamide) showing part of the ladder pattern produced by the strategy diagrammed in figure 8.

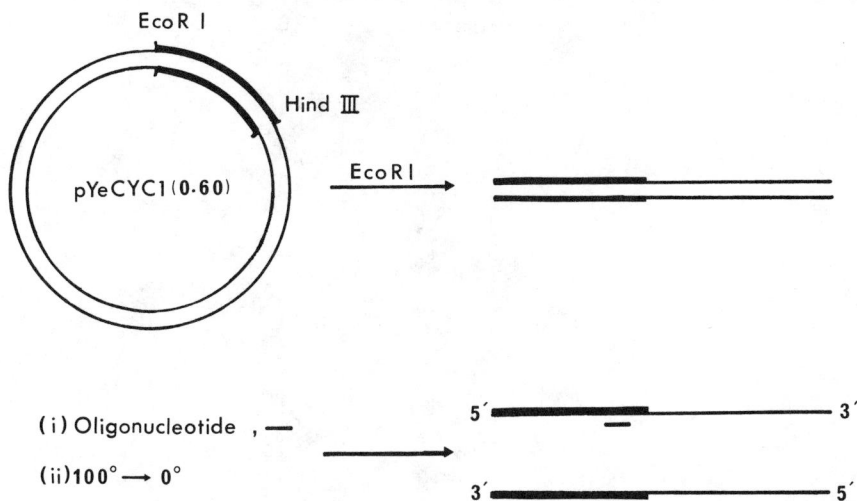

Fig. 10. Strategy for use of a synthetic oligodeoxyribonucleotide as a primer for enzymatic terminator sequence determination using template from double-stranded DNA (Smith et al., 1979).

this is the case for the cytochrome c genes of higher organisms. Apart from the additional amino acids at the N-terminus, iso-2-cytochrome c differs from the 108 amino acids of iso-1-cytochrome c at seventeen positions. In the N-terminal region of the genes, there are an even larger number of differences; silent differences in the wobble positions. Consequently, the very high conservation of the DNA sequences in the C-terminal region of the genes is particularly striking, and presumably of biological significance. In prokaryotes such conservation in otherwise divergent DNA sequences is always associated with dual functions (Sanger et al., 1978; Godson et al., 1978). These dual functions can be a pair of overlapping genes, an origin of DNA replication superimposed on a coding region, or a promoter superimposed on a coding region. Additional dual functions can easily be hypothesized. It will be of great interest if the DNA at the C-terminus of these genes has such a dual function. One further comment on this region of the cytochrome c genes is appropriate. Part of the sequence codes for a tract of amino acids that is conserved in all cytochromes c (Borden and Margoliash, 1976). It has been assumed that the conservation of amino acid sequence implies essential functions for the amino acids. Conservation of DNA sequence provides an alternate explanation.

The sequences before the 5'-ends of the coding regions for the yeast cytochromes c, though not identical, are very similar in structure. As has been discussed elsewhere in detail, this structure is most compatible with

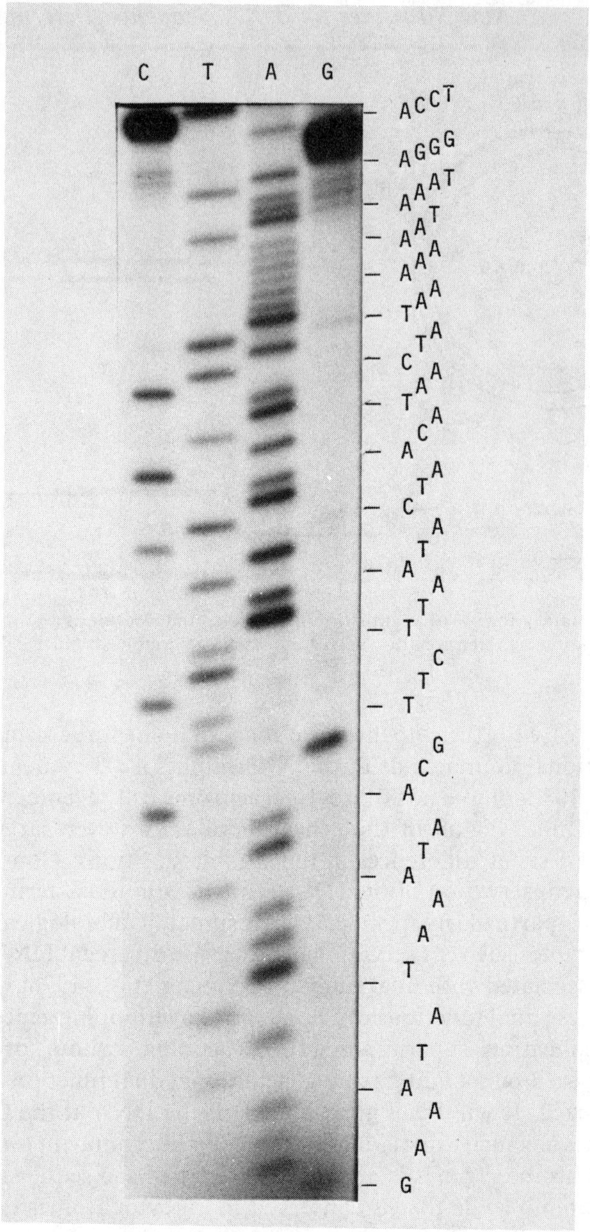

Fig. 11. Gel electrophoresis (12% polyacrylamide) showing part of a ladder pattern produced by the strategy diagrammed in figure 10. The primer was $d(pA_5GA_3)$ using pYeCYC1(0.6) DNA as template (Smith et al., 1979).

the Kosak model of initiation of eukaryote protein synthesis (Smith et al., 1979; Kozak, 1978). Beyond the 3′-end of the genes, there are T-rich sequences that could well be transcription termination signals (Smith et al., 1979, Montgomery et al., 1979).

SEQUENCE OF MUTANTS AT THE SUP-4 LOCUS OF YEAST

An alternate approach to defining the function of nucleotides in a DNA sequence is to make mutants defective in the gene of interest and then determine the sequence of the mutant DNAs. The SUP-4 locus of yeast is an appropriate target for such a study. Since it codes for a suppressor tRNA (Goodman et al., 1977), it is easy to screen for point mutants defective in that function (Kurjan et al., 1979). The DNA from the SUP4 locus of the mutants can be isolated fairly readily as a recombinant in a bacterial plasmid vector (Kurjan et al., 1979). The technical point of interest is that use of the enzymatic terminator method for DNA sequence determination with a synthetic oligodeoxyribonucleotide primer provides the most rapid route to the sequence of the mutant DNAs; the strategy is the one described earlier (fig. 10). The results of these sequence determinations are shown in figure 14. The most striking observation is that all mutations map within the boundaries of the structural gene. This contrasts with a study of point mutations in the gene of a prokaryote suppressor tRNA some of which map in the promoter region on the 5′-side of the structural gene (Berman and Landy, 1979). The location of the mutations in SUP4, together with other studies on eukaryote t-RNA and 5 S genes (Telford et al., 1979; Brown, 1979) is supportive of a model where the structural gene also functions as the promoter. Such a dual functionality is a convincing explanation of the conservation of sequence in independent copies of the same tRNA genes in eukaryotes (Goodman et al., 1977).

The SUP4 gene contains an intervening sequence, and one of the mutant changes is within this sequence (fig. 14). It will be of interest to determine whether the resultant defect in gene expression is in transcription and/or processing. Assuming that the mutant DNA is transcribed to give a mutant RNA (fig. 15), it is possible to draw a model, dependent on a change in secondary structure, that could explain a defect in processing (fig. 16). Another important consequence of the ready availability of the mutant DNA sequences is that it allows the possibility of correlation of DNA sequence changes with the recombinant genetic map (fig. 14). The excellent and detailed agreement between the two maps illustrates the power of fine structural genetic mapping in yeast.

```
          FnuEI                MboII                              PstI HinfI          HhaI
5'-GATCTTGATACATACGTTGGTATTTACTTTGAAGACTGCTGCATTGTGATGTACGAAAAGAAAGGGAAATAACGACAACTGCGAGGACTGAACCTGCCGGGCAAAGCCCAA
3'-CTAGAACTATGTATGCAACCATAAATGAAACTTCTGACGACGTAACACATACATGCTTTCTTTCCCTTTATTGCTGTTGACGTCCTGAGCTTGACGGCCCGTTTCGGGTT
                                                                          TaqI  FnuDII
                                         60

                 HaeIII                                      EcoRII                              HinfI
AAGATTTCTAATCTTCGCCTTAACCACTCGGCCAAGTTGCCTTAATCTCTTGAGTTTGTGTAAGGACAACAACACTGTCCACCTGGCATGGAATATTACCATTATTCAGATTCGGGCTT
TTCTAAAGATTAGAAAGCGGAATTGGTGAGCCGGTTCAACGGAATTAGAGAACTCAAACACATTCCGTTGTGACAGGTGACCGTACCCTTATAATGTAATAAGGTCTAAGCCCGAA   240
                                              180                                              TaqI  FnuDII
UUCUAAAGAUUAGAAAGCGGAAUUGGUGAGCCGGUUCAACGG-5'

             AluI   HpaI                                        HphI
GTTATTAGGACTTTTTACCGGCTTGATTTACCGTCATTACCTATCCTACACACTATTACATCACAACTACACTTGGTGAAGAAATCACAGAGCCTTGGATTGTTTAACAATTATATGTTTA
CAATAATCCTGAAAAATGGCCGAACTAAATGGCAGTAATGGATAGGATGTGATATGGTAGTTGATGTGAACCACTTCTTTAGTGTCTCGGAACCTAAACATTGTTAATATACAAAT   360
                                              300                                         FnuEI BamHI
                                                                                          MboII
                                                                                          TaqI

        AsuI MnlI                                   HphI                                    AluI
CGCAATTTAATCTTGGGAAAGCAGGGCCTCATTTTGTGGACATAGCACGGTTTGTCACCCTCGTCGTGCAAAAAATATTTCGACTTCTACTCTTTTGAAGTTCGTGATCCTTC
GCGTTAAATTAGAACCCTTTCGTCCCGGAGTAAACACCGTATCGTGCCAACAGTGGGAGCAGCACGTTTTTATAAAGCTGAAGATGAGAAAAACTTCAAACAACTAGCCTAGGGAAG   480
                                                                                            FnuEI
                      HaeIII                                                                  MnlI

       TaqI    HhaI
TAAGGAAAAGTCGAATGCTGTCATTGCCGATATAAAGGCCGAGAACGATAGTGACTGATATTAAGTTATTGTCATCATAAGTAATTTTTCTCTGATATCGTAGCCCTAGTTTTTA
ATTCCTTTTCAGCTTACGACAGTAACGCGTATATTTCCCGTCTTGCTACACTAACATAATTCAATAACAGTAGTATATTCATTAAAAAGAGACTATAGACATCGGGATCAAAAAT   600
                     AluI                       MboII                                       AluI
                                                                                           540

      AluI                    MnlI
ATATTTTCGTAAAATAGCTCAAAAAAATTTCTGTTTAAAATTCTTCTGGGTATATACACATGTTCCCTGCTAAATTAAAAGAAAAAAGGAACCGAAGAAGAGGCAATTTATAAAGTT
TATAAAGCATTTTATCGAGTTTTTTTAAAGACAAATTTAAGAAGACCCATATATGTGTACAAGGGACGATTTAATTTTCTTTTTTCCTTGGCTTCTTCCGTTAAATATTTTCAA   720
                     MnlI                                                                      660

       AluI                                              XbaI                     FnuDII       AluI
ATCACTCCGGCATATTGAGCGGAAAAAATCCGTCAAAAAAAGTTCTGTAAGTCGTCATAAGTGACTATGGTATCAAACACCAATTAACGACGACGTACCGATAATGAAGGCAACTT
TAGTAGGAGCCGTATAACTCGCCTTTTTTAGGCAGTTTTTCAAGACATTCAGCAGTATTCACTGATACCATAGTTTGTGGTTAATTGCTGCATGGCTATTACTTCCGTTGAA
                                                             FnuEI                            780

      AluI                               BglII MboII                    Fnu4HI MboII                   TaqI  Fnu4HI     AluI
CGTAGCTGAAACAGAACGAAGTTAGGATGGATGAAATTCTAATCAGATCTTGGAGCTTCCTTGAAGGAGGATGGCAGCTTCTTCACCCATGGCCGCAGAAATGATAGTGACCATCAAATCCTTACCTTCATCGAAGCAGTTG
GCATCGACTTTGTCTTGCTTCAATCTTTATTCTGCTTCCCTTTATAGGCAGTTTTTTATTAGTCTTAGAAGATGTTAGAAATGGTATTTTCTTTTTGCAGTGCTTTTAGGAATGGAATGGATGAAGTAGCTTCTCGAAC
                                            FnuEI                   AluI  HphI                                      AluI
```

```
        HphI                    HindIII                                                      HphI
CATGCTGTCACCCAACCTACCTTCTGGTCGTTGACATCGTCTTTGGTTCACCGTTCATGGTCATCAAGGACAAGTAGCCATCATCAATATCCAACAATTGGTATTCGCTTCTCTTGAC  1200
GTACGACAGTGGGGTTGGATGGAAGACCACGCAACTGTAGCAGAAACCAAAGTGCAACTGTAGCAGTTCCTGTTCATCGGTAGTAGTTATAGGTTGTTAACCATAAGCGAAGAGAACTG
                                                              AsuI                                             HpaII
                                         MboII                                                         HgaI
                                                                                              MnlI
                                                                                                 AvaII
AAATGGAACTTCCAAGTGTGAGTGGATGGAGACAAATCTTCCAACTTCTTACCAGTGAAGAATATCAAGGTAACCAAATGGACTTGGCGTGACCGTGCTTACCGCTCTTGGAAGTGGA  1320
TTTACCTTGAAGGTTCAACTCACCTACCTCTGTTTAGAAGGTTGAAGAATGGTCACTTCTATAGTTCCATTGGTTTACCTGGAACCGCACTGGCAGAACCTTCACCT

                                                                                              MboII 1440
             AccI                                                                                   HgaI
CATGTCGACAATCTTACATGGTCTACCTTTGATGACAACGAGAAACCATTCTTTCTCAAGGCAGAACATTGCATTGGGTAGGTGGCGGAGGCACCAGCGTCAGCATTTCAAAGGTTGTGTT
GTACAGCTGTTAGAATGTACCAGATGGAAACTACTGTTGCTCTTTGGTAAGAAAGAGTTCCGTCTTGTAACGTAACCATCCACCGCCTCCGTGGTCGCAGTCGTAAAGTTTCCAACACAA
 TaqI                                                                                                                1380
MboII
                                                                                                             MboII
CTTCGTCAGACAGTGTTTAGTGTGTGAATGAATAGGTGTATGTTTTCTTTTGCAGACAATTAGGAACAAGGTAAGGAGGTAAGTGTAGAATAAGAATTAAAAAGAAGAGAACAAG
GAAGCAGTCTGTCACAAATCACACACTTACTTTATCCACATACAAAAGAAAACGTCTGATTAATCCTTGTTCCATTCCCTTGATTCCACATCTTATTCTTAATTTTTCTTCTCTTGTTC  1560

                                                                                                           MboII
TTGAAAGGCAAGTGAAATTCAAGAAAAAGTCAATTGAAGTACAGTAAATTGACCTGAATATATCTGAATTATCTGAGTTCCGAGCAACAATGAGTTACCGAAGAGAACAATGGAAAACTT
AACTTTCCGTCAACTTTAAAGTCTTTTTCAGTTAACTTCATGTTAACTGGACTTATATGACTTCAAGGTCGTTGTTACTCAATGGCTTCTCTTGTTACCTTATCCTTATCCTTATCCTTGAA  1680
MboI                                        XhoI
                                                  HphI        GCTC
                                                     TaqI                          MnlI
TGAACGAAGAAGGAAGGCAGGAAGGAGGAAGAAAATTTTAGGCCAGAACAATAGGCAACAATAGGCACCAAGCAACGAACAGAACAAGTGAAAAAAACGAAAAAAAACACAGAAAAGAAT
ACTTGCTTCTTCCTTCCGTCCTTCCTTTCTTTTTTAAAATCCGGTCTTGTTATCCGTTGTTATCCGTTGTTGTCGTTGCTTGTTCACTTTTTTTTTTTGTCTTTTCTTA
                                                                                                              1800

GCAGAAAGTTGTAAACTGAATGAAAATCTGTGAGACCAGAGAAAAAAATAAAAAAAAAAAGAGGACGAAACAAAAAAGTGAAAAAAATGAAAAAATTTTTTGGAAAA
CGTCTTTCAACATTTGACTTTTTTTTTTTTTCGCTCCTTTCTTTTTTTTTTTATTTTTTTTTTTTTTTTCCTCCTGCTTTGTTTTTTTTTCACTTTTTTACTTTAAAAAACCTTT
                                                       HinfI                                Smal    FnuEI 1920
                                                                                          HpaII
CCAAGAAATGAATTATATTTCCCGTGTGAGACGACATCGTCGAATAACGTCAGGTAACATTTATGTAACTCAATTCCTACCTGAATCTAAAATTCCCGGGAGCAAGATCAAGATGT
GGTTCTTTACTTAATATAAAGGGCACACTCTGCTGTAGCAGCTTATTGCAGTCCATTGTAAATACATTGAGTTAAGGATGGACTTAGATTTTAAGGGCCCTCGTTCAGTTCTACA
 HphI FnuEI HpaII        HaeIII          MboII  HhaI AluI              FnuEI   FnuDII   XhoI
                                                                              FnuDII
TTTACCGATCTTTCCGGCTCTCTTTGCCGGGGTTTACGGACGATGACCGAGACGGCCCAGCTCATTTGGCCAGCCTTGGTTGGTGATCAAGATCAGGTAGCAATCCTCCGAG
AAAGTGGCTAGAAAGGCCAGAGAAACCGGCCCAAATGCCTGCTACTGGCTCTGCGGAGAGTAAACCGGTCGCAACCACGTCCCTAGTTAGGCATCCGGTAGGAGCGTC  TaqI
                                                              BclI      AvaIII
FnuEI  EcoRII                                                                        TaqI
CAGATCCGCCAGGCGCGTGTATAGCGGAAATGGCCAGGCAACTTTAGTGCTGACACATACAGGCCATATATATATGTGCGACGACACATGCATATGGCATGCGTCTGTCTGTATGT
GTCTAGGCGGTCCCACATATCGCCTTTACCGGTCCGTTGAAATCACGACGTGTATGTCCGGTATATATATATACACACGCTGCTGTGTACGTATATACCGTACGAGACAGACATACA
          HpaII                                                                          FnuEI          HgiAI
                                                                                                          2280
```

```
                    MboII                                    AsuI
ATATAAAACTCTTGTTTCTTTTCTCTAAATATTCTTTCCTTATACATTAGGTCCTTGTAGCATAAATTACTATACTTCTATAGACACGCAAACACAAATACACACTAAATTAA
TATATTTTGAGAACAAAGAAGAAAGAGATTTATAGAAGAAATATGTAATTCCAGGAAACATCGTATTTAATGATGAAGAATCTGCGTTTGTGTTATGTGTTGTGATTTAATT    2400
                                                                       AvaII

     MetThrGluPheLysAlaGlySerAlaLysLysGlyAlaThrLeuPheLysPheLysArgCysLeuGlnCysHisThrValGluLysGlyGlyProHisLysLysValGlyProAsnLeuHis
         EcoRI HaeIII                                    AccI                                                              AsuI
TAATGACTGAATTCAAGGCCGGTTCTGCTAAGAAAGGTGCTACACTTTTCAAGACTAGATGTCTACAATGCCACACCGTGGAAAAGGTGGCCCACATAAGGTTGGTCCAAACTTGCATG
ATTACTGACTTAAGTTCCGGCCAAGACAATTCTTTCCACGATGTGAAAGTTCTGATCTACAGATGTTACGGTGTGGCACCTTTTCCACCGGGTGTATTCCAACCAGGTTGACGTAC    2520
               HpaII                                                                 HaeIII                  AvaII

 GlyIlePheGlyAlaArgHisSerGlyGlnAlaGlyTyrSerTyrThrAspAlaAsnIleLysLysAsnValLeuTrpAspGluAsnAsnMetSerGluTyrLeuThrAsnProLysLys
               AluI
GTATCTTTGGCAGACACTCTGGTCAGCTGAAGGGTATTCCTACACAGATCCAATATCAAGAAAAAACGTGTGTGGGGACGAAAATAACATGTCAGAGTACTTCAGAGTCTGACTAACCCAAAGAAAT
CATAGAAACCGTCTGTGAGACCAGTTCGACTTCCATAAGGATGTGTCTAGGTTATAGTTCTTTTTGCACAACACCTGCTTTTATTGTACAGTCTCATGAACTGATTGGGTTTCTTTA    2640

 TyrIleProGlyThrLysMetAlaPheGlyGlyLeuLysGluLysGlnLysAspArgAsnAspLeuIleThrTyrLeuLysLysAlaCysGlu
          EcoRII HaeIII                       MboII                                              AsuI                   TaqI
ATATTCCTGGTACCAAGATGGCCTTTGGTGGGTGAGAAGGAAAAAGACGACTTAATTACCTACTTGAAAAAGCCTGTGAGTAAACAGGCCCCTTTTCCCTTTGTCGCATATCA
TATAAGGACCATGGTTCTACCGGAAACCACCCAACTTCTTCTTTTGTCTTTGAATTAATGGATGAACTTTTTCGGACACTCATTTGTCCGGGGAAAAGGAAACAGCTATAGT    2760
   KpnI                                                                                                        HaeIII

                                              AsuI
TGTAATTAGTTATGTCACGCTTACATTCACGCCTCCCCCCCAATCCGCTCTAACCGAAAAGGAAAGGAGTTAGACAACCTGAAGTCTAGGTCCCTATTTATTTTTATTAGTTATGTTAG
ACATTAATCAATACAGTGCGAATGGTAAGTGCCGGAGGGGGGGTTAGGCGAGATTGGCTTTCCTTCCTCAATCTGTTGGACTTCAGATCCAGGGATAAAATAAAAAAATATCAATACAATC    2880
                                     HgaI                                                      AvaII
                                                                                                     HgaI
TATTAAGAACGTTATTTATTTCAAATTTTTCTTTTTTTCTGTACAGACGGCGTGTACCGCATGTAACATTATCTGAAAACCTTGCTTGAGAAGGTTTGGGACGCTCGAAGGCTTTAA
ATAATTCTTGCAATAAATATAAGTTTAAAAGAAAAAAGACATGTCTGCGCACATGCGTACATTGTAATATGACTTTTGAACGAACTCTTCCAAAACCCTGCAGGTCCGAAATT    3000
                                                    FnuDII  2940
                                                                                                         TaqI

  HindIII
  TTTGCAAGCTT-3'
  AAACGTTCGAA-5'
  AluI
```

Fig. 12. DNA sequence at the CYC1 locus of *Saccharomyces cerevisiae*. The sequence was determined in part by the strategies diagrammed in figures 8 and 10 and in part by the chemical cleavage method (Smith et al., 1979; Leung, Montgomery, et al., 1979). The gene for serine t-RNA (Page and Hall, 1979) and that of iso-1-cytochrome c are indicated by the sequences of their gene products.

AAAGGCACAACACATATATATCGTTGTTGAAGCTCGAGAAGATTAGATCAGAATAG

TTCTCTTTTTGTTGAGGTTGAAACAAAATCAAAGACTTCACAAGAGATCACATACAAGCATTTATTCACATTACTTTAAGTAACTTCAGTAAACTACAATCAATTACATCAATAAACAAAC
TAAAACTCTGTTTTCCTTCTTTTTCTAAATATTCTTTCCTTATACATTAGGTCCTTTGTAGCATAAATTACTATACTTCTATAGACACGCAAACAAATACAACAAATAAATTAATA
GCATATATATGTGTGCGACGACACATGATCCATATATGCATGTGTCTCTGTATGTATA

ATG GCT AAA GAA AGT ACG GGT GGT GGA TTC AAA AAG CCA GGC TCT GCA CCT AAA AAC ATC AAC AAC GTC AAA AAC AAC GAT AGT AGT GAT ATG TCC GAG TAC TTG ACG TAC TTG ACT TTA ATT ACT GGT GAA AAG CCA GGT CAG AAG GTA TAT TCT CAT ACA GAT ATTA CTG
Met Ala Lys Glu Ser Thr Gly Gly Gly Phe Lys Lys Pro Gly Ser Ala Pro Lys Asn Ile Asn Asn Val Lys Asn Val Lys Asn Asp Ser Met Ser Glu Glu Asn Asn Asp Glu Lys Tyr Ser His Thr Asp Ile Val

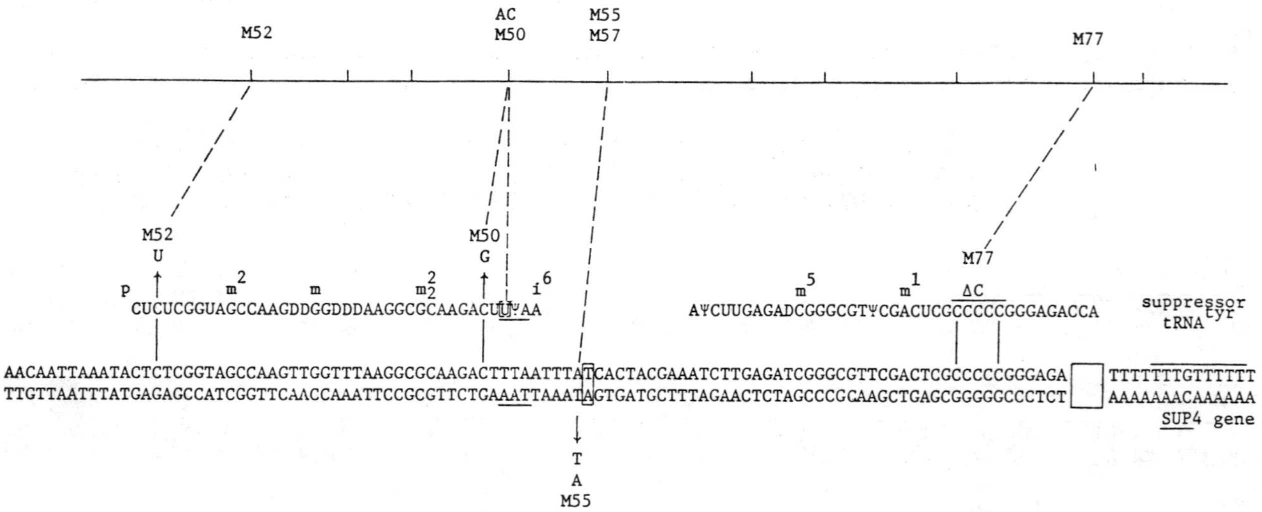

Fig. 14. The sequence of the SUP4 gene, the sequence of the suppressor tRNA^{tyr}, the alterations found in mutants M52, M50, M55, and M57 together with their alignment with the genetic map as determined by meiotic recombination frequencies (Kurjan et al.,1979). The sequences of the mutants were determined using the enzymatic terminator method with the synthetic oligodeoxyribonucleotide d (pAAA-AACAAAA) as primer (corresponding to overlined sequence at the 3' end of the gene). The anticodon (AC) is underlined, and the base pair within the box varies between different t-RNA^{tyr} genes.

```
                          A
                           C OH
                           C
                           A
                  pC · G
                   U · A
      M52   U ← C · G
                   U · G
                   C · G
                   G · C   ΔC      M77
                   G · C┐           
                 U       └ →             C   1
                         C C C G C   U  C  A m
              A          · · · · ·              G
                     2   G G G C G  T  ψ   C
        D  D  G  A  m              - 
     D     A    A C C G        m5C
   Gm            · · ·              D
   G                               A
     D          A G G C            G
       D  D  A                     A
            A      2G
             m2   C · G
                  A · U
                  A · U
                  G · C
                  A · ψ
      M50   G ← C       A
             U         i6A
            [U]       A
                   ψ
```

Mature <u>SUP4</u> tRNA^tyr

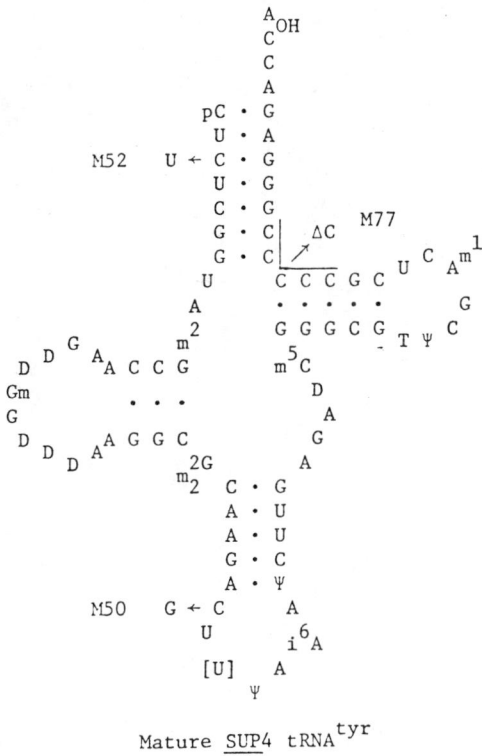

Fig. 15. Structure of mature SUP4 tRNA *tyr* with the positions of the nucleotide changes in mutant M50, M52, and M77; M77 is completely defective, but M50 and M52 have residual suppressor activity (Kurjan et al., 1979).

CONCLUSIONS

This article has sought to identify some salient points in the development of the present-day rapid DNA sequencing methodologies. It describes some novel strategies for the use of these methods in the determination of parvovirus and yeast DNA sequences. The DNA sequences in turn have provided new insights into the organization and function of eukaryote DNAs.

ACKNOWLEDGMENTS

Research in the author's laboratory was supported by the Medical Research Council of Canada through the provision of a research grant and a Career Investigatorship. The enthusiasm and hard work of Caroline R. Astell, Shirley Gillam, Patricia Jahnke, David Y. Leung, and Anne C. P.

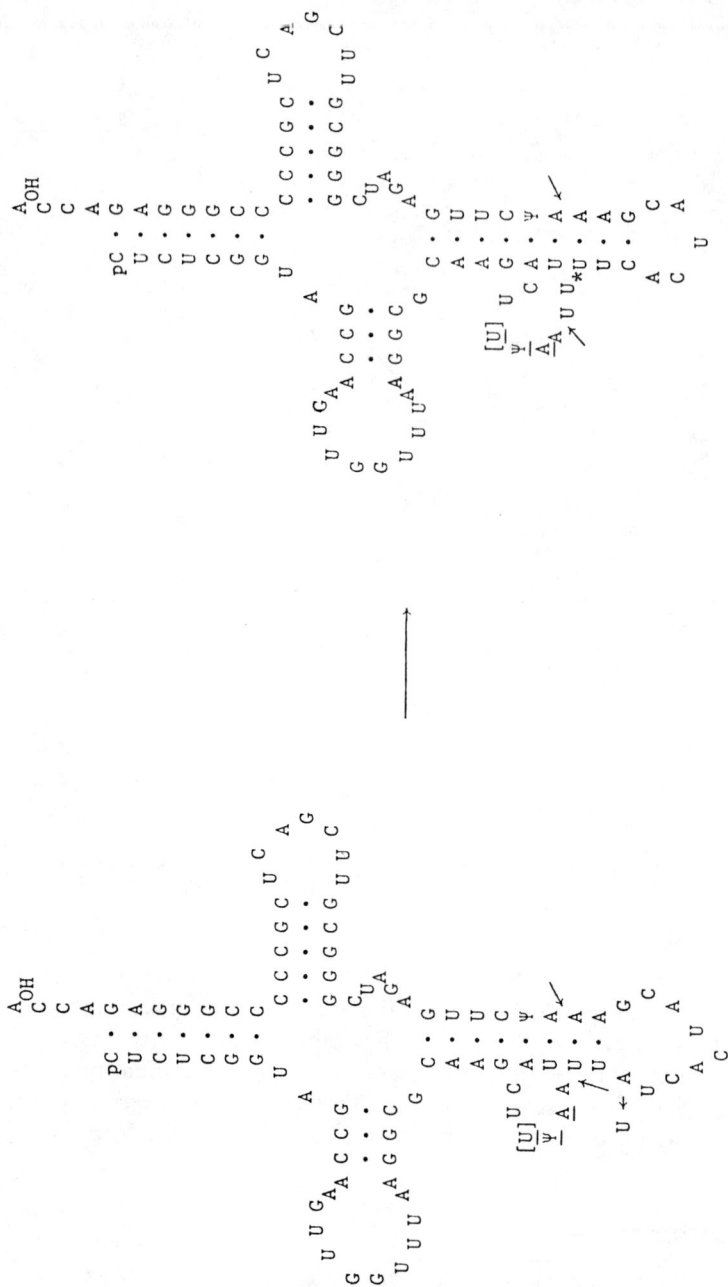

Fig. 16. Possible structures of the 92 nucleotide precursors for the tRNAs of SUP4 and mutants M55 and M57 (indpendent isolates with the same nucleotide change). It is hypothesized that the mutant precursor can shift to the form shown on the right (changes nucleotide indicated with an asterisk). This changes the positions of the points at which cleavage occurs (Abelson, 1979) during t-RNA maturation and, hence, results in a structure that may be resistant to cleavage.

SUP4
92 Nucleotide precursor

Mutants M55 & M57
92 Nucleotide precursor

Lui were essential ingredients in DNA sequence experiments described above. The project on parvovirus DNAs is a collaboration with Marie B. Chow and David C. Ward, Yale University, and that on yeast DNA is a collaboration with Gerard Faye, Benjamin D. Hall, Janet Kurjan, Donna L. Montgomery, and Peter Shalit of the University of Washington.

REFERENCES CITED

Astell, C. R., M. Smith, M. B. Chow, and D. C. Ward. 1979a. Structure of the 3′-hairpin termini of four rodent parvovirus genomes: nucleotide sequence homology at the origins of DNA replication. Cell 17:691–703.

Astell, C. R., M. Smith, M. B. Chow, and D. C. Ward. 1979b. Sequence of the 3′-terminus of the genome from Kilham rat virus, a non-defective parvovirus. Virology 96:669–74.

Astell, C. R., M. Smith, M. B. Chow, and D. C. Ward. 1979c. Unpublished results.

Berman, M. L., and A. Landy. Personal communication.

Berns, K. I., and W. W. Hanswirth. 1978. Parvovirus DNA structure and replications. *In* D. C. Ward and P. Tattersall (eds.). Replication of mammalian parvovirus, pp. 13–32. Cold Spring Harbor Laboratory, Cold Spring Harbor, New York.

Borden, D., and E. Margoliash. 1976. Amino acid sequences of proteins-eukaryote cytochromes. *In* G. D. Fasman (ed.), Handbook of biochemistry and molecular biology, 3:268–79. Chemical Rubber Company, Cleveland.

Bos, J. L., K. A. Osinga, G. Van der Horst, P. Borst. 1979. Nucleotide sequence of the mitochondrial structural genes for cysteine-tRNA and histidine-tRNA of yeast. Nucl. Acids Res. 6:3255–66.

Brosius, J., M. L. Palmer, P. J. Kennedy, and H. F. Noller. 1978. Complete nucleotide sequence of a 16S ribosomal RNA gene from *Escherichia coli*. Proc. Nat. Acad. Sci. USA 75:4801–5.

Brown, D. D. 1979. Personal communication.

Brown, N. L., and M. Smith. 1977. The sequence of a region of bacteriophage ϕX174 DNA coding for parts of genes A and B. J. Mol. Biol. 116:1–28.

Brownlee, G. G., and F. Sanger. 1967. Nucleotide sequences from the low molecular weight ribosomal RNA of *Escherichia coli*. J. Mol. Biol. 23:337–53.

Brownlee, G. G., and F. Sanger. 1969. Chromatography of ^{32}P-labelled oligonucleotides on thin layers of DEAE-cellulose. Eur. J. Biochem. 11:395–99.

Burton, K., and G. B. Petersen. 1960. The frequencies of certain sequences of nucleotides in deoxyribonucleic acid. Biochem. J. 75:17–27.

Ehresmann, C., P. Stiegler, P. Carbon, and J. P. Ebel. 1977. Recent progress in the determination of the primary sequence of the 16S RNA of *Escherichia coli*. FEBS Lett. 84:337 41.

Englund, P. T. 1972. The 3′-terminal nucleotide sequences of T7 DNA. J. Mol. Biol. 66:209 24.

Fellner, P. 1974. Structure of the 16S and 23S ribosomal RNAs. *In* M. Nomura, A. Tissieres, and P. Lengyel (eds.), Ribosomes, pp. 169–91. Cold Spring Harbor Laboratory, Cold Spring Harbor, New York.

Fiers, W., R. Coutreras, F. Duerwick, G. Haegeman, D. Iserentant, J. Merregaert, W. Min Jou, F. Molemans, A. Raeymaekers, A. Van den Berghe, G. Volckaert, and M. Ysebaert. 1976. Complete nucleotide sequence of bacteriophage MS2 RNA: primary and secondary structure of the replicase gene. Nature 260:500–507.

Gillam, S., F. Rottman, P. Jahnke, and M. Smith. 1977. Enzymatic synthesis of oligonucleotide of defined sequence: synthesis of a segment of yeast iso-1-cytochrome c gene. Proc. Nat. Acad. Sci. USA 74:96–100.

Godson, G. N., B. G. Barrell, R. Staden, and J. C. Fiddes. 1978. Nucleotide sequence of bacteriophage G4 DNA. Nature 276:236–47.

Goodman, H. M., M. V. Olson, and B. D. Hall. 1977. Nucleotide sequence of a mutant eukaryotic gene: the yeast tyrosine-inserting ochre suppressor *SUP4-o*. Proc. Nat. Acad. Sci. USA 74:5453–57.

Hall, R. H. 1971. The modified nucleosides in nucleic acids. Columbia University Press, New York.

Hensgens, L. A. M., L. A. Grivell, P. Borst, and J. L. Bos. 1979. Nucleotide sequence of the mitochondrial structural gene for subunit 9 of yeast ATPase complex. Proc. Nat. Acad. Sci. USA 76:1663–67.

Hirs, C. H. W., and S. N. Timasheff, (eds.). 1977. Enzyme structure (Part E). Methods in Enzymology 47:1–668.

Holley, R. W. 1968. Experimental approaches to the determination of nucleotide sequences of large oligonucleotides and small nucleic acids. Prog. Nucl. Acid Res. Mol. Biol. 8:37–47.

Korn, L. J., C. L. Queen, and M. N. Wegman. 1977. Computer analysis of nucleic acid regulatory sequences. Proc. Nat. Acad. Sci. USA 74:4401–5.

Kozak, M. 1978. How do eukaryote ribosomes select initiation regions in messenger RNA? Cell 15:1109–23.

Kurjan, J., S. Gillam, M. Smith, and B. D. Hall. 1979. Unpublished results.

Leung, D. W., A. C. P. Lui, H. Merilees, B. C. McBride, and M. Smith. 1979. A restriction enzyme from *Fusobacterium nucleatum* 4H which recognizes GCNGC. Nucl. Acids Res. 6:17–25.

Leung, D. W., D. L. Montgomery, B. D. Hall, and M. Smith. 1979. Unpublished results.

Ling, V. 1971. Partial digestion of ^{32}P-fd DNA with T4 endonuclease IV. FEBS Lett. 19:50–54.

Ling, V. 1972. Pyrimidine sequences from the DNA of bacteriophages fd, f1, and ϕX174. Proc. Nat. Acad. Sci. USA 69:742–46.

Lui, A. C. P., B. C. McBride, G. F. Vovis, and M. Smith. 1979. Site specific endonucleases from *Fusobacterium nucleatum*. Nucl. Acids Res. 6:1–15.

Maat, J., and A. J. H. Smith. 1978. A method for sequencing restriction fragments with dideoxynucleoside triphosphates. Nucleic Acids Res. 5:4537–45.

Macino, G., and A. Tzagoloff. 1979. Assembly of the mitochondrial membrane system. The DNA sequence of a mitochondrial ATPase gene in *Saccharomyces cerevisiae*. J. Biol. Chem. 254:4617–23.

Marotta, C. A., B. G. Forget, M. Cohen-Solal, and S. M. Weissman. 1976. Nucleotide sequence analysis of coding and non-coding regions of human β-globin mRNA. Prog. Nucl. Acid Res. Mol. Biol. 19:165–75.

Maxam, A. M., and W. Gilbert. 1977. A new method for sequencing DNA. Proc. Nat. Acad. Sci. USA 72:560–64.

Maxam, A. M., and W. Gilbert. 1979. Sequencing end-labeled DNA with base-specific chemical cleavages. Methods in enzymology 65 (in press).

McCallum, D., and M. Smith. 1977. Computer processing of DNA sequence data. J. Mol. Biol. 116:28–30.

McGeoch, D. J. 1979. Structure of the gene N: gene NS intercistronic junction in the genome of Visicular Stomatitis virus. Cell 17:673–81.

Mills, D. R., and F. R. Kramer. 1979. Structure-independent nucleotide sequence analysis. Proc. Nat. Acad. Sci. USA 76:2232–35.

Montgomery, D. L., B. D. Hall, S. Gillam, and M. Smith. 1978. Identification and isolation of the yeast cytochrome c gene. Cell 14:673–80.

Montgomery, D. L., D. W. Leung, M. Smith, P. Shalit, G. Faye, and B. D. Hall. 1979. Unpublished results.

Ohmori, H., J. Tomizawa, and A. M. Maxam. 1978. Detection of 5-methyl-cytosine in DNA sequences. Nucl. Acid. Res. 5:1479–86.

Page, G., and B. D. Hall. 1979. Personal communication.

Peattie, D. A. 1979. Direct chemical method for sequencing RNA. Proc. Nat. Acad. Sci. USA 76:1760–64.

Platt, T. 1978. Regulation of gene expression in the tryptophan operon of *Escherichia coli. In* J. H. Miller and W. S. Reznikoff (eds.), The operon, pp. 263–302. Cold Spring Harbor Laboratory, Cold Spring Harbor, New York.

Proudfoot, N. J., C. C. Cheng, and G. G. Brownlee. 1976. Sequence analysis of eukaryote mRNA. Prog. Nucl. Acid Res. Mol. Biol. 19:123–34.

Rae, P. M. M., and R. E. Steele. 1979. Absence of cytosine methylation of C-C-G-G and G-C-G-C sites in the rDNA coding regions and intervening sequences of Drosophila and the rDNA of other higher insects. Nucl. Acids. Res. 6:2987–95.

Reznikoff, W. S., and J. N. Abelson. 1978. The *lac* promoter. *In* J. H. Miller and W. S. Reznikoff (eds.), The operon, pp. 221–43. Cold Spring Harbor Laboratory, Cold Spring Harbor, New York.

Sadowski, P. D., and I. Bakyta. 1972. T4 endonuclease IV. J. Biol. Chem. 247:405–12.

Sanger, F. 1959. Chemistry of insulin. Science 129:1340–44.

Sanger, F., and C. A. Coulson. 1975. A rapid method for determining sequences in DNA by primed synthesis with DNA polymerase. J. Mol. Biol. 94:441–48.

Sanger, F., S. Nicklen, and A. R. Coulson. 1977. DNA sequencing with chain-terminating inhibitors. Proc. Nat. Acad. Sci. USA 74:5463–67.

Sanger, F., and A. R. Coulson. 1978. The use of thin acrylamide gels for DNA sequencing. FEBS Lett. 87:107–10.

Sanger, F., A. R. Coulson, T. Friedmann, G. M. Air, B. G. Barrell, N. L. Brown, J. C. Fiddes, C. A. Hutchison, III, P. M. Slocombe, and M. Smith. 1978. The nucleotide sequence of bacteriophage ϕX174. J. Mol. Biol. 125:225–46.

Schoen, A. E., R. G. Cooks, J. L. Wiebers. 1979. Modified bases characterized in intact DNA by mass-analyzed ion kinetic energy spectrophotometry. Science 203:1249–51.

Smith, A. J. H. 1979a. The use of exonuclease III for preparing single stranded DNA for use as a template in the chain terminator sequencing method. Nucl. Acids Res. 6:831–48.

Smith, A. J. H. 1979b. DNA sequence analysis by primed synthesis. Methods in enzymology 65: (in press).

Smith, H. O. 1979. Nucleotide sequence specificity of restriction endonucleases. Science 205:455–62.

Smith, M., D. W. Leung, S. Gillam, C. R. Astell, D. L. Montgomery, and B. D. Hall. 1979. Sequence of the gene for iso-1-cytochrome c in *Saccharomyces cerevisiae*. Cell 16:753–61.

Staden, R. 1977. Sequence data handling by computer. Nucl. Acids. Res. 4:4037–51.

Staden, R. 1978. Further procedures for sequence analysis by computer. Nucl. Acids Res. 5:1013–15.

Staden, R. 1979. A strategy of DNA sequencing employing computer programs. Nucl. Acids Res. 6:2601–10.

Stewart, J. W., and F. Sherman. 1974. Yeast frameshift mutations indneitied by sequence changes in iso-1-cytochrome c. *In* L. Prakash, F. Sherman, M. W. Miller, C. W. Lawrence, and H. W. Taber (eds.), Molecular and environmental aspects of mutagenesis, pp. 102–27. Charles C. Thomas, Sprinfield, Illinois.

Tal, J., D. Ron, P. Tattersall, S. Bratosin, and Y. Alonk. 1979. About 30% of minute virus of mice RNA is spliced out following polyadenylation. Nature 279:649–51.

Tattersall, P. 1978. Parvovirus protein structure and virion maturation. *In* D. C. Ward and P. Tattersall (eds.), Replication of mammalian parvoviruses, pp. 53–72. Cold Spring Harbor Laboratory, Cold Spring Harbor, New York.

Tattersall, P., and D. Ward. 1978. The parvoviruses: an introduction. *In* D. C. Ward and P. Tattersall (eds.), Replication of mammalian parvoviruses, pp. 3–12. Cold Spring Harbor Laboratory, Cold Spring Harbor, New York.

Telford, J. L., A. Kressmann, R. A. Koski, R. Grosschedl, F. Muller, S. G. Clarkson, and M. L. Birnstiel. 1979. Delimitation of a promoter for RNA polymerase III by means of a functional test. Proc. Nat. Acad. Sci. USA 76:2590–94.

Waalwijk, C., and R. A. Flavell. 1978. MspI, an isoschizomer of Hpa II which cleaves both unmethylated and methylated Hpa II sites. Nucl. Acids Res. 5, 3231–36.

Weissmann, C., M. A. Billeter, H. M. Goodman, J. Hindley, and H. Weber. 1973. Structure and function of phage RNA. Ann. Rev. Biochem. 42:303–28.

Woese, C. R., G. E. Fox, L. Zablen, T. Uchida, L. Bonen, K. Pechman, B. J. Lewis, and D. Stahl. 1975. Conservation of primary structure in 16S ribosomal RNA. Nature 254:83–86.

Wu, R., and A. D. Kaiser. 1968. Structure and base sequence in the cohesive ends of bacteriophage lambda DNA. J. Mol. Biol. 35:523–37.

Wu, R., C. P. Bahl, and S. S. Narang. 1978. Synthetic oligodeoxyribonucleotides for analyses of DNA structure and function. Prog. Nucl. Acid Res. Mol. Biol. 21:102–41.

Zabin, I., and A. V. Fowler. 1978. β-Galactosidase, the lactose permease protein and thiogalactoside transacetylase. *In* J. H. Miller and W. S. Reznikoff (eds.), The operon, pp. 89–121. Cold Spring Harbor Laboratory, Cold Spring Harbor, New York.

Zimmern, D., and P. Kaesberg. 1978. 3′-Terminal nucleotide sequence of encephalomyocarditis virus RNA determined by reverse transcriptase and chain terminating inhibitors. Proc. Nat. Acad. Sci. USA 75:4257–61.

LYDIA VILLA-KOMAROFF

The Synthesis of Eukaryotic
Proteins in Prokaryotic Cells

4

INTRODUCTION

The study of eukaryotic genes has been greatly facilitated by the ability to isolate and to amplify individual eukaryotic genes in bacteria. In some cases the bacterial cell can synthesize the protein encoded by the inserted eukaryotic DNA. The ability to generate large amounts of specific eukaryotic proteins has exciting implications for studies of structure and function, for the generation of specific antigens, and for the generation of proteins for therapeutic uses. In addition, the expression of foreign polypeptide chains may provide information on the normal mechanisms of transcriptional and translational control in bacterial cells. Some of these applications depend not only on producing the correct polypeptide chain, but also on producing biologically functional proteins.

To achieve the expression of a eukaryotic sequence in a prokaryotic cell, one must be able to obtain the sequence encoding the protein of interest, one must have a method to detect the protein and the RNA transcript of the eukaryotic sequences, and the polypeptide encoded by those sequences must be stable in the prokaryotic cell.

In some cases the production of biologically active eukaryotic gene products in prokaryotic cells has been demonstrated by the ability of cloned eukaryotic DNA to provide functions missing or deficient in the bacterial host. In these cases one obtains the sequence encoding the protein and detects the synthesis of the protein in the same step. This method has been used to detect the production of eukaryotic enzymes that have bacterial homologs. When this method is used, care must be taken to

Department of Microbiology, University of Massachusetts Medical School, Worcester, Massachusetts 01605 ensure that the "complementation" observed is not due to the reversion of the bacterial host. In most cases a deletion of the appropriate function must be used. Scrupulous care must be taken to preclude the inadvertant cloning of bacterial DNA that could complement the deleted function.

To select a fragment of eukaryotic DNA that is able to complement a particular mutation in *Escherichia coli*, the efficiency of the cloning system must be high enough to ensure that the number of hybrid clones obtained is large enough to contain all the genomic sequences of the organism under study. The DNA encoding the desired function must also be transcribed into functional mRNA in *E. coli*. In experiments utilizing simple eukaryotes such as yeast or neurospora, both criteria can be met using chromosomal DNA. The size of these genomes is small enough so that the entire genome can be represented by about 5×10^3 independent transformants if the cloned segments are about 10^4 base pairs in size. In the case of higher eukaryotes, about 5×10^5 transformants must be generated in order to have a 99% probability of containing the entire genome (Clarke and Carbon, 1976). Methods are now available for rapidly generating large numbers of transformants (Maniatis et al., 1978), or the complexity of the DNA to be cloned can be reduced by selecting a fraction of the DNA that has the sequence of interest (e.g., Tilghman et al., 1977), so the large size of most eukaryotic genomes no longer precludes the use of chromosomal sequences for the expression of functions that can complement mutations in *E. coli*. However, the coding region of many higher eukaryotic genes is interrupted by noncoding sequences (Brack and Tonegawa, 1977; Breathnach et al., 1977; Jeffreys and Flavell, 1977; Lai et al., 1978; Tilghman et al., 1978). Since there is no evidence that bacteria contain the RNA processing enzymes necessary to remove introns from the primary transcript, mature mRNA could not be formed in bacteria from a DNA sequence containing an intron. The use of double stranded copy DNA (ds-cDNA) circumvents both of these problems: it allows one to greatly enrich for the sequence of interest by utilizing RNA from tissue where that protein is produced, and it provides a DNA sequence containing no introns in the structural coding region (for a review see Efstratiadis and Villa-Komaroff, 1979).

In cases where the expression of a eukaryotic gene does not confer a selectable phenotype on the host, the detection of the sequences encoding the protein of interest, the detection of the synthesis of the protein, and the demonstration that the protein is biologically active must be done separately.

The most straightforward way to identify clones containing the wanted sequence is to hybridize mRNA (or cDNA) to DNA in the bacterial colonies (Grunstein and Hogness, 1975). However, in many cases, the mRNA cannot be purified in large enough amounts to be used as a probe. In these cases the DNA in the clones can sometimes be screened by using hybrid-arrested translation, a method based on the principal that RNA in the form of an RNA:DNA hybrid does not direct protein synthesis (Paterson et al., 1977).

If the amino acid sequence of the protein is known and if the protein is small, a sequence encoding the known polypeptide chain can be chemically synthesized (Itakura et al., 1975). In this case the successful synthesis and cloning can be confirmed by determining the nucleotide sequence of the cloned DNA (Maxam and Gilbert, 1977; Sanger et al., 1977). (In the case where the amino acid sequence is not known, it can be deduced from the nucleic acid sequence.)

Once the DNA sequence encoding the protein has been identified, it must be inserted into a phage or plasmid vector in such a way that it is likely to be expressed, and an assay must be available to detect the synthesis of the protein. The synthesis of proteins can be demonstrated by showing that the inserted DNA directs the synthesis of a polypeptide with the physical or immunological characteristics of the protein in question.

Although the detection of antigenic determinants suggests that at least part of the molecule has assumed its correct tertiary structure, an immunoreactive molecule is not necessarily biologically active. The biological activity of a eukaryotic molecule synthesized in prokaryotic cells must be demonstrated by showing that the protein produced in bacterial cells is capable of producing the same biological effects as the protein obtained from its normal source.

The eukaryotic proteins that have been produced in bacterial cells include enzymes, polypeptide hormones, viral polypeptides, and the proteins ovalbumin and actin (see table 1). Three basic strategies for producing functional proteins in bacterial cells have been employed. First, the insertion of chromosomal DNA encoding the protein of interest such that transcription is unlikely to be under the control of bacterial sequences. Second, the insertion of the structural sequence for the protein of interest into a bacterial operon such that transcription is under control of the bacterial sequences and a hybrid protein is produced. Third, the insertion of the structural sequence for the protein of interest directly adjacent to the control elements of a bacterial operon such that transcription of the eukaryotic sequences is under control of the bacterial elements but the eukaryotic protein contains no bacterial amino acids. In some cases the coding sequences have been inserted into a bacterial gene encoding a secretory protein so that the product is secreted into the periplasmic space.

All these strategies have been used successfully; however, no single strategy is equally successful with all eukaryotic sequences.

CLONING AND EXPRESSION OF EUKARYOTIC ENZYMES

Although some early evidence indicated that eukaryotic DNA could be transcribed in prokaryotic cells (Morrow et al., 1974; Chang et al., 1975; Kedes et al., 1975; Meagher et al., 1977), the first indications that eukaryotic sequences could be correctly transcribed and translated came from studies utilizing cloned chromosomal DNA from simple eukaryotes. In these studies a fragment of yeast or neurospora DNA complemented *E. coli* mutants lacking enzymatic activities.

The initial report of the functional expression of eukaryotic DNA in *E. coli* was made by Struhl and colleagues (1976). Chromosomal yeast DNA was digested with the restriction endonuclease Eco R1 and ligated to the R1 arms of λgt, a phage vector deleted for the gene necessary for integration (Thomas et al., 1974). The recombinant phage was integrated into the bacterial chromosome of the histidine auxotroph hisB463 by utilizing the integration functions of a helper phage. This bacterial host lacks the activity for the enzyme imidazole glycerol phosphate (IGP) dehydratase. The investigators showed that the reversion frequency of the mutation was less than 10^{-11} and could not be induced by ultraviolet irradiation, ethyl methane sulfonate, nitrosoguanidine, or ICR191 and is therefore most likely a deletion. When yeast DNA was inserted into the phage and integrated into the bacterial chromosome, the frequency of complementation was 10^{-8}. If the bacterial host was cured of the prophage, histidine was once again required for growth. The phages used for these experiments had a temperature sensitive repressor, so stable lysogeny and a his$^+$ phenotype should occur only at temperatures less than 35° if the complementing function is encoded on the phage DNA. None of the his$^+$ colonies obtained were able to grow on rich medium at 42°. The authors concluded that the DNA inserted into the vector phage was able to complement the lesion in hisB463. The authors then demonstrated that an Eco R1 fragment of identical size to the DNA inserted in the hybrid phage and complementary to it could be detected in yeast DNA by the blotting method of Southern (1975). They then concluded that yeast DNA was able to complement the lesion in hisB463. They also concluded that transcription of the eukaryotic sequences was initiated in the inserted eukaryotic DNA because the orientation of the eukaryotic DNA in the phage vector did not affect the ability of the DNA to complement the missing function.

In a subsequent study Struhl and Davis (1977) demonstrated that the cloned yeast DNA in the recombinant phage λgtSchis contained the structural gene for the enzyme IGP dehydratase. Both genetic and biochemical evidence supported this conclusion. The Eco R1 fragment equivalent to the fragment cloned in the initial study was isolated from mutants of yeast lacking IGP hydratase activity and cloned. These yeast DNA fragments could not complement the lesion in hisB463. However, his⁺ recombinants could be obtained from phage crosses between several of the noncomplementing hybrid phages. Complementation of the hisB mutant was shown not to be due to suppression because IGP dehydratase activity could be found in *E. coli* containing the yeast fragment even if the entire histidine locus had been deleted from the cells. The enzymatic activity found in the lysogenized bacterial cells had properties similar to those of the wild type yeast enzyme and not to the homologous *E. coli* enzyme.

A slightly different approach was taken by Ratzkin and Carbon (1977). Randomly sheared DNA from yeast was joined to the plasmid Col E1 by the poly(dA · dT) connector method (Lobban and Kaiser, 1973). The yeast DNA to be cloned was randomly sheared rather than cleaved with a restriction endonuclease in order to ensure that any gene of interest remained intact in at least a portion of the DNA fragments generated. The hybrid DNA was then used to transform *E. coli* to colicin resistance. All of the transformants were pooled, and hybrid plasmid DNA was isolated and used to transform various auxotrophic mutants. One of the mutants used was hisB463. Three plasmids were identified that converted hisB463 to his⁺. The three plasmids contained a segment of yeast DNA of similar size and with a single Eco R1 restriction site. The inserted DNA was shown to have sequences in common with those in λgtSchis by the demonstration that the addition of single-stranded DNA from λgtSchis markedly increased the rate of reassociation of pYehis2 DNA.

The pool of hybrid plasmids was also used to transform JA199, a strain of *E. coli* that contains a lesion in the leuB gene, which encodes β-isopropylmalate dehydrogenase. After transformation leu⁺ colonies appeared at a frequency of 10^{-6}. Restriction analysis of the plasmid DNA isolated from some of the leu⁺ transformants revealed four different patterns of R1 restriction fragments. DNA from the four types of hybrid plasmids were used to transform three different leuB mutants. DNA from one of the types, pYeleu10, could complement all of the leuB mutants, but could not complement mutations in other genes of the leu operon. This plasmid had been isolated from the fastest-growing leu⁺ transformant. The other three plasmids could only complement the point mutation present in JA199. This result suggests that these plasmids were suppressing the mutation in JA199.

Because well-characterized deletions of the leuB region in *E. coli* were not available, the investigators utilized leu deletion mutants in *Salmonella typhimurium* to provide evidence that the DNA in pYeleu10 contained the structural gene for an enzyme that could replace β-isopropylmalate dehydrogenase. The pYeleu10 DNA was transferred to the *Salmonella* mutants by F-mediated transfer (Clarke and Carbon, 1976). The plasmid DNA could complement deletions of the leuB gene, but not deletions in other genes of the leu operon. No biochemical characterization of the complementing activity was done in this study.

The studies of Struhl and colleagues and of Ratzkin and Carbon clearly demonstrate that functional eukaryotic proteins can be produced in bacterial cells. Other studies quickly followed in which complementation of *E. coli* auxotrophic mutants by cloned fragments of yeast or neurospora DNA was reported (see table 1).

The cloning and expression of the structural gene for catabolic dehydroquinase from *Neurospora crassa* demonstrated that the subunit assembly of eukaryotic enzymes can occur in prokaryotic cells, since the active form of this enzyme is a multimer of 20 identical subunits of molecular weight 10,000 (Vapnek et al., 1977). This is probably a general phenomenon since β-galactosidase, an enzyme made up of four subunits of molecular weight 135,000, has been successfully expressed in *E. coli* containing a DNA fragment from the yeast *Kluyveromyces lactis* (Dickson and Markin, 1978). Both of these studies used immunological characterization in addition to biochemical characterization to demonstrate that the complementing activity was the eukaryotic enzyme homologous to the *E. coli* enzyme.

The first report of the successful expression of an enzyme from a higher eukaryote was made by Chang and her collaborators (1978) when they reported the expression of mouse dihydrofolate reductase (DHFR). This enzyme, which catalyzes the conversion of dihydrofolic acid to tetrahydrofolic acid, has a much lower affinity for the antimetabolic drug trimethoprim than does the corresponding bacterial enzyme (Burchall and Hitching, 1965). Therefore, a bacterial cell able to synthesize this enzyme in a functional form would be resistant to levels of trimethoprim that ordinarily would inhibit its growth.

Sequences encoding DHFR were initially obtained by making ds-cDNA to partially purified mRNA isolated from cultured methotrexate-resistant mouse cells. The double-stranded cDNA was inserted into the Pst restriction endonuclease site of the plasmid pBR322 by the oligo (dG·dC) joining method (Boyer et al., 1977). This plasmid is a popular cloning vehicle because it is amplifiable, it encodes resistance to both tetracycline and ampicillin (Bolivar et al., 1977), and its entire sequence is known (Sutcliffe,

1978). The annealed molecules were used to transform *E. coli* strain $_x$1776 as required by the NIH guidelines for recombinant DNA research. This strain of *E. coli* is resistant to a high level of trimethoprim because of its thy$^-$ mutation, so the direct selection scheme could not be used. The investigators therefore identified the transformants containing sequences encoding DHFR by probing the DNA in the tetracycline-resistant colonies with cDNA made to highly purified DHFR mRNA. At this point $_x$2282, a thy$^+$ varient of $_x$1776, was approved for EK2 use, so the direct selection scheme could be used. Plasmid DNA was isolated from pools of the $_x$1776 transformants and used to transform $_x$2282. This strain was also directly transformed with freshly prepared ds-cDNA annealed to pBR322. Colonies that expressed increased resistance to trimethoprim were obtained in both sets of transformations. Partial sequence analysis of the cloned DNA provided direct verification that the sequences encoded DHFR, since it corresponded to the amino acid sequence reported for the mouse enzyme. The enzyme was also shown to have the biochemical and immunological properties of the mammalian enzyme. Some of the clones that had immunologically detectable DHFR were not resistant to trimethoprim, clearly demonstrating that immunoreactive molecules are not always biologically active. The level of expression of DHFR varied in independently derived transformants. The variation in expression appeared to depend at least in part on the length of the d(G · C) tail that connected the coding sequences, although there is no simple relationship between length of the tail and expression. Deletion of 3' nontranslated regions of the eukaryotic mRNA also appeared to affect the level of expression.

In the studies utilizing yeast or neurospora chromosomal DNA, the DNA was inserted into the phage or plasmid vectors such that no bacterial control elements were located adjacent to the inserted DNA. Thus, it is not surprising that initiation of transcription occurs within the inserted DNA in these cases. It is not clear whether these transcriptional signals are authentic yeast signals recognized by the *E. coli* enzymes or whether a sequence that fortuitously resembles an *E. coli* sequence is present in those sequences that have been successfully expressed.

The sequences encoding the mammalian enzyme were inserted into the gene encoding β-lactamase with the expectation that transcription would be controlled by the β-lactamase promoter. However, expression occurred even in cases where the coding region was out of phase with the β-lactamase coding sequence or in the incorrect orientation. The structure of one of the plasmids that led to the expression of DHFR as 0.01% of the soluble protein is β-lactamase $_{(1-184)}$–d(G)$_{11}$-ATG-DHFR. In this plasmid the DHFR sequence is out of phase with the β-lactamase sequence. A protein

TABLE 1

Eukaryotic Proteins Synthesized in Prokaryotic Cells

Protein	Organism	Type of DNA Cloned	Cloning Method	Evidence for Expression	Promoter	Product	Clone Designation	References
Enzymes								
Imidazole glycerol phosphate dehydratase	*Saccharomyces cerevisiae*	Chromosomal	Ligation of R1 fragments to R1 arms to λgt	D complementation[1]; genetic[2] biochemical[1]	In inserted DNA	Active enzyme	gt-Schis	Struhl et al. (1976); Struhl, Davis (1977)
Imidazole glycerol phosphate dehydratase	*Saccharomyces cerevisiae*	Chromosomal	Poly d(A)·d(T) joining of random shear DNA to plasmid ColE1	S + D complementation	In inserted DNA	Active enzyme	pYehis	Ratzkin, Carbon (1977); Clark, Carbon (1976)
β-isopropylmalate dehydrogenase	*Saccharomyces cerevisiae*	Chromosomal	Poly d(A)·d(T) joining of random shear DNA to plasmid ColE1	D complementation	In inserted DNA	Active enzyme	pYeleu	Ratzkin, Carbon (1977); Clark, Carbon (1976)
Catabolic dehydroquinase (5-dehydroquinate hydrolyase)	*Neurospora crassa*	Chromosomal	Ligation of R1-Hind III fragments to R1-Hind III cut pBR322	Complementation; biochemical; physical[4]	In inserted DNA	Active enzyme	pVK	Vapnek et al. (1977); Alton et al (1978)
Argininosuccinate lyase	*Saccharomyces cerevisiae*	Chromosomal	Poly d(A)·d(T) joining of random shear DNA to ColE1		In inserted DNA	Active enzyme	pYearg	Clark, Carbon (1976)
β-galactosidase	*Kluyveromyces lactis*	Chromosomal	Ligation of R1 partial digestion fragments to R1 cut pBR322	D complementation; physical; biochemical; immunological[?]	Probably in inserted DNA	Active enzyme	pK1	Dickson, Markin (1978)
Tryptophan synthetase	*Saccharomyces cerevisiae*	Chromosomal	Recloning of Bam fragments from ColE1 (dA·dT) yeast DNA hybrids into pBR313	D complementation; genetic	IS2 inserted into cloned yeast fragment	Active enzyme	pYetrp 5	Waltz, et al (1978)
Dihydrofolate reductase	AT-3000 mouse cells	ds-cDNA	Oligo d(G)·d(C) joining of ds-cDNA to Pst linearized pBR322	Biochemical; immunological; complementations;[?] physical	β-lactamase: possibly in inserted DNA in some cases	Active enzyme	pDHFR	Chang et al (1978)
Orotine-5'-phosphatase	*Saccharomyces cerevisiae*	Chromosomal	Poly d(A)·d(T) joining of random shear DNA to pMB9	D complementation	Probably in inserted DNA	Active enzyme	pDHFR	Bach et al. (1979)
Galactokinase	*Saccharomyces cerevisiae*	Chromosomal	Ligation of BglIII cut DNA to Bam cut pBR322	D complementation	Probably in inserted DNA	Active enzyme	GB	Schell, Wilson (1979)
Galactokinase	*Saccharomyces cerevisiae*	Chromosomal	Ligation of HindIII cut DNA to HindIII cut pBR322	D complementation; physical	Probably in inserted DNA	Active enzyme	pJD	Citron et al. (1979)

Hormones

Protein	Source	DNA	Construction	Detection	Fused to	Fusion type	Plasmid	Reference
Somatostatin	Human	Synthetic DNA	Ligation of R_1-Bam fragment to pBR322 bearing β-galactosidase gene	Immunological; biological assay	β-galactosidase	Intracellular fusion protein	pSOM	Itakura et al. (1977)
Proinsulin	Rat	ds-cDNA	Oligo d(G)·d(C) joining of ds-cDNA to Pst linearized pBR322	Immunological; biological assay	β-lactamase	Secreted fusion protein	pI	Villa-Komaroff et al. (1978)
A and B chains of insulin	Human	Synthetic DNA	As in somatostatin	Immunological; physical	β-galactosidase	intracellular fusion protein	pIA pIB	Goeddel et al. (1979b)
Growth hormone	Rat	ds-cDNA	Poly d(A)·d(T) joining to pMB9 followed by restriction and recloning in pBR322	Immunological; physical	β-lactamase	Intracellular fusion protein	pE-RGH	Seeburg et al. (1978)
Growth hormone	Human	ds-cDNA	Addition of synthetic linkers to ds-cDNA, connection to pBR322 restriction, and recloning into ptrpEDS-1	Immunological; physical	Trytophan	Intracellular fusion protein	ptrpED50	Martial et al. (1979)
Growth hormone	Human	ds-cDNA and synthetic DNA	Oligo d(G)d(C) connection of ds-cDNA to pBR322 followed by restriction, connection of synthetic DNA, and recloning	Immunological; physical	β-galactosidase	Not a fusion protein	pHGH	Goeddel et al. (1979a)

Viral Peptides

Protein	Source	DNA	Construction	Detection	Fused to	Fusion type	Plasmid	Reference
Hepatitis B core antigen		Viral DNA	Oligo d(G)·d(C) joining of DNA to Pst linearized pBR322	Immunological	β-lactamase	Not a fusion protein	pHVB	Pasek et al. (in press)
SV-40 small t antigen		Viral DNA	Combination of restriction cleavage and digestion with S_1 nuclease	Immunological; physical	β-galactosidase	Not a fusion protein	pTR	Roberts et al. (1979a)

Other Peptides

Protein	Source	DNA	Construction	Detection	Fused to	Fusion type	Plasmid	Reference
Ovalbumin	Chicken	ds-cDNA	Use of R_1 linkers to attach Taq fragment of cloned ds-cDNA to the lac UVs promoter plasmid p0P203	Immunological; physical	β-galactosidase	Fusion protein	pUC1001	Fraser, Bruce (1978)
Ovalbumin	Chicken	ds-cDNA	Blunt-end ligation of a Hha fragment of cloned ds-cDNA to the lacUV5 promoter plasmid pOMPO	Immunological; physical	β-galactosidase	Fusion protein	pOMP	Mercereau-Pujalon et al. (1978)
Actin	*Dictyostelium discoideum*	Chromosomal	Poly d(A)·d(T) joining of random shear DNA to pMB9	Physical; biochemical	?	Not a fusion protein	M6 KHI0 pDd actin 2	Kindle, Firtel (1978)

1. D-complementation means deletion mutants were used; S-complementation means point mutants were used.

2. Refers to the use of yeast mutants.

3. Refers to properties of the enzyme such as heat stability, pH optimum, Km for substrate cation dependence.   Different studies studied different properties.

4. Refers to properties such as sedimentation rate, migration in polyacrylamide gels.

5. Refers to the use of antibodies specific for the eukaryotic protein in question.

6. In this case, the inserted DNA does not complement a mutation in the host but rather confers a new property, i.e., increased resistance to trimethoprim.

of the same size as DHFR can be detected in cells containing this plasmid. The authors suggest that a ribosome binding site has been constructed by the Pst-oligo–d(G)–ATG sequence. One of the plasmids that allowed expression of DHFR has the DNA inserted in the incorrect orientation with respect to β-lactamase, suggesting that transcription is occurring from a sequence in the distal portion of the β-lactamase gene.

CLONING AND EXPRESSION OF MAMMALIAN HORMONES

The expression of a mammalian hormone in a bacterial cell would not be expected to confer a selectable phenotype upon a bacterium, so an alternative method must be used to detect the synthesis of these proteins. Immunological reactivity with antibody made against the hormone of interest has been the assay of choice (see table 1). The DNA sequences encoding the hormones have been synthesized either chemically or enzymatically using mRNA as the template. In many cases the chromosomal sequences encoding these proteins do contain introns (Lomedico et al., 1979; Fiddes et al., 1979), so that the chromosomal sequences cannot be expected to be expressed in *E. coli*. The biological activity of the hormones produced in the bacterial cells has been demonstrated only in a few cases.

The first hormone to be produced in *E. coli* was human somatostatin (Itakura et al., 1977). Somatostatin, a peptide made up of 14 amino acids, inhibits the secretion of a number of hormones, including growth hormone, insulin, and glucogon. Since the amino acid sequence of the peptide was known, a short DNA fragment could be designed that encoded those 14 amino acids. In order to optimize the expression of this sequence, the investigators designed the sequence with the following features. First, the amino acid codons known to be favored in *E. coli* for the expression of the MS2 genome were used where possible. Second, the fragments were designed to avoid inter- and intramolecular interactions that might interfere with the assembly of the overlapping fragments that were planned. Third, dG·dC rich sequences followed by dA·dT rich sequences were avoided since such sequences might terminate transcription. Fourth, a methionine codon was inserted preceding the normal NH$_2$-terminal amino acid of somatostatin; and two nonsense codons were inserted following the COOH-terminal codon. A feature of the sequence that facilitated its insertion into the plasmid vector pBR322 was the presence of single-stranded cohesive termini for the Eco R1 and the Bam H1 restriction endonucleases at the 5′ termini.

The triester method was used to synthesize eight fragments (Itakura et al., 1975), which were then annealed, ligated together, and inserted into one of two plasmids derived from pBR322. Each of these plasmids has an Eco R1 site at a different region of the β-galactosidase gene. Insertion of the synthetic gene into either of these R1 sites brings the expression of the gene under control of the lac operon. After insertion of the somatostatin fragment into these plasmids, transcription and translation should result in a somatostatin peptide preceded either by ten amino acids (pSOM1 series) or by virtually the entire β-galactosidase subunit structure (pSOM11 series). Since there are no methionine residues in the 14 amino acids of somatostatin, the hormone can be released from the hybrid polypeptide by cyanogen bromide cleavage.

The recombinant plasmid pSOM1 was constructed first. Nucleotide sequence analysis was used to confirm the insertion of the proper sequence in the proper orientation to produce a somatostatin peptide. Cyanogen bromide-treated extracts from the cells containing this plasmid were assayed by a radioimmune assay for the presence of somatostatin. None could be detected. Because reconstruction experiments showed that somatostatin was very rapidly degraded in bacterial extracts, the investigators concluded that the failure to detect somatostatin was due to intracellular degradation of the peptide. They therefore decided to try to protect the peptide by attaching it to a large polypeptide, e.g., the subunit of β-galactosidase. The synthetic fragment was attached to the second plasmid described above. Cyanogen bromide-treated extracts from a number of independent transformants were analyzed for somatostatin activity. In this case four clones had easily detectable amounts of radioimmune activity. In most cases no radioimmune activity was detected if the extracts were not treated with cyanogen bromide. This was not unexpected since the antiserum used required a free NH_2-terminal alanine. These clones as well as several that did not have any activity were analyzed by restriction endonucleases. Those clones that were expressing a peptide with somatostatin antigenic determinants had the appropriate orientation of the fragment while the others had the opposite orientation.

The synthesis of the hybrid peptide was shown to be under the control of the lac operon by showing that the level of somatostatin radioimmune activity increased if the cells were treated with an inducer of the lac operon. The total yield of somatostatin varied from 0.001 to 0.03% of the total cell protein. The overproduction of incomplete and inactive β-galactosidase appeared to confer a disadvantage for cell growth since deletions of the lac region occurred with a high frequency: after 15 generations about one-half of the cells in culture were no longer constitutive for the synthesis of β-galactosidase.

The cyanogen bromide-treated extracts of the cells containing somatostatin radioimmune activity inhibited the release of growth hormone from rat pituitary cells, whereas extracts from cells with no radioimmune activity had no effect.

The chemical synthesis of a somatostatin gene was an elegant and impressive accomplishment; however, this approach is limited to peptides that are small and whose amino acid sequence is known. A more generally applicable technique is the use of mRNA encoding the protein of interest as a template for the synthesis of ds-cDNA. This approach was used by Villa-Komaroff and her colleagues (1978) to clone and express one of the rat insulin genes.

Although mature insulin contains two chains of 20 and 30 amino acids, it is initially synthesized as part of a longer chain of 109 amino acids called preproinsulin (Chan et al., 1976; Steiner et al., 1972). A hydrophobic leader sequence is cleaved off of the nascent molecule, presumably as the polypeptide chain moves through the endoplasmic reticulum (Blobel and Dobberstein, 1975), producing a proinsulin molecule. The proinsulin molecule folds into its proper tertiary structure and the C peptide is cleaved from its middle. If the two chains of insulin are denatured, they do not easily or efficiently reassemble correctly (Humbel et al., 1972). The rat and the mouse contain two distinct genes for insulin, unlike most other vertebrates, which contain only one (Clark and Steiner, 1969).

An X ray-induced transplantable rat beta cell tumor (Chick et al., 1977) was used as the source of preproinsulin mRNA. The total poly(A) containing RNA of the cells was used as template to synthesize ds-cDNA. The ds-cDNA was inserted into the Pst site of the plasmid pBR322 by the oligo(dG·dC) joining procedure and used to transform the *E. coli* strain $_x$1776. Because the mRNA encoding preproinsulin was only a small fraction of the total mRNA, purified mRNA was not available to screen the transformants. (Subsequent analysis of the mRNA indicated that the insulin mRNA constituted 0.3% of the total mRNA.) However, Ullrich and his collaborators (1977) had cloned and sequenced ds-cDNA corresponding to parts of the rat insulin genes. They reported that treatment of cDNA with Hae III, a restriction endonuclease that will cleave single-stranded DNA (Horiuchi and Zinder, 1975), generated an 80 base fragment greatly enriched for insulin sequences. The investigators therefore cleaved cDNA made to insulinoma mRNA with Hae III and isolated an 80 base fragment. This fragment was used to screen one-third of the transformants that had been obtained. About 20% of the colonies hybridized to this probe, but restriction analysis of a few of the transformants which hybridized most strongly to the probe indicated that the cloned sequences did not correspond to insulin. Some of the positive clones were rescreened using the hybrid arrested translation method (Paterson et al., 1977). This method is based on the fact that mRNA in the form of an RNA:DNA molecule will not direct cell-free protein synthesis. Aliquots of the insulinoma mRNA were incubated with linearized plasmid DNA from nine clones under conditions that favored DNA:RNA hybridization (Casey and Davidson, 1977), and a radioimmune assay was used to look for the specific inhibition of insulin synthesis. DNA from one of the clones inhibited both total protein synthesis and the synthesis of im-munoprecipitable material. Direct sequence analysis confirmed the presence of insulin sequences in this clone. The inserted DNA encoded the entire preproinsulin chain except for the first two amino acid residues of the reported preregion. The insulin sequence was excised from the plasmid with Pst and used to probe the rest of the transformants. Forty-eight of 1,745 clones contained sequences homologous to the sequence in clone pI19.

The ds-cDNA had been inserted into the Pst site of pBR322 because this site is in the coding region for β-lactamase, the enzyme that confers resistance against penicillin. This enzyme is a periplasmic enzyme. Insertion of the insulin sequences should cause expression of the insulin sequence as a hybrid protein transported outside of the cell, if the inserted DNA were in the correct orientation and connected to the coding region of the plasmid DNA with a $d(G) \cdot d(C)$ tail that put the triplets of the insulin insertion in phase with the coding region of the plasmid. The hybrid protein should contain 159 amino acids of penicillinase connected by a glycine bridge to a portion of the preproinsulin sequences. The 48 transformants containing insulin sequences were screened with a sandwich radioimmune assay (Broome and Gilbert, 1978) for the presence of a hybrid protein with both penicillinase and insulin antigenic determinants. One of the clones contained such a protein, and this protein was shown to be present in the periplasm of the cells. The DNA in the clone was sequenced and shown to contain the sequence of proinsulin inserted in the correct orientation. In this clone amino acid 182 of penicillinase is connected by six glycine residues to the fourth amino acid of proinsulin.

In their initial report the investigators thought that the penicillinase portion of the hybrid molecule retained enzymatic activity because many of the transformants containing insulinoma sequences were resistant to ampicillin. Subsequent investigation has indicated that these cells most likely contained plasmids, in addition to the recombinant plasmids, that had lost the insertion and had regenerated an intact penicillinase gene (Villa-Komaroff et al., 1979; Villa-Komaroff and Stahl, unpub. obs.).

Because the detection of antigenic determinants does not necessarily imply biological activity, the insulin portion of the hybrid protein produced in the bacterial cells was assayed for its ability to stimulate the uptake of glucose by fat cells (Naber et al., in preparation). The assay measures the conversion of $1\text{-}^{14}C$-glucose to $^{14}CO_2$ (Renold et al., 1960). Although this conversion is stimulated by both insulin and proinsulin, proinsulin is only 2 to 10% as active as insulin in this bioassay (Rubenstein et al., 1972). Trypsinization of proinsulin removes the connecting C peptide and generates a molecule with the same activity as insulin. Because the oxidation of glucose can be stimulated nonspecifically by a variety of materials, the hybrid protein was partially purified on an antibody affinity column, then treated with trypsin and analyzed. The trypsin-treated material had biological activity equal to or greater than that predicted by radioimmune assay of the material. The biological activity could be suppressed by anti-insulin serum, and no activity was detected in cells that had no radioimmune activity.

The combination of chemical synthesis and enzymatic synthesis of DNA fragments is perhaps the most versatile approach. This is illustrated by the elegant work of Goeddel and his coworkers in the production of human growth hormone (1979a).

Human growth hormone, a protein of 191 amino acids, is synthesized in the anterior lobe of the pituitary. HGH is initially synthesized with a 24 amino acid signal sequence (Bancroft, 1973). The investigators knew from their previous work (Goodman et al., 1979) that sites for the restriction enzyme HaeIII were present in the 3' noncoding region of the mRNA and in the sequence encoding amino acids 23 and 24 of HGH. They therefore synthesized ds-cDNA using as template total poly(A) containing RNA from pituitary. The ds-cDNA was digested with the restriction endonuclease HaeIII, and a fragment of 550 base pairs was isolated by gel electrophoresis. This fragment was then inserted into the Pst site of pBR322 with the oligo(dG·dC) joining method. This method will not only regenerate the Pst site on either side of the insertion but will also regenerate the HaeIII sites in the ds-cDNA. The hybrid DNA was used to transform $_x$1776. The DNA in the transformants was screened in the colony hybridization assay (Grunstein and Hogness, 1975) by using radiolabeled cDNA to somatomamotropin mRNA (Shine et al., 1977). This probe is nearly identical to human growth hormone in nucleotide sequence (Goodman et al., 1979). DNA in one of the positive clones was sequenced and shown to correspond to the 550 base pair fragment encoding most of HGH.

Coding sequences for the first 24 amino acids of HGH were chemically synthesized as 12 overlapping segments that contained an Eco R1 site and a methionine codon preceding the first HGH codon, a HaeIII site between amino acids 23 and 24, and a HindIII site after the codon for amino acid 24. This synthetic fragment was inserted into pBR322 cut with both R1 and HindIII. The sequence of the cloned fragment was verified by direct sequence analysis.

A DNA fragment containing the entire HGH sequence was produced by ligating the two separately cloned sequences together. The judicious use of restriction enzymes allowed the investigators to select for the fragment with the correct alignment of the two pieces. The fragment was inserted into a plasmid vector containing two lac promoters. Again, the use of restriction enzymes insured the insertion of the fragment in the proper orientation with respect to the promoters. The initial construction resulted in a hybrid plasmid (pHGH107) where the lac ribosome binding site was separated from the initiation codon by 11 base pairs. In the naturally occuring lac operon, there are 7 base pairs between the ribosome binding site and the initiation codon, so the investigators also constructed a plasmid pHGH197-1) where the exact lac sequence was restored.

Extracts of $_x$1776 containing the two plasmids were tested for HGH expression by radioimmune assay. HGH could be detected in bacterial extracts from cells containing either pHGH107 or pHGH107-1; however, the introduction of four extra base pairs resulted in the increased, rather than decreased, expression. Five polypeptides could be precipitated with anti-HGH serum: one of them comigrated with authentic growth hormone on SDS-polyacrylamide gel electrophoresis; the other four were smaller. To verify that the synthesis of the HGH was under control of the lac operon, a strain of bacteria that overproduces the lac repressor was transformed with pHGH107. In these cells no HGH antigenic activity could be found unless an inducer of the lac operon was added. The peptide produced in the bacterial cells has not yet been assayed for biological activity.

CLONING AND EXPRESSION OF VIRAL POLYPEPTIDES

Hormones produced in bacterial cells must be biologically active (or able to be converted to a biologically active state) in order to be of use therapeutically. But viral polypeptides could be used for the production of vaccines if they simply contain antigenic determinants that are present on a virus particle.

Hepatitis B virus (HBV) infection is widespread. Between 3 and 5% of healthy blood donors in Western Europe and the U.S.A. show serological evidence of past infection, and about 0.1% are chronically infected. In many Asian and African countries, the majority of the adult population has been infected, and 5 to 10% are chronically infected (Szmuness et al., 1978). Although most infections are subclinical and are followed by complete recovery, a significant proportion result in more serious disease.

Plasma from some blood donors and from patients infected with HBV contains particles that are probably infectious virions. These particles consist of an outer envelope containing the HBV surface antigen and an inner core containing the core antigen, a double-stranded circular DNA molecule with a gap in one strand, and a DNA polymerase activity that can fill this gap. Passively or actively acquired antibody to the surface antigen of HBV confers some immunity to subsequent infection, indicating that a vaccine might be of use in preventing the disease. However, HBV cannot be grown in tissue culture and normally infects only humans and apes, so both the preparation of vaccines to test and the study of the virus have been limited.

DNA from the virus particles has been cloned by several groups (Burrell et al., 1979; Charnay et al., 1979; Sninsky et al., 1979), and the expression of the core antigen has been reported (Burrell et al., 1979; Pasek et al., 1979). The expression of core antigen was detected in bacterial cells that had been transformed with restriction fragments of HBV DNA that had been inserted into the Pst site of pBR322 by the oligo(dG·dC) joining method. All of the recombinants examined that produced the core antigen have the HBV DNA inserted in phase with the β-lactamase gene, and all begin within one or two bases of the same sequence. This unique attachment was not expected because the original fragments came from DNA digested with different enzymes. The investigators expected that the core antigen would be expressed as a hybrid protein containing both β-lactamase and all or part of the core antigen. This was not the case. Instead, in the 6 producing strains, the β-lactamase sequence is fused by 5 to 8 glycine codons to a peptide sequence that terminates 25 amino acid codons farther on. Three nucleotides after the stop codon there is an initiation codon and a sequence encoding a polypeptide of 183 residues. Termination codons are abundant in the other two possible translation phases throughout this region. The investigators conclude that the core antigen is translated *de novo* from a hybrid transcript.

To establish the feasibility of vaccine production from viral antigens produced in bacterial cells, sterile extracts of the bacterial cells were mixed with Freund's adjuvant and injected into rabbits. Additional injections of the same sample were given 2 and 5 weeks after the initial injection. Two animals were injected with crude extracts, and two were injected with partially purified material. Sera from all four rabbits contained antibodies against the core antigen of the virus.

MAXIMIZING THE EXPRESSION OF CLONED SEQUENCES

Many of the potential uses of eukaryotic proteins made in prokaryotic cells depends on the ability to make these proteins in large amounts. The amount of a particular protein made in a cell varies in the different systems studied. For example, up to 1.5% of the total cellular protein is ovalbumin when the structural sequences for this protein are inserted into the lac operon (Fraser and Bruce, 1978); on the other hand, only 100 molecules per cell of insulin were made when these sequences were inserted under the control of β-lactamase (Villa-Komaroff et al., 1978). This difference cannot be accounted for on the basis of the intrinsic differences between the two promoters. If the β-lactamase promoter were as effective as it could be, we would expect 2×10^5 molecules of insulin per cell. Many factors could account for the difference in the amount of protein synthesized. Specific considerations include the stability of the protein within the bacterial cell and the stability of the RNA transcript encoding the protein. A more general consideration is the presence of any bacterial function that interferes with transcription or translation of the eukaryotic sequences. The structure of the hybrid gene is clearly important since high rates of expression depend on efficient transcription and translation, which depend in turn on the presence of a strong promoter and a properly positioned ribosome binding site.

The cloning of somatostatin demonstrated the importance of the stability of the protein within the bacterial cell. As mentioned earlier, the expression of this protein was not detected unless it was protected by the presence of a large polypeptide. Insulin is also susceptible to bacterial proteases in mixing experiments (Goldberg, pers. comm.). However, unlike somatostatin, the presence of a bacterial protein covalently linked to insulin does not appear necessary to protect it from degradation. Plasmids have been constructed in which all of the β-lactamase sequences are deleted and insulin is still detected in approximately the same amount as long as the protein is secreted to the periplasmic space. If the polypeptide cannot be transported to the periplasm, much less protein can be detected (Talmadge, pers. comm.).

The state of the cells may also affect the stability of the foreign protein. Villa-Komaroff and her collaborators (1978) found that no insulin could be detected if the cells were grown in rich broth. Goeddel and his collaborators (1979a) found that the amount of growth hormone detected decreased if the cells were assayed in stationary phase instead of log phase.

The stability of the RNA encoding the foreign protein also affects the amount of the protein that can be synthesized. Hautala and her coworkers (1979) have shown that mRNA encoding the eukaryotic enzyme catabolic dehydroquinase is stabilized in bacterial cells that are deficient in polynucleotide phosphorylase. In these cells the half-life of this mRNA is 2.8 times greater than it is in wild-type cells. This results in a 20- to 100-fold increase in the amount of catabolic dehydroquinase produced in the bacterial cells. The half-life of prokaryotic mRNAs encoded by the plasmid vector does not increase, and the expression of the *E. coli* enzyme does not increase in this mutant; however, the copy number of the plasmid does increase in the polynucleotide phosphorylase-deficient cells. When these cells were transformed with a plasmid containing the insulin sequences under the control of the β-lactamase gene but with most of the β-lactamase sequences deleted, the amount of insulin detected increased to 5,000 molecules per cell (Talmadge, pers. comm). The synthetic insulin chains synthesized by Goeddel and his colleagues (1979b) are synthesized in large amounts in the bacterial cell. The difference between the amount of protein produced in the two cases may in part reflect the instability of the eukaryotic mRNA sequence since in this case a similar polypeptide chain is encoded by two different sequences, a synthetic one and the natural one. Clearly not all eukaryotic sequences are particularly unstable. Human and rat growth hormone as well as ovalbumin have been produced in bacterial cells in reasonable amounts (Seeburg et al., 1978; Martial et al., 1979; Fraser and Bruce, 1978; Mercereau-Puijalon et al., 1978).

There may be other functions in the host cell that interfere with the production of the eukaryotic protein. Vapnek and coworkers (1977) noted that they obtained different levels of complementation in different strains of *E. coli* containing the mutation that they wished to complement. The plasmid encoded the information that was responsible for the complementation, but the extent of complementation depended on information encoded by the bacterial chromosome. The bacterial function was not characterized in this case.

Some of the functions that interfere with the production of a eukaryotic protein may be encoded on the cloned eukaryotic sequence. Clark and Carbon transformed arg⁻ cells with hybrid plasmids containing yeast DNA that they knew complemented argH deletions and selected for faster growing variants. These variants were shown to contain higher levels of the enzyme argininosuccinate lyase. In this case the alterations occurred not in the bacterial genome but in the inserted yeast DNA. These alterations included deletions, insertions, and rearrangements. In one case the rearrangement responsible for the increased expression of a eukaryotic gene was shown to be the insertion of IS2 into the yeast DNA (Waltz et al., 1978). IS2 is a bacterial insertion sequence that is able to act as a negative effector in one orientation and as a gene activator in the opposite orientation (for review see Kleckner, 1977).

The distance between the promoter and the ribosome binding site is a critical factor in the efficiency of expression of an inserted eukaryotic gene. A ribosome binding site consists of the translational start codon, AUG (or GUG) and a sequence complementary to bases on the 3' end of the 16S ribosomal RNA (Shine and Delgarno, 1975). The Shine and Delgarno sequences (SD sequences) have been found in most of the prokaryotic mRNAs examined. They consist of 3 to 9 bases positioned 3 to 12 bases from the AUG (for review see Steitz, 1978). Variation in the distances between a strong promoter and the eukaryotic sequences can be obtained by using the homopolymer connector method as in the cloning of insulin (Villa-Komaroff et al., 1978) or dihydrofolate reductase (Chang et al., 1978). A limited variation in this separation can be obtained by removing the overlapping ends of a restriction cut between the promoter and the ribosome binding site and religating the blunt ends (Goedell et al., 1979a). Several vectors have been constructed that either contain strong promoters or result in an increased copy number of a plasmid vector (Bernard et al., 1979; Rao and Rogers, 1978; Uhlin et al., 1979). Although these vectors will be useful, a general method is needed for positioning a promoter at any distance from any gene and for varying the distance between the SD sequence and the start codon. Such a method has been developed by Roberts and colleagues (1979b). The promoter used for this work is a cloned fragment containing two lac promoters (Backman et al., 1976; Backman and Ptashne, 1978). This fragment does not encode a translational start, but it does contain an SD sequence. The use of this fragment should allow the construction of hybrid genes that will result in the expression of the eukaryotic protein with no amino acids in the polypeptide chain encoded by a bacterial gene. The basic approach is to clone a gene such that a unique restriction site is located within 100 base pairs of the 5' end of the gene. The hybrid plasmid is then linearized with that restriction enzyme, and varying amounts of DNA are removed by digesting the DNA first with exonuclease III (under conditions where the Exo III removes 8 to 10 bases per minute per end), then with the single-strand-specific exonuclease S1. The lac promoter fragment is then inserted into the plasmid. This procedure results in a set of plasmids containing the promoter at varying distances from the gene.

The investigators first used this method to examine the effect of the distance between the promoter and the gene on the expression of λ cro protein. They found that the level of cro protein produced was strongly dependent on the structure of the hybrid gene, but no general rule describing the optimal structure could be deduced. For example, one plasmid (pTR213) resulted in the synthesis of about 190,000 monomers of cro per cell (1.6% of the soluble protein). Another plasmid (pTR199) contains 3 less base pairs than pTR213 and directs the synthesis of only 0.1 of the amount protein as pTR213. A third plasmid (pTR214) contains 8 base pairs less than pTR213 but directs the synthesis of about the same amount of protein as pTR213.

Roberts and colleagues (1979a) used this method to produce the small t antigen of the virus SV-40. Viral DNA containing the small t gene was cloned, and the lac fragment was inserted as described above. The sequence of the region of the fusion between the bacterial gene and the viral gene was determined for plasmids isolated from clones which expressed varying amounts of small t. The expression of small t was most efficiently directed by plasmids in which the SD sequence of the lac gene is positioned close to the ATG of the viral sequence. The most efficient expression was obtained from a plasmid where the SD sequence is 9 base pairs from the ATG. A plasmid where 11 base pairs separate the SD sequence from the ATG directs the synthesis of about one-half the amount of small t. The presence of 17 base pairs between the two components of the hybrid ribosome binding site results in the greatly decreased synthesis of small t. If the first two bases of the initiation codon are deleted, no synthesis of small t can be detected.

REFERENCES

Alton, N. K., J. A. Hautala, N. H. Giles, S. R. Kushner, and D. Vapnek. 1978. Transcription and translation in *E. coli* of hybrid plasmids containing the catabolic dehydroquinase gene from *Neurospora crassa*. Gene 4:241–59.

Bach, M.-L., F. Lacroute, and D. Botstein. 1979. Evidence for transcriptional regulation of orotidine-5-phosphate decarboxylase in yeast by hybridization of mRNA to the yeast structural gene cloned in *Escherichia coli*. Proc. Nat. Acad. Sci. USA 76:386–91.

Backman, K., M. Ptashne, and W. Gilbert. 1976. Construction of plasmids carrying the cl gene of bacteriophage λ. Proc. Nat. Acad. Sci. USA 73:4174–78.

Backman, K., and M. Ptashne. 1978. Maximizing gene expression on a plasmid using recombination in vitro. Cell 13:65–71.

Bancroft, F. C. 1973. Intracellular location of newly synthesized growth hormone. Exp. Cell Res. 79:275–78.

Bernard, H.-V., E. Remaut, M. V. Hershfield, H. K. Das, D. R. Helinski, C. Yanofsky, and N. Franklin. 1979. Construction of plasmid cloning vehicles that promote gene expression from the bacteriophage lambda PL promoter. Gene 5:59–76.

Blobel, G., and B. Dobberstein. 1975. Transfer of proteins across membranes. 1. The presence of proteolytically processed and unprocessed nascent immunoglobulin light chains on membrane-bound ribosomes of murine myeloma. J. Cell Biol. 67:835–51.

Bolivar, F., R. L. Rodriguez, P. J. Greene, M. C. Betlach, H. L. Heyneker, H. W. Boyer, J. H. Crossa, and S. Falkow. 1977. Construction and characterization of new cloning vehicles. II. A multipurpose cloning system. Gene 2:95–113.

Boyer, H. W., M. Betlack, F. Bolivar, R. L. Rodriguez, H. L. Heyneker, J. Shine, and H. M. Goodman. 1977. The construction of molecular cloning vehicles. *In* R. F. Beers, Jr., and E. G. Bassett (eds.), Recombinant molecules impact on science and society, pp. 9-20. Proc. 10th Miles Symp, Raven, New York.

Brack, C., and S. Tonegawa. 1977. Variable and constant parts of the immunoglobulin light chain gene of a mouse myeloma cell are 1250 nontranslated bases apart. Proc. Nat. Acad. Sci. USA 74:5652–56.

Breathnach, R., J. L. Mandel, and P. Chambon. 1977. Ovalbumin gene is split in chicken DNA. Nature 270:314–19.

Broome, S., and W. Gilbert. 1978. Immunological screening method to detect specific translation products. Proc. Nat. Acad. Sci. USA 75:2746–49.

Burchall, J. J., and G. H. Hitching. 1965. Inhibitor binding analysis of dihydrofolate reductase from various species. Molec. Pharmacol. 1:126–36.

Burrell, C. J., P. MacKay, P. J. Greenway, P. H. Hofschneider, and K. Murray. 1979. Expression in *Escherichia coli* of hepatitis B virus DNA sequences cloned in plasmid pBR322. Nature 279:43–47.

Casey, J., and H. Davidson. 1977. Rates of formation and thermal stabilities of RNA:DNA duplexes at high concentrations of formamide. Nucl. Acids Res. 4:1539–52.

Chan, S. J., P. Keim, and D. F. Steiner. 1976. Cell-free synthesis of rat preproinsulins: characterization and partial amino acid sequence determination. Proc. Nat. Acad. Sci. USA 73:1964–68.

Chang, A. C. Y., J. H. Nunberg, R. J. Kaufman, H. A. Erlich, R. T. Schmike, and S. N. Cohen. 1978. Phenotypic expression in *E. coli* of a DNA sequence coding for mouse dihydrofolate reductase. Nature 275:617–24.

Chang, A. C. Y., R. A. Lansman, and D. A. Clayton. 1975. Studies of mouse mitochondrial DNA in *Escherichia coli*: structure and function of the eukaryotic-prokaryotic chimeric plasmids. Cell 6:231–44.

Charnay, P., C. Pourcel, A. Louise, A. Fritsch, and P. Tiollais. 1979. Cloning in *Escherichia coli* and physical structure of hepatitis B virion DNA. Proc. Nat. Acad. Sci. USA 76:2222-2226.

Chick, W. L., S. Warren, R. N. Chute, A. A. Like, V. Lauris, and K. C. Kitchen. 1977. A transplantable insulinoma in the rat. Proc. Nat. Acad. Sci. USA 74:628–32.

Citron, B. A., M. Feiss, and J. E. Donelson. 1979. Expression of the yeast galactokinase gene in *Escherichia coli*. Gene 6:251–64.

Clark, J. L., and D. F. Steiner. 1969. Insulin biosynthesis in the rat: demonstration of two proinsulins. Proc. Nat. Acad. Sci. USA 62:278–85.

Clarke, L., and J. Carbon. 1976. A colony bank containing synthetic Col El hybrid plasmids representative of the entire *E. coli* genome. Cell 9:91–99.

Dickson, R. C., and J. S. Markin. 1978. Molecular cloning and expression in *E. coli* of a yeast gene coding for β-galactosidase. Cell 15:123–31.

Efstratiadis, A., and L. Villa-Komaroff. 1979. Cloning of double-stranded cDNA. *In* A. Hollaender and J. Setlow (eds.), Genetic engineering: Principles and methods, 1:15–36. Plenum, New York.

Fiddes, J. C., P. H. Seeburg, F. N. DeNoto, R. A. Hallewell, J. D. Baxter, and H. M. Goodman. 1979. Structure of genes for human growth hormone and chorionic somatommatropin. Proc. Nat. Acad. Sci. USA 76:4294–98.

Fraser, T. H., and B. J. Bruce. 1978. Chicken ovalbumin is synthesized and secreted by *Escherichia coli*. Proc. Nat. Acad. Sci. USA 75:5936–40.

Goeddel, D. V., H. L. Heyneker, T. Hogumi, R. Arentzen, K. Itakura, D. G. Yansura, M. Ross, G. Miozzari, R. Crea, and P. H. Seeburg. 1979a. Direct expression in *Escherichia coli* of a DNA sequence coding for human growth hormone. Nature 281:544–48.

Goeddel, D. V., D. G. Kleid, F. Bolivar, H. L. Heyneker, D. G. Yansura, R. Crea, T. Hirose, A. Kraszewski, K. Itakura, and A. D. Riggs. 1979b. Expression in *Escherichia coli* of chemically synthesized genes for human insulin. Proc. Nat. Acad. Sci. USA 76:106–11.

Goodman, H. M. et al. 1979. *In* Specific eukaryotic genes, pp. 179–90. J. Engberg, H. Klenow, and V. Leick (eds.), Munskagaard, Copenhagen.

Grunstein, M., and D. S. Hogness. 1975. Colony hybridization: method for the isolation of cloned DNAs that contain a specific gene. Proc. Nat. Acad. Sci. USA 72:3961–65.

Hautala, J. A., C. L. Bassett, N. Giles, and S. R. Kushner. 1979. Increased expression of a eukaryotic gene in *Escherichia coli* through stabilization of its messenger RNA. Proc. Nat. Acad. Sci. USA 76:5774–78.

Horiuchi, K., and N. D. Zinder. 1975. Site specific cleavage of single-stranded DNA by a Hemophilus restriction enzyme. Proc. Nat. Acad. Sci. USA 72:2555–58.

Humbel, R. E., H. R. Bossard, and H. Zahn. 1972. Chemistry of insulin. *In* D. F. Steiner and N. Frienkel (eds.), The handbook of physiology, 1:111–32. Williams and Wilkins, Baltimore.

Itakura, K., T. Hirose, R. Crea, A. D. Riggs, H. L. Heyneker, F. Bolivar, and H. W. Boyer. 1977. Expression in *Escherichia coli* of a chemically synthesized gene for the hormone somatostatin. Science 198:1056–63.

Itakura, K., N. Katagiri, S. A. Narang, C. P. Bahl, K. J. Marians and R. Wu. 1975. Chemical synthesis and sequence studies of deoxyribooligonucleotides which constitute the duplex sequence of the lactose operator of *Escherichia coli*. J. Biol. Chem. 250:4592–4600.

Jeffreys, A. J., and R. N. Flavell. 1977. The rabbit β-globin gene contains a large insert in the coding sequence. Cell 12:1097–1108.

Kedes, L. H., A. C. Y. Chang, D. Housman, and S. N. Cohen. 1975. Isolation of histone genes from unfractioned sea urchin DNA by subculture cloning in *E. coli*. Nature 255:533–538.

Kindle, K. and R. A. Firtel, 1978. Identification and analysis of Dictyostelium actin genes, a family of moderately repeated genes. Cell 15:763–78.

Kleckner, N. 1977. Translocatable elements in prokaryotes. Cell 11:11–23.

Lai, E. C., S. L. C. Woo, A. Dugaiczyk, J. F. Catterall, and B. W. O'Malley. 1978. The ovalbumin gene structural sequences in native chicken DNA are not continuous. Proc. Natl. Acad. Sci. USA 75:2205–9.

Lobban, P. E., and A. D. Kaiser. 1973. Enzymatic end-to-end joining of DNA molecules. J. Mol. Biol. 78:453–71.

Lomedico, P., N. Rosenthal, A. Efstratiadis, W. Gilbert, R. Kolodner, and R. Tizard. 1979. The structure and evolution of the two nonallelic rat proinsulin genes. Cell 18:545–58.

Maniatis, T., R. C. Hardison, E. Lacey, J. Lauer, C. O'Connell, D. Quon, Gek Kee Sim, and A. Efstratiadis. 1978. The isolation of structural genes from libraries of eukaryotic DNA. Cell 15:687–701.

Martial, J. A., R. A. Hallewell, J. D. Baxter, and H. M. Goodman, 1979. Human growth hormone: complementary DNA cloning and expression in bacteria. Science 215:612–17.

Maxam, A., and W. Gilbert. 1977. A new method for sequencing DNA. Proc. Nat. Acad. Sci. USA 74:560–64.

Meagher, R. G., R. C. Tait, M. Betlach, and H. W. Boyer. 1977. Protein expression in *E. coli* minicells by recombinant plasmids. Cell 10:521–36.

Mercereau-Puijalon, O., A. Royal, B. Cami, A. Garapin, A. Krust, G. Gannon, and P. Kourilsky. 1978. Synthesis of an ovalbumin-like protein by *Escherichia coli* K12 harbouring a recombinant plasmid. Nature 275:505–10.

Morrow, J. F., S. N. Cohen, A. C. Y. Chang, H. W. Boyer, H. M. Goodman, and R. B. Helling. 1974. Replication and transcription of eukaryotic DNA in *Escherichia coli*. Proc. Nat. Acad. Sci. USA 71:1743–47.

Naber, S., L. Villa-Komaroff, S. Broome, W. Gilbert, and W. C. Chick. Biologic activity of proinsulin synthesized by bacteria (in preparation).

Pasek, M., T. Goto, W. Gilbert, B. Zink, H. Schaller, P. MacKay, G. Leadbetter, and K. Murray. 1979. Hepatitis B virus genes and their expression in *E. coli*. Nature 282:575–79.

Paterson, B. M., B. E. Roberts, and E. L. Kuff. 1977. Structural gene identification and mapping by DNA· mRNA hybrid arrested cell-free translation. Proc. Nat. Acad. Sci. USA 74:4370–79.

Rao, R. N., and S. G. Rogers. 1978. A thermoinducible phage. Col E1 plasmid chimera for the overproduction of gene products from cloned DNA segments. Gene 3:247–63.

Ratzkin, B., and J. Carbon. 1977. Functional expression of cloned yeast DNA in *Escherichia coli*. Proc. Nat. Acad. Sci. USA 74:487–91.

Renold, A. E., D. R. Martin, Y. M. Dagenais, J. Steinke, R. J. Nickerson, and M. S. Sheps. 1960. Measurement of small quantities of insulin-like activity using rat adipose tissue. 1. A proposed procedure. J. Clin. Invest. 39:1487–98.

Roberts, T. M., I. Bikel, R. R. Yocum, D. M. Livingston, and M. Ptashne. 1979a. Synthesis of simian virus 40 t antigen in *Eschericia coli*. Proc. Nat. Acad. Sci. USA 76:5596–5600.

Roberts, T. M., R. Kacick, and M. Ptashne. 1979b. A general method for maximizing the expression of a cloned gene. Proc. Nat. Acad. Sci. USA 76:761–64.

Rubenstein, A. H., F. Melani, and D. F. Steiner. 1972. Circulating proinsulin: immunology, measurement, and biological activity. *In* R. O. Greep, E. B. Astwood, D. F. Steiner, N. Freinkl, and S. R. Geiger (eds.), Handbook of physiology. Section 7: Endocrinology, vol. 1, Endocrine pancreas, p. 515. American Physiol. Soc., Washington, D.C.

Sanger, F., S. Nicklen, and A. R. Coulson. 1977. DNA sequencing with chain-terminating inhibitors. Proc. Nat. Acad. Sci. USA 74:5463–67.

Schell, M. A., and B. D. Wilson. 1979. Cloning and expression of the yeast galactokinase gene in an *Escherichia coli* plasmid. Gene 5:291–313.

Seeburg, P. H., J. Shine, J. A. Martial, R. K. Ivarie, J. A. Morris, A. Ullrich, J. D. Baxter, and H. M. Goodman. 1978. Synthesis of growth hormone by bacteria. Nature 276:795–98.

Shine, J., and L. Dalgarno. 1975. Determinants of cistron specificity in bacterial ribosomes. Nature 254:34–38.

Shine, J., P. H. Seeburg, J. A. Martial, J. D. Baxter, and H. M. Goodman. 1977. Construction and analysis of recombinant DNA for human chorionic somatomammotropin. Nature 270:494–99.

Sninsky, J. J., A. Siddiqui, W. S. Robinson, and S. N. Cohen. 1979. Cloning and endonuclease mapping of the hepatitis B viral genome. Nature 279:346–48.

Southern, E. 1975. Detection of specific sequences among DNA fragments separated by gel electrophoresis. J. Mol. Biol. 98:503–17.

Steiner, D. F., W. Kemmler, J. L. Clark, P. E. Oyer, and A. H. Rubenstein. 1972. The biosynthesis of insulin. In R. O. Greep, E. B. Astwood, D. F. Steiner, N. Freinkel, and S. R. Geiger (eds.), Handbook of physiology. Section 7, Endocrinology, vol. 1: Endocrine pancreas, p. 175. Amer. Physiol. Soc., Washington, D.C.

Steitz, J. 1978. Genetic signals and nucleotide sequences in messenger RNA. In R. Goldberger (ed.), Biological regulation and development. 1:349–99. Plenum, New York.

Struhl, K., and R. W. Davis. 1977. Production of a functional eukaryotic enzyme in *Escherichia coli*: cloning and expression of the yeast structural gene for imidazoleglycerolphosphate dehydratase (his3). Proc. Nat. Acad. Sci. USA 74:5255–59.

Struhl, K., and J. R. Cameron, and R. D. Davis. 1976. Functional genetic expression of eukaryotic DNA in *Escherichia coli*. Proc. Nat. Acad. Sci. USA 73:1471–75.

Sutcliffe, J. G. 1978. The complete nucleotide sequence of the *Escherichia coli* plasmid pBR322. Cold Spring Harbor Symp. Quant. Biol. 43:77–90.

Szmuness, W., E. J. Harley, H. Ikram, and C. E. Stevens. 1978. In Viral hepatitis. ed. G. N. Vyas, S. N. Cohen, and R. Schmid. Sociodemographic aspects of epidemiology of Hepatitis B. Franklin Institute Press, Philadelphia.

Thomas, M., J. R. Cameron, and R. W. Davis. 1974. Viable molecular hybrids of bacteriophage lambda and eukaryotic DNA. Proc. Nat. Acad. Sci. USA 71:4579–83.

Tilghman, S., D. C. Tiemeier, F. Polsky, J. G. Seidman, A. Leder, L. W. Enquist, B. Norman, and P. Leder. 1977. Cloning specific segments of the mammalian genome: bacteriophage λ containing mouse globin and surrounding gene sequences. Proc. Nat. Acad. Sci. USA 74:4406–10.

Tilghman, S. M., D. C. Tiemeier, J. G. Seidman, B. M. Peterlin, M. Sullivan, J. V. Maizel, and P. Leder. 1978. Intervening sequence of DNA identified in the structural portion of a mouse β-globin gene. Proc. Nat. Acad. Sci. USA 75:725–29.

Ullrich, A., J. Shine, J. Chirgwin, R. Pictet, E. Tischer, W. J. Rutter, and H. M. Goodman. 1977. Rat insulin genes: construction of plasmids containing the coding sequences. Science 196:1313–19.

Uhlin, B. C., S. Molin, P. Gustafsson, and K. Nordstrom. 1979. Plasmids with temperature-dependent copy number for amplification of cloned genes and their products. Gene 6:91–116.

Valenzuela, P., R. Gray, M. Quiroga, J. Zaldivar, H. M. Goodman, and W. Rutter. 1979. Nucleotide sequence of the gene coding for the major protein of hepatitis B virus surface antigen. Nature 281:815–19.

Vapnek, D., J. A. Hautala, J. W. Jacobson, N. H. Giles, and S. R. Kushner. 1977. Expression in *Escherichia coli* K-12 of the structural gene for catabolic dehydroquinase of *Neurospora crassa*. Proc. Nat. Acad. Sci. USA 74:3508–12.

Villa-Komaroff, L., A. Efstratiadis, S. Broome, P. Lomedico, R. Tizard, S. P. Naber, W. L.

Chick, and W. Gilbert. 1978. A bacterial clone synthesizing proinsulin. Proc. Nat. Acad. Sci. USA 75:3727–31

Villa-Komaroff, L., S. P. Naber, S. Broome, A. Efstratiadis, P. Lomedico, R. Tizard, W. L. Chick, and W. Gilbert. 1980. The synthesis of proinsulin in bacteria. *In* E. G. Bassett (ed.), Polypeptide hormones. (In press.)

Walz, A., B. Ratzkin, and J. Carbon. 1978. Control of expression of a cloned yeast (*Saccharomyces cerevisiae*) gene (trp5) by a bacterial insertion element (1S2). Proc. Nat. Acad. Sci. USA 75:6172–76.

DONALD H. DEAN

Cloning in and of *Bacillus*

5

From the inception of recombinant DNA technology, there has been an interest in cloning *Bacillus* genes and in developing molecular cloning systems for *Bacillus subtilis*. At first this effort was aimed at developing a "safer cloning system" (Appendix A of the 1976 Edition of the NIH Guidelines for Recombinant DNA Research); but with the decreasing fears associated with genetic engineering, interest has turned to exploiting the unique properties of bacilli.

Bacillus subtilis is a gram positive spore-forming aerobe which has been genetically well characterized. Its genome has been demonstrated to be a complete circle by autoradiography (Wake, 1973) and by genetic transduction (Lepesant-Kejzlarova et al., 1974). Two genetic maps have been published that are comprehensive and accurate (Young and Wilson, 1976; Hoch, 1978). The *Bacillus* Genetic Stock Center was established in 1978 at the Ohio State University to collect, maintain, and distribute cultures from an extensive collection of genetic mutants. A computer-assisted map of the *B. subtilis* genome prepared by the *Bacillus* Genetic Stock Center is shown in figure 1.

Bacillus subtilis or closely related strains are used commercially in Japan for the production of edible curd, "natto," and worldwide in the production of many antibiotics (e.g., Gramicidin and Bacitracin [Shoji, 1978; Katz and Demain, 1977]) and enzymes (e.g., α-amylase and penicillinases [Priest, 1977]). The benefit of cloning industrially important genes is obvious. Alpha-amylase, which is made and excreted by *B. subtilis, B. amyloli-*

Department of Microbiology and Department of Genetics, Ohio State University, Columbus, Ohio 43210.

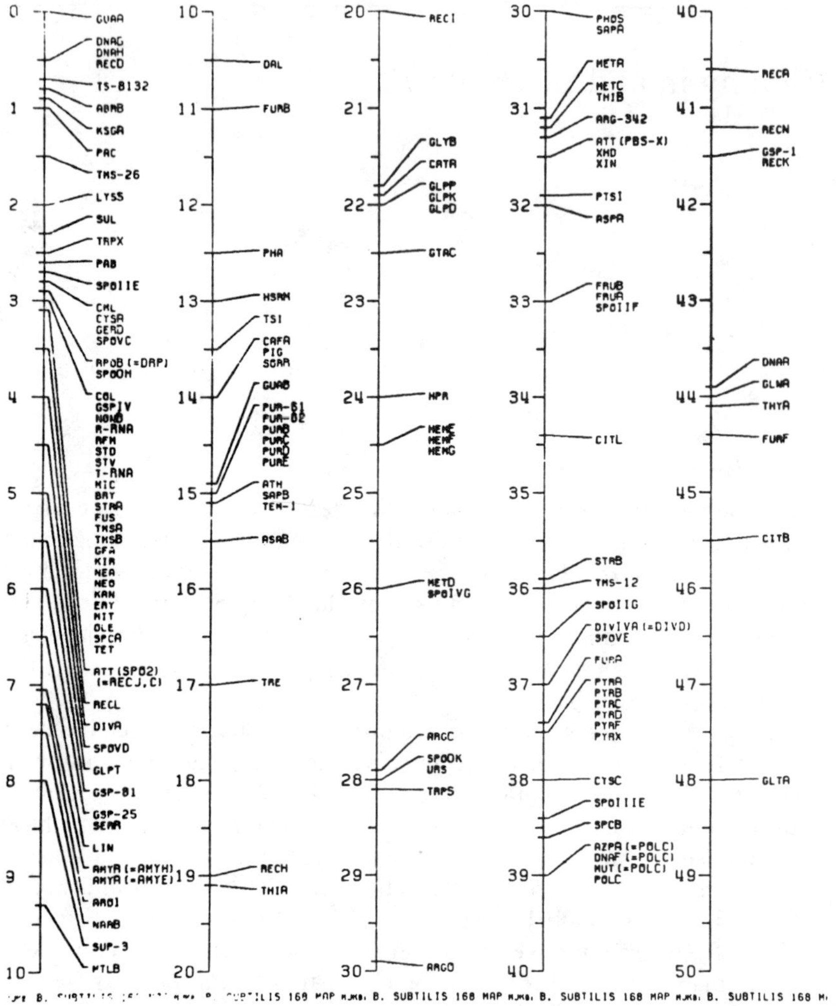

Fig. 1. Computer-generated genetic map of *Bacillus subtilis* 168. Ten equal sections are drawn as they would appear in a clockwise direction on the chromosome. The map positions

quefaciens, and *B. licheniformis,* is used commercially for sizing paper and the conversion of starch to glucose and syrups. The production strains currently preferred are natural isolates of *B. licheniformis* that make a heat-stable α-amylase (fermentation temperatures usually exceed 50°C unless costly cooling is employed). The present goal of engineering genes of commercial importance is to amplify production and to combine desirable

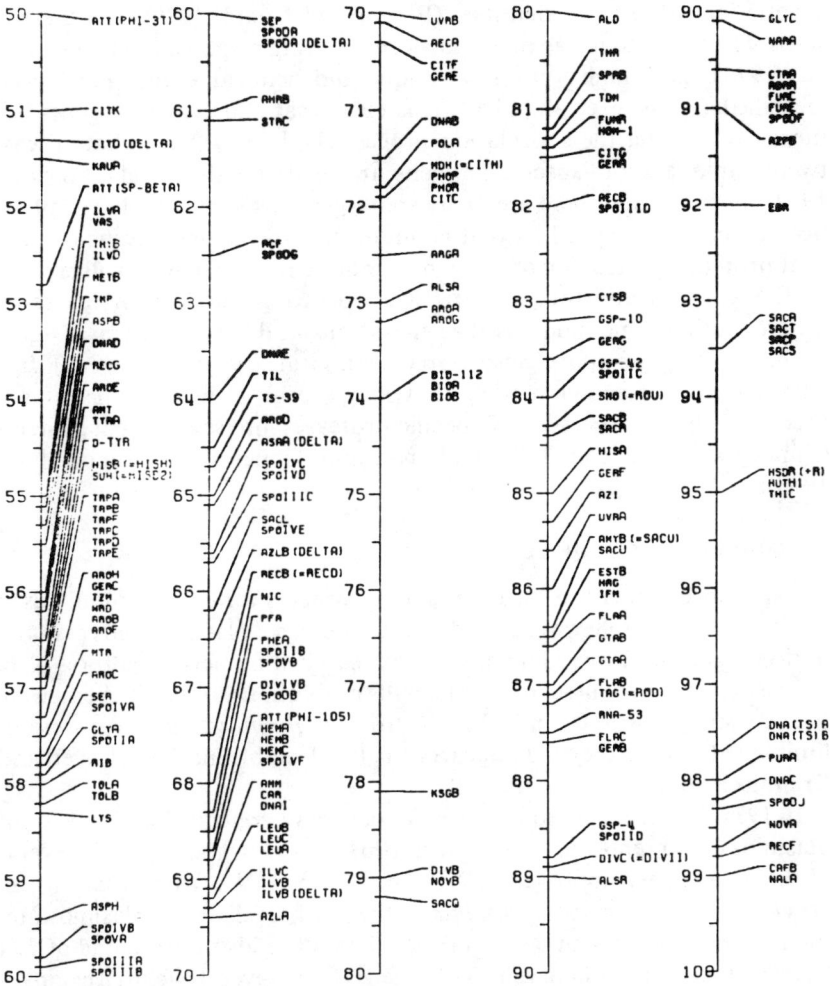

50 — ATT (PHI-3T) 60 — SEP / SPO0A / SPO0A (DELTA)   70 — UVRB / RECA / CITF GENE   80 — ALD   90 — GLYC / NARA

51 — CITK 61 — AHRB / STRC   71 — DNAB / POLA   81 — THR / SPRB / TDH / FUMR HOM-1 / CITG GERA   91 — CTRA / ADMA / FURC / FURE / SPO0F / AZPB

— CITD (DELTA) / KAUA / ATT (SP-BETA)

52 — ILVA / VAS 62 — RCF / SPO0G   72 — MDH (=CITH) / PHOP / PHOR / CITC   82 — RECB / SPO111D   92 — EBR

— TH:B / ILVD / METB

53 — TKP / ASPB / DNAD / RECG / AROE 63 —   73 — ARGA / ALSA / ARDA / AROG   83 — CYSB / GSP-10 / GERG / GSP-42 SPO11C   93 — SACR / SACT / SACP / SACS

54 — AHT / TYRA / D-TYR / HISB (=HISH1) / SUH (=HISC2) 64 — DNAE / TIL / TS-39 / ARGD / ASAA (DELTA)   74 — BID-112 / BIOA / BIOB   84 — SHO (=ROU) / SACB / SACA / HISA   94 —

55 — TRPA / TRPB / TRPF / TRPC / TRPD / TRPE 65 — SPO1VC / SPO1VD / SPO111C / SACL / SPO1VE / AZLB (DELTA)   75 —   85 — GERF / AZI / UVRA / AMYB (=SACU) / SACU   95 — HSDA (+R) / HUTH1 / THIC

56 — AROH / GERC / TZM / HRD / AROB / AROF 66 — RECB (=RECD) / NIC / PFA   76 —   86 — ESTB / HRG / IFM / FLAA / GTAB / GTAA

57 — AROC / SER / SPO1VA / GLYA / SPO11A / RIB 67 — PHEA / SPO11B / SPOVB / DIV1VB / SPO0B / ATT (PHI-105) / HEMA / HEMB / HEMC / SPO1VF   77 —   87 — FLAB / TAG (=ROD) / RNA-53 / FLRC / GERB   97 — DNA(TS) A / DNA(TS) B / PURA

58 — TOLA / TOLB / LYS 68 — AMM / CAA / DNA1   78 — KSGB   88 —   98 — DNAC / SPO0J / NOVA / RECF

— LEUB / LEUC / LEUA   — GSP-4 SPO11D / DIVC (=DIV11)

59 — 69 — ILVC / ILVB / ILVB (DELTA) / AZLA   79 — DIVB / NOVB / SACQ   89 — ALSR   99 — CARB / NALA

— ASPH / SPO1VB / SPOVA

60 — SPO111A / SPO111B 70 —   80 —   90 —   100 —

B. SUBTILIS 168 MAP   B. SUBTILIS 168 MAP   B. SUBTILIS 168 MAP   B. SUBTILIS 168 MAP   B. SUBTILIS 168

are only rough approximation on this edition of the map. (a) Markers from 1 to 50, (b) facing page markers from 50 to 100.

characteristics such as heat stability, faster growth rate of host, and less expensive growth requirements. The recent attempts in cloning the α-amylase gene will be discussed later in this chapter.

Recombinant DNA techniques will also aid academic studies in *Bacillus*. Bacterial sporulation is a morphogenic process consisting of at least seven stages, distinguishable in the electron microscope. Thirty-five separate genetic loci have been identified (Piggot and Coote, 1976), and there is statistical evidence for as many as 44 genes being responsible (Hranueli et al., 1974). Although very few unique biochemical events have been identified (Hanson et al., 1970), it is clear that spore-specific genes are turned on during the sporulation process (Doi, 1977a) and that at least twelve unique spore-specific antigens are made (Walker and Thomson, 1972). Even though there are many spore-gene products that have not yet been detected, one product that is unquestionably spore-specific is spore coat protein. The cloning of this gene, reported in this chapter, will provide a DNA substrate for testing two operating models of bacterial differentiation. One model is that alterations in RNA polymerase (sigma exchange) turn on spore genes (Greenleaf, et al., 1973; Doi, 1977b). It is also proposed (second model) that spore coat protein is made as a precursor that is processed by specific proteases into several coat proteins (Pandey and Aronson, 1979; Goldman and Tipper 1978; Munoz et al., 1978).

PLASMID CLONING VEHICLES FOR BACILLUS

Early development of plasmid cloning systems for the commonly used strain, *B. subtilis* 168, was delayed by the absence of plasmids and episomes in this organism. Lovett and coworkers searched numerous cultures of *B. subtilis* and *pumilus* in search of plasmids usable in *B. subtilis*. Several were found, namely pPL10, and pPL576 from *B. pumilus*, but these were of limited value since they did not carry easily selectable markers (Lovett and Bramucci, 1975).

In 1977, Ehrlich showed that *Staphylococcus aureus* antibiotic-resistant plasmids could replicate in *B. subtilis* (Ehrlich, 1977), and currently several of these have been approved by NIH for cloning DNA from *B. subtilis, B. amyloliquefaciens*, and *B. pumilus* at the HV1 level. These plasmids and their properties are included in table 1. The plasmids pUB110 and pC194 have been used as cloning vehicles for genes from several bacilli (Keggins et al., 1978; G7yczan and Dubnau, 1978; Ehrlich, 1978).

A natural recombination event between pC194 and pS194 resulted in a plasmid vehicle, pSC194 (=pSA2100), with one selectable marker (CM)land one (SM) that is inactivated by *Eco* RI (Löfdahl et al., 1978a, b; Gryczan et al., 1978). More recent findings that pC194 is a transposition element (Dubnau and Novick, personal communication; Martin and Dean, personal observations) may reduce the value of pSC194 and pC194 as cloning systems. The erythromycin-resistant plasmid pE194 has been fully characterized and explored as a cloning vehicle (Weisblum et al.,

Table 1

Staphylococcus Aureus Plasmids
Approved For
Cloning in *Bacillus Subtillis*

In November 1977, the NIH reviewed data which were submitted in conjunction with requests for approval to use certain *Staphylococcus aureus* plasmids for cloning experiments in *Bacillus subtilis*. Under present NIH Guidelines, approval has been given for the use of *Staphylococcus aureus* plasmid vectors to clone DNA from *Bacillus pumilus*, *Bacillus licheniformis* and *Bacillus subtillis*, and any phage or plasmids that replicate in these species, in *Bacillus subtilis* under P1 containment conditions.

Approval of these systems as HV1 systems under the proposed revised Guidelines, or extension of these systems to clone DNA from other bacterial species, must await information on the survival of the host in its natural environment, and possible use of asporogenic strains.

Plasmid	Phenotype Conferred	Literature References
pUB110	Kan^r/Neo^r	2
pSA0501	Str^r	3
pC194	Cm^r	3, 4
pUB112	Cm^r	1, 3, 4
pSA2100	Str^r Cm^r	3
pC221	Cm^r	4
pC223	Cm^r	4
pT127	Tet^r	4

Phenotype abbreviations are: Kanomycin-Kan^r, Neomycin-Neo^r, Streptomycin-Str^r, Chloramphenicol-Cm^r, Tetracycline-Tet^r

Literature References

1. Chopra, I., P.M. Bennett, and R.W. Lacey, A variety of staphylococcal plasmids present as multiple copies. J. Gen. Microbiol. 79: 343-345, 1973.

2. Lacey, R.W., and I. Chopra, Genetic studies of a multiresistant strain of *Staphylococcus aureus*. J. Med. Microbiol. 7: 285-297, 1974.

3. Iordanescu, S. Recombinant plasmid obtained from two different, compatible staphylococcal plasmids. J. Bacteriol. 124: 597-297. 1974.

4. Ehrlich, S.D., Replication and expression of plasmids from *Staphylococcus aureus* in *Bacillus subtilis*, Proc. Natl. Acad. Sci. USA 74: 1680-82. 1977

1978). Other plasmids from *S. aureus* have been reported that may also prove useful for cloning in *Bacillus* (Kono et al., 1978a, b; Wilson and Baldwin, 1978).

More recent efforts at isolating plasmids from bacilli have been successful (Tenaka et al., 1977; Tanaka and Koshikawa, 1977; LeHegarat and Anagnostopoulos, 1977; and Kreft et al., 1978). These plasmids and their properties are shown in table 2. Among these, several have been reported as molecular cloning vehicles (Horinouchi et al., 1977; Tanaka and Sakagunchi, 1978; Kreft et al., 1978), as will be discussed later in this chapter.

Hybrid or combination plasmids between *E. coli* plasmids and plasmids

TABLE 2

BACILLUS PLASMIDS USEFUL FOR RECOMBINANT DNA RESEARCH

PLASMID	MOLECULAR WEIGHT	BACTERIAL ORIGIN	SELECTABLE MARKERS	RESTRICTION ENZYME SITES				REFERENCES
				EcoRI	HindIII	Bam	BSU	
pBC16	4.25kb	*B. cereus*	Tc	1	0	0	0	Kreft et al., 1978
pBC16-1	2.7kb	*B. cereus*	Tc	1	6	0	0	Kreft et al., 1978
pBS161	8.2kb	*B. subtilis–B. cereus*	Tc					Kreft et al., 1978
pBS161-1	3.65kb	Constructed	Tc	0	1	0	0	Kreft et al., 1978
pAT1060	5.4md	*B. subtilis*		1	5	0	2	Horinouchi et al., 1977
pLS11	5.4Md	*B. subtilis*		1	5	1	ND*	Tonaka et *al.*, 1977
pLS12	5.4Md	*B. subtilis*		1	5	1	ND	Tonaka et al., 1977
pLS101-103	4.2	*B. natto*		1	ND	1	ND	Tonaka et al., 1978
pGY31	7.6	*B. subtilis*		1	ND	ND	ND	LeHagarat and Anagnostopoulos, 1977
pGY32	3.6	*B. subtilis*		1	ND	ND	ND	LeHagarat and Anagnostopoulos, 1977
pGY7	5.0	*B. subtilis*		2	ND	ND	ND	LeHagarat and Anagnostopoulos, 1977

* ND: not determined.

(from *Bacillus* or *Staphylococcus*) that replicate in *B. subtilis* have proved extremely useful in Japan, Germany, and France (Ehrlich, 1978; Horinouchi et al., 1977; Kreft et al., 1978). With these plasmids, selection for desired genes may take place in a heterologous genetic background (*E. coli*) to prevent recombination. The recombinant DNA may then be cloned back into *B. subtilis* for genetic analysis. Until recently these studies have been prohibited in the United States by the NIH Guidelines for Recombinant DNA Research. Interspecies cloning between two nonpathogenic bacteria is classified as P2, HV1. An HV1 cloning host has been approved for *B. subtilis*.

The results of interspecies cloning between *B. subtilis* and *E. coli* have revealed some interesting observations. Plasmid-borne antiobiotic-resistant genes from *Bacillus* (Kreft et al., 1978) or *Staphylococcus* plasmids (Ehrlich, 1978; Rapoport et al., 1978) are expressed in *E. coli*, but plasmid-borne antiobiotic-resistant markers from *E. coli* plasmids (Ampicillin, kanamycin, and chloramphenicol) are not expressed in *B. subtilis*. This would indicate either that the *B. subtilis* RNA polymerases are more sophisticated than the *E. coli* counterpart or that *B. subtilis* promoters are less restrictive. There is little information on the latter possibility, but for the former there is abounding circumstantial evidence. *E. coli* makes the same set of proteins throughout its cell cycle (Lutkenhaus et al., 1979) whereas *B. subtilis* has a very complex and changing protein pattern (Linn and Losick, 1976). *B. subtilis* is also known to have significant alterations in its RNA polymerase and several new RNA polymerase binding components during its life cycle (Doi, 1977b).

Similar evidence of the permissive nature of *E. coli* RNA polymerase is also seen in expression of mRNA of *B. subtilis* plasmids in cell-free extracts of *E. coli* (Horinouchi et al., 1979) and the expression of *Bacillus* phage SP01 in *E. coli* minicells (see abstract of Amann and Reeve, this volume).

BACTERIOPHAGE CLONING VEHICLES FOR BACILLUS

Phage cloning systems have particular advantages over their plasmid counterparts. Phages repackage their recombinant DNA into easily isolated particles. *In vitro* repackaging increases the efficiency of cloning over transfection or transformation. A clone bank developed by phages is easily maintained as a lysate as opposed to a bank obtained by plasmids, which requires the maintenance of in excess of 2,000 individual colony clones. Phages amplify their DNA and its product during infection, and, for example, a value of 500 fold amplification of cloned product has been reported (Panasenko et al., 1977).

Within the last year two phage cloning systems have been reported for *B. subtilis*. Kawamura and colleagues (1979) found an ingeneous way to select specialized transducing phages carrying restriction fragments of *B. subtilis* DNA. Bacterial DNA was digested with *Eco* RI (actually *B. subtilis* DNA from the defective prophage PBSX was used as the substrate) and ligated to a mixture of fragments from an *Eco* RI digest of temperate phage ρ11. This bacteriophage has about 21 *Eco* RI fragments (Dean et al., 1976). The ligation mixture was transformed into a ρ11 lysogen, and a desired marker was selected. The recombination step is highly likely to be between the fragments of phage DNA and the prophage, thereby resulting in a recombinant prophage with the selected marker cloned into it. Upon induction of the lysogens and subsequent infection of new cells, a high frequency of the lysogens formed are found to possess the selected marker. The phage ρ11 has a large genome and can accommodate much additional DNA without loss of viability. A similar approach was made with another temperate *Bacillus* phage, ϕ105, but this smaller-sized genome prevents construction of helper independent phages.

In my laboratory we have developed temperate phage–lytic phage cloning vehicles by standard recombination DNA techniques. Our approach was to search new and available phages for those with a single restriction enzyme site. We found that ρ14 had a single *Sal* I site and a single *Bgl* II site (Dean et al., 1978). After extensive physical characterization (Perkins et al., 1978); (Rudinski and Dean, 1979) and comparison to other phages by host range immunity and serology (Dean et al., 1978), it was found that ρ14 was 90% related to a well-characterized temperate phage ϕ105. The close relationship of ρ14 to ϕ105 greatly aided the development of ρ14 as a cloning vehicle. In order to accommodate large pieces of DNA, a set of ρ14 deletion mutants have been isolated by a rapid technique (Kroyer and Dean, 1979). Analysis of these deletions by heteroduplex and restriction analysis (Kroyer et al., 1979) has shown that both clear plaque (ϕdoc phages) and turbid plaque phages (ϕdo phages) have characteristics of cloning vehicles theoretically capable of carrying about 2×10^6 daltons of DNA into their *Sal* I site (Dean et al., 1979, Recombinant DNA Technical Bulletin 2:9–14). To date we have only been able to detect 0.5×10^6 dalton inserts however. The clear plaque mutants are under development as HV-1 cloning systems for *B. subtilis*, while the turbid plaque phages are available as a intraspecies vehicle for several species of *Bacillus*. The temperate vehicle will allow single gene-pair complementation for the analysis of genetic regulation of sporulation genes. The physical map of ρ14 and the ϕdo and ϕdoc deletions are shown in figure 2. Figure 3 demonstrates that ϕdo7 can carry additional DNA at the *Sal* I site.

SalGI
BglII

φ105

DI:291 DI:1c

Δt7 Δc36

φ105
EcoRI
SacI   D   I   E   (J)G   B   H F   C
                                            D E   C
Kpnl
Smal   D   C   E   G¹   G²   F D B   E

ρ14

ρ14
EcoRI
SacI   C   D   A   F D¹   G H   E
                                        D²   C
Kpnl
Smal   C   B   F G E   D   B   C   E
                                                    D

kbp   5   10   15   20   25   30   35
Mdal   5   10   15   20   25

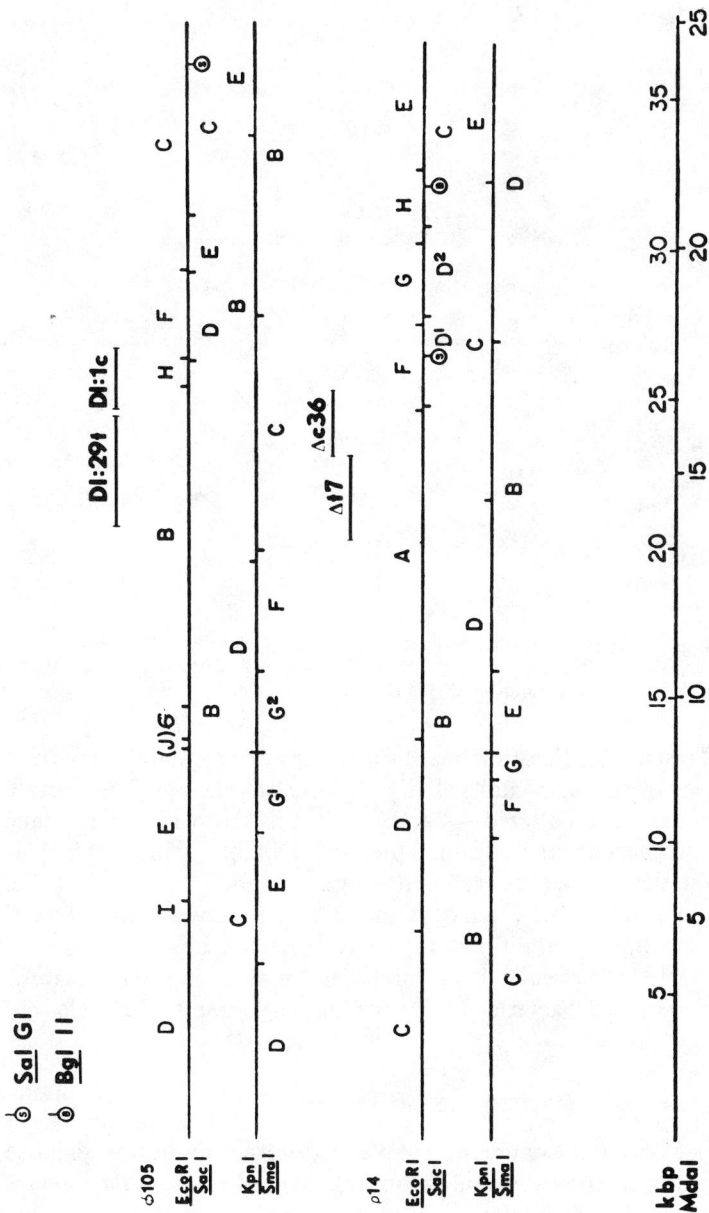

Fig. 2. The physical map of temperate *B. subtilis* phages φ105 and ρ14. Bars above the maps indicate selected deletion mutants; the φ105 turbid plaque deletion mutant is DI:29t, and the clear plaque deletion mutant is DI:1C; the turbid plaque deletion mutant of ρ14 is φdo7; the deleted region is indicated by △t7, and the clear plaque deletion mutant of ρ14 is φdoc36, indicated by △C36.

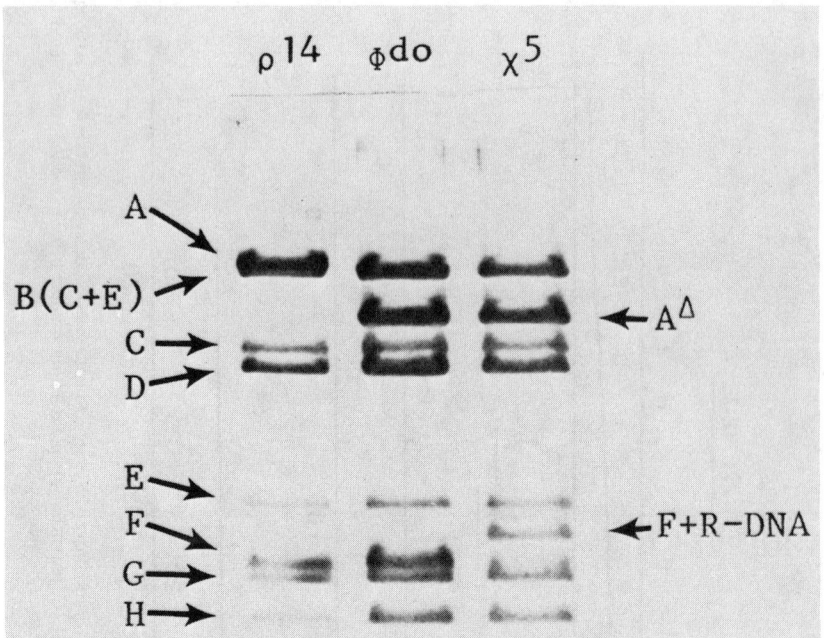

Fig. 3.   Agarose gel (0.7%) electrophoresis patterns of ρ14; the cloning vehicle, φdo7; and a chimeric phage, x5, carrying recombinant DNA (R-DNA). The letters A-H are *Eco* RI restriction bands (see figure 2), and band B is the union of the ends, C and E).

The *Eco* RI restriction digestion of ρ14 shows the normal migration pattern on a 0.7% agarose horizontal gel (fig. 3, lane 1). The φdo7 temperate deletion mutant has an altered A band that is shifted down, A (fig. 3, lane 2). When φdo7 accepts an additional piece of DNA, the chimeric phage is called **x**. One such clone, **x**5, is shown in figure 3, lane 3. The cloning site, *Sal* I, is within the *Eco* RI F band. Figure 3, lane 3, shows a shift in the F ban demonstrating that additional DNA has been cloned. A clone bank of *B. subtilis* DNA is currently being prepared by attaching *Sal* I adaptor molecules to sheared bacterial DNA and ligating these to *Sal* I cleaved φdo7.

CLONING OF BACILLUS SUBTILIS GENES

In a similar vein to the expression of plasmid-borne antibiotic resistance markers mentioned above, several chromosomal markers from *Bacillus* are expressed in *E. coli* when cloned on plasmids or phages. A thymidylate synthetase gene, *thy*P, from *Bacillus* phage φ3T (Duncan et al., 1977; Ehrlich et al., 1978), *trp*C$_2$ (Keggins et al., 1978), *leu* (Mahler and Halvorson, 1977; Rapoport et al., 1978, in press), and *his* A, *gly* B, and *thr* (Rapoport et al., 1978, in press) have all been cloned into *E. coli*. The markers *pyr* and *leu* from *B. subtilis* also have been introduced into *E. coli* through λgt and shown to express there (Chi et al., 1978).

In the case of the *B. subtilis leu* gene, cloned into *E. coli* on plasmid pMB9, the cloned DNA was apparently lost upon return to *B. subtilis* (Mahler and Halvorson, 1977). Although *B. subtilis* is not thought to have a restriction modification system (Ehrlich et al., 1976; Trautner et al., 1974), this effect was believed to be due to internal nucleases (Mahler and Halvorson, 1977). More recently it has been shown that recombinant DNA cloned in *E. coli* may be returned to *B. subtilis* only if a covalently closed circular form of recombinant plasmid is used for standard genetic transformation (Rapoport et al., 1979, in press). Use of recombination-deficient *B. subtilis* also prevents loss of the cloned fragment through recombination.

Segall and Losick (1977) have cloned into *E. coli* a 4.4 k base piece of *B. subtilis* DNA that is expressed during sporulation. The fragment of DNA was selected from *Eco* RI digested *B. subtilis* DNA by hybridization to a particular sporulation-specific RNA of 0.4 k bases. This use of cloning demonstrated a particular benefit of cloning *B. subtilis* DNA into *E. coli* to study sporulation.

Another example of the use of recombinant DNA in the study of sporulation is the recent cloning in the author's laboratory of the spore coat protein (SCP) gene. William Martin used the sandwich radio-immune assay of Broome and Gilbert (1978) to detect *E. coli* chimeras making SCP. Figure 4 shows the autoradiograms revealing positive clones making SCP. Screening of the plasmids retrieved from the clones shows inserts of 0.1 to 4.0 M daltons. Preliminary evidence reveals that the *E. coli* clones are not making SCP as it is found on the mature spore. It appears that the clones make a higher molecular weight protein. This finding agrees nicely with the proposal (Pandey and Aronson, 1978; Munoz et al., 1978) that SCP is made as a precursor and processed into the lower molecular weight coats by specific proteases. Further analysis of the protein made by the clone and the precursor made by sporulating cells by standard protein characterization will be needed to definitively prove that these clones contain the SCP gene.

As mentioned in the Introduction, another area of great interest in *Bacillus* research is industrial production of enzymes and antibiotics. Richard Perro in the author's laboratory has been studying *E. coli* clones carrying *B. subtilis* DNA coding for α-amylase. Figure 5 shows a plate with clones that make α-amylase (wild-type *E. coli* does not make α-amylase).

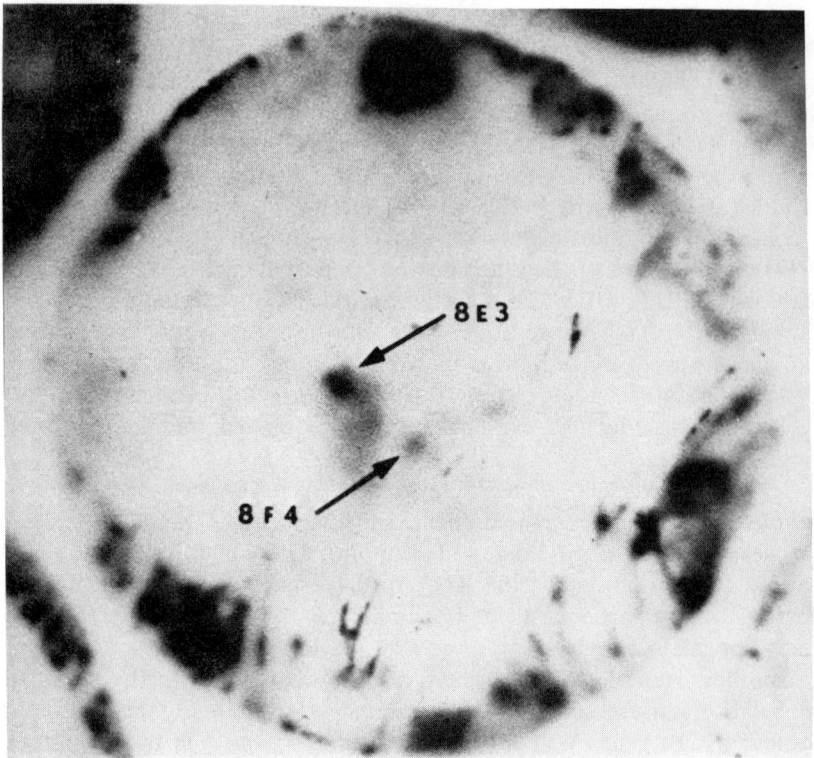

Fig. 4.   Sandwich radio-immune assay of *E. coli* clones carrying *B. subtilis* DNA. The clones expressing *Bacillus* spore coat protein are recognized by ^{125}I-labeled antisera to spore coat protein.

These clones have been grown on plates containing starch, and, because *E. coli* does not normally excrete enzymes, they have been lysed with chloroform. After 30 min., to allow defusion of the α-amylase, the plates are flooded with iodine, which stains the starch. Clones have a halo zone around their colonies, and iodine does not bind within the colony. Control colonies do not have a halo zone, and they do bind iodine. In future experiments α-amylase genes will be inserted into a *B. subtilis* plasmid or phage vehicle and returned to *B. subtilis*, where studies on excretion and amplified expression may be contained.

CONCLUSION

This chapter has related the need of, and purpose for, developing *B. subtelis* as a cloning system. Both phage and plasmid cloning vehicles

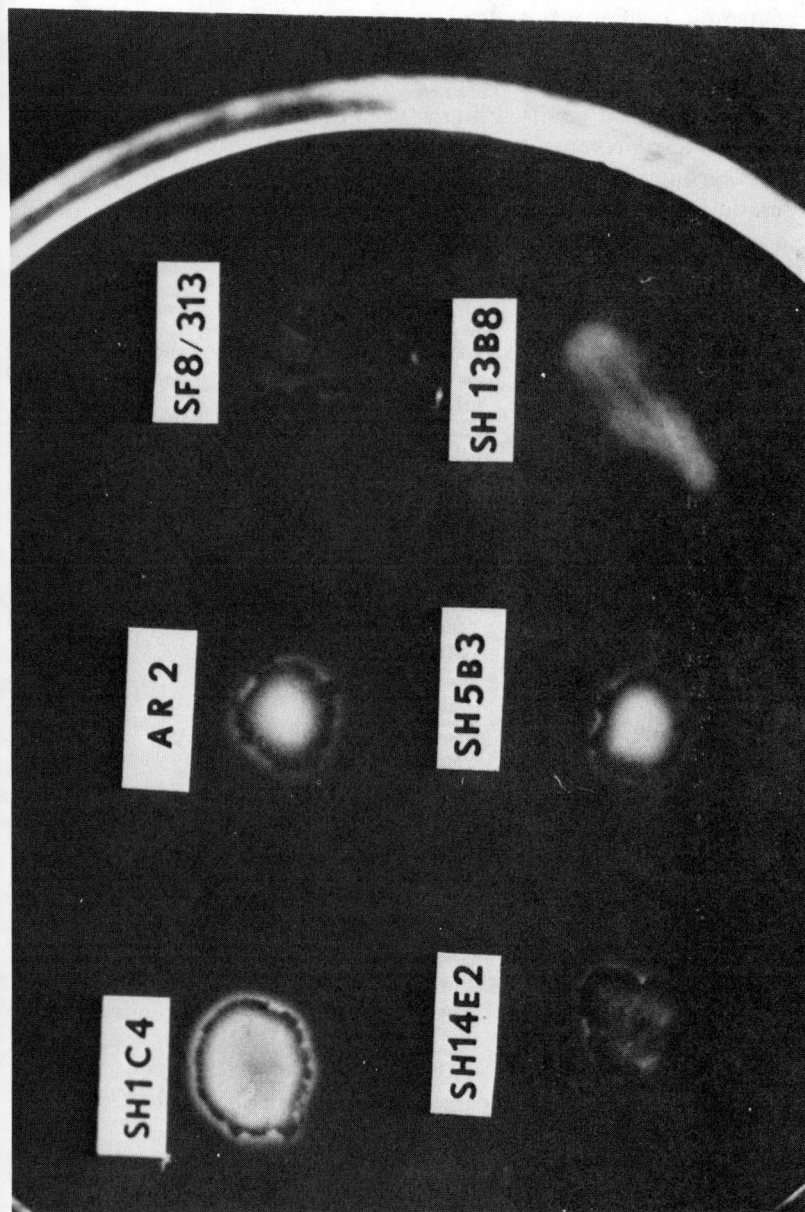

Fig. 5. Starch-iodine plate assay of *E. Coli* clones expressing α-amylase. The clones SH 1 C4, AR2, SH 13 E2, SH 5 Be, and SH 13 B8 degrade the starch in the plate and are not stained with iodine. SF 8/313, the parent strain, SF8, infected with the plasmid cloning vehicle, pBR313 but no recombinant DNA, shows no starch degradation.

systems are now available, and advancements in the fields of sporulation or industrial production by genetic engineer are now beginning. Cloning of *Bacillus* DNA into *E. coli* has proved valuable since *Bacillus* DNA is expressed in *E. coli* with equity. Cloning *Bacillus* DNA into a heterologous genetic species also prevents recombination and provides an antigenically neutral background. With this approach, significant genes in the pathway of sporulation and the industrial production of enzymes have been isolated and reported here for the first time.

ACKNOWLEDGMENTS

I wish to thank James M. Kroyer, John B. Perkins, and Mark S. Rudinski for their work on the development of the ϕdo and ϕdoc phages; William Martin for his work on cloning the spore coat protein; and F. Richard Perro for his work on cloning α-amylase. Research on these projects is supported by NIH Grant GM 26172. I also thank Michael Kaelbling, programmer, and Scott Martin, curator, of the *Bacillus Genetic Stock Center (BGSC)* for preparation of the *B. subtilis* map. The BGSC is supported by NSF Grant DEB 7809339.

REFERENCES CITED

Broome, S., and W. Gilbert. 1978. Immunological screening method to detect specific translation products. Proc. Nat. Acad. Sci. USA 75;2746-49.

Chi, N.-Y. W., S. D. Ehrlich, and J. Lederberg. 1978. Functional expression of two *Bacillus subtilis* chromosomal genes in *Escherichia coli*. J. Bacteriol. 133:816-21.

Dean, D. H., J. C. Orrego, K. W. Hutchison, and H. O. Halvorson. 1976. New temperate bacteriophage for *Bacillus subtilis*, ρ11, J. Virol. 20:509-19.

Dean, D. H., J. B. Perkins, and C. D. Zarley. 1978. Potential temperate bacteriophage molecular vehicle for *Bacillus subtilis*. *In* G. Chambliss and J. C. Vary (eds.), Spores VII, pp. 144-49. American Society for Microbiology, Washington, D.C.

Doi, R. H. 1977a. Role of ribonucleic acid polymerase in gene selection in procaryotes. Bacteriol. Rev. 41:568-94.

Doi, R. H. 1977b. Genetic control of sporulation. Ann. Rev. Genet, 11:29-48.

Duncan, C. H., G. A. Wilson, and F. E. Young. 1977. Transformation of *Bacillus subtilis* and *Escherichia coli* by a hybrid plasmid pCD1. Gene 1:153-67.

Ehrlich, S. D. 1977. Replication and expression of plasmids from *Staphylococcus aureus* in *Bacillus subtilis*. Proc. Nat. Acad. Sci. USA 74:1680-82.

Ehrlich, S. D. 1978. DNA cloning in *Bacillus subtilis*. Proc. Nat. Acad. Sci. USA 75:1433-36.

Ehrlich, S. D., H. Bursztyn-Pettegrew, I. Stroynowski, and J. Lederberg. 1976. Expression of the thymidylate synthetase gene of the *Bacillus subtilis* bacteriophage Phi 3-T in *Escherichia coli*. Proc. Nat. Acad. Sci. USA 73:4145-49.

Goldman, R. C., and D. J. Tipper. 1978. *Bacillus subtilis* spore coats: complexity and purification of a unique polypeptide component. J. Bacteriol. 135:1091–1106.

Greenleaf, A. L., T. G. Linn, and R. Losick. 1973. Isolation of a new RNA polymerase-binding protein from sporulating *Bacillus subtilis*. Proc. Nat. Acad. Sci. USA 70:490–94.

Gryczan, T. J., S. Contente, and D. Dubnau. 1973. Characterization of *Staphylococcus aureus* plasmids introduced by transformation into *Bacillus subtilis*. J. Bacteriol. 134:318–29.

Gryczan, T. J., and D. Dubnau. 1978. Construction and properties of chimeric plasmids in *Bacillus subtilis*. Proc. Nat. Acad. Sci. USA 75:1428–32.

Hanson, R. S., J. A. Peterson, and A. A. Yousten. 1970. Unique biochemical events in bacterial sporulation. Ann. Rev. Microbiol. 24:53–90.

Hoch, J. A. 1978. Developmental genetics at the beginning of a new era. *In* G. Chambliss and J. C. Vary (eds.), Spores VII, pp. 119-21. American Society for Microbiology, Washington, D.C.

Horinouchi, S., T. Uozumi, T. Hoshino, A. Ozaki, S. Nakajima, T. Bepper, and K. Arima. 1977. Molecular cloning and in vitro transcription of *Bacillus subtilis* plasmid in *Escherichia coli* Molec. Gen. Genet. 157:175–82.

Horanueli, D., P. J. Piggot, and J. Mandelstam. 1974. Statistical estimate of the total number of operons specific for *Bacillus subtilis* sporulation. J. Bacteriol. 119:684–90.

Katz, E., and A. L. Demain. 1977. The peptide antibiotics of *Bacillus*: chemistry biogenesis and possible functions. Bacteriol. Rev. 41:449–74.

Kawamura, F., H. Saito, and Y. Ikeda. 1979. A method for construction of specialized transducing phage ρ11 of *Bacillus subtilis*. Gene 5:87–91.

Keggins, K. M., P. S. Lovett, and E. J. Duvall. 1978. Molecular cloning of genetically active fragments of *Bacillus* DNA in *Bacillus subtilis* and properties of the vector plasmid pUB110. Proc. Nat. Acad. Sci. USA 75:1423–27.

Kono, M., M. Sasatsu, H. Hamashima, and T. Yamakawa. 1978. Transformation of *Bacillus subtilis* with staphylococcus plasmid DNA. Microbios Lett. 5:55–59.

Kono, M., M. Sasatsu, H. Hamashima, and T. Yamakawa. 1978. Transformation of *Bacillus subtilis* with staphylococcal plasmid (pTP-2) DNA resistant to penicillin and tetracycline. Microbios Lett. 6:67–76.

Kreft, J., K. Bernhard, and W. Goebel. 1978. Recombinant plasmids capable of replication in *B. subtilis* and *E. coli*. Molec. Gen. Genet. 162:59–67.

LeHégarat, J.-C., and C. Anagnostopoulos. 1977. Detection and characterization of natural occuring plasmids in *Bacillus subtilis*. Molec. Gen. Genet. 157:167–74.

Lepesant-Keijzlarova, J., J.-A. Lepesant, J. Walle, A. Billault, and R. Dedonder. 1975. Revision of the linkage map of *Bacillus subtilis* 168: indications for circularity of the chromosome. J. Bacteriol. 121:823–34.

Linn, T., and R. Losick. 1976. The program of protein synthesis during sporulation in *Bacillus subtilis*. Cell 8:103–14.

Löfdahl, S., J.-E. Sjöström, and L. Philipson. 1978a. Characterization of small plasmids from *Staphylococcus aureus*. Gene 3:149–59.

Löfdahl, S., J.-E. Sjöström and L. Philipson. 1978b. A vector for recombinant DNA in *Staphylococcus aureus*. Gene 3:161–72.

Lovett, P. S., and M. G. Bramucci. 1975. Plasmid deoxyribonucleic acid in *Bacillus subtilis* and *Bacillus pumilus*. J. Bacteriol. 124:484–90.

Lutkenhaus, J. F., A. Moore, M. Masters, and W. D. Donachie. 1979. Individual proteins are synthesized continuously throughout the *Escherichia coli* cell cycle. J. Bacteriol. 138:352–60.

Mahler, I., and H. O. Halvorson. 1977. Transformation of *Escherichia coli* and *Bacillus subtilis* with a hybrid plasmid molecule. J. Bacteriol. 131:374–77.

Munoz, L., Y. Sadaie, and R. H. Doi. 1978. Spore coat protein of *Bacillus subtilis*, structure and purcursor synthesis. J. Biol. Chem. 253:6694–701.

Panasenko, S. M., J. R. Cameron, R. W. Davis, and I. R. Lehman. 1977. Five hundred fold overproduction of DNA ligase after induction of a hybrid lambda lysogen constructed *in vitro*. Science 196:188–89.

Pandey, N. K., and A. I. Aronson. 1979. Properties of the *Bacillus subtilis* spore coat. J. Bacteriol. 137:1208–18.

Piggot, P. J., and J. G. Coote. 1976. Genetic aspects of bacterial endospore formation. Bacteriol. Rev. 40:908–62.

Priest, F. G. 1977. Extracellular enzyme synthesis in the genus *Bacillus*. Bacteriol. Rev. 41:711–53.

Rudinski, M. S., and D. H. Dean. 1979. Evolutionary considerations of related *Bacillus subtilis* temperate phages ϕ105, ρ14, ρ10 and ρ6 as revealed by heteroduplex analysis. Virology 99:57–65.

Segall, J., and R. Losick. 1977. Cloned *Bacillus subtilis* DNA containing a gene that is activated early during sporulation. Cell 11:751–61.

Shoji, J. 1978. Recent chemical studies on peptide antibiotics from the genus *Bacillus*. Adv. Appl. Microbiol. 24:187–214.

Tanaka, T., and T. Koshikawa. 1977. Isolation and characterization of four types of plasmids from *Bacillus subtilis* (natto). J. Bacteriol. 131:699–701.

Tanaka, T., M. Kuroda, and K. Sakaguchi. 1977. Isolation and characterization of four plasmids from *Bacillus subtilis*. J. Bacteriol. 129:1487–94.

Tanaka, T., and K. Sakaguchi. 1978. Construction of a recombinant plasmid composed of *B. subtilis* leucine genes and a *B. subtilis* (natto) plasmid: its use as cloning vehicle in *B. subtilis* 168. Molec. Gen. Genet. 165:269–76.

Trautner, T. A., B. Pawlek, S. Bron, and C. Anagnostopoulos. 1974. Restriction and modification in *Bacillus subtilis*: biological aspects. Mol. Gen. Genet. 131:181–91.

Wake, R. G. 1973. Circularity of the *Bacillus subtilis* chromosome and further studies on its bidirectional replication. J. Mol. Biol. 77:569–75.

Walker, P. D., and R. D. Thomson. 1972. Immunology of spores and sporeforms. *In* H. O. Halvorson, R. Hanson, and L. L. Campbell (eds.), Spores V, pp. 321–39. American Society for Microbiology. Washington, D.C.

Weisblum, B., M. Y. Graham, T. Gryczan, and D. Dubnau. 1979. Regulation of plasmid copy number control: isolation and characterization of high copy number (cop) mutants of plasmid pE194. J. Bacteriol. 137:635–43.

Wilson, C. R., and J. N. Baldwin. 1978. Characterization and construction of molecular cloning vehicles within *Staphylococcus aureus*. J. Bacteriol. 136:402–13.

Young, F. E., and G. A. Wilson. 1972. Genetics of *Bacillus subtilis* and other gram-positive sporulating bacilli. *In* H. O. Halvorson, R. Hanson, and L. L. Campbell (eds.), Spores V, pp. 77–106. American Society for Microbiology. Washington, D.C.

The Origin of Deletions in Yeast as Revealed by Transformation

6

INTRODUCTION

Deletion of a large segment of DNA represents a drastic mutational change. Unlike duplications, base-pair changes, transpositions, translocations, and inversions, gross deletions result in the loss of DNA sequences that, in most instances, cannot be recovered. Whereas base substitution mutations and frameshift mutations can revert to give a functional gene, mutations resulting in the loss of several genes or even a portion of a gene cannot. Deletions could occur during DNA replication, recombination, or repair of lesions induced by radiation or chemicals. Since deletions of unique sequences represent irretrievable losses of function, it seems likely that organisms have evolved mechanisms that prevent the wholesale loss of genetic material during DNA replication and repair.

Under some circumstances deletion of genes may provide an important evolutionary avenue. For example, gene duplication and deletion could serve as a regulatory mechanism permitting large changes in the amounts of a gene product. In some environments a large amount of a particular gene product may be required, and amplification of the gene that codes for this product would confer a selective advantage. When the amplified set of sequences is no longer required, deletion of the excess information could make room for amplification of yet other sequences that might be required to meet new environmental or developmental conditions. This sequence of events, amplification followed by deletion of the amplified DNA, has been described in bacteria (Anderson and Roth, 1977; Davies and Rownd, 1972;

Department of Biochemistry, Molecular and Cell Biology, Cornell University, Ithaca, New York 14853

Hashimoto and Rownd, 1975), and some evidence for such events exists in eukaryotes (Tartoff, 1975). Deletion of DNA sequences may also serve as a defense mechanism against deleterious elements like viruses. Thus, an equilibrium must exist between the forces that compel an organism to conserve sequences required for unique functions and those that catalyze gene deletion.

This paper will examine the occurrence of deletions in the yeast *Saccharomyces cerevisiae*. Until recently, deletions in this organism could be obtained and characterized only by a classical genetic approach. Deletions were defined by two genetic criteria: reversion and recombination. Deletions are stable, failing to revert to wild type at measurable frequencies. In addition they fail to recombine with two or more point mutations that recombine with each other. These criteria are not definitive since they allow for the possible confusion of deletions and inversions. Therefore, many investigators classified mutations fulfilling these criteria as "multi-site." Recently, the use of specific, cloned DNA sequences as probes in Southern hybridization has permitted a number of mutations that fulfill these genetic criteria to be identified as physical deletions of specific DNA sequences. The survey in this paper shows that, with a few exceptions, there is a low incidence of deletions in yeast. In some regions of the genome, deletions represent a high proportion of the mutational events. In these "deletion prone" regions, there is reason to suspect the presence of duplicated DNA sequences. In fact, the analysis of duplications created by transformation indicates that they give rise to deletions of the duplicated segment at a high frequency.

GENETIC SYSTEMS IN WHICH DELETIONS HAVE BEEN IDENTIFIED

Deletions have been uncovered in a number of gene systems of yeast. The most advantageous systems for the detection of deletions are those where large numbers of mutants can be generated and analyzed easily. The systems described in this paper are those in which the genetics is developed sufficiently to permit the detection of gross deletions by genetic fine structure analysis and by reversion studies.

Adenine Mutants

The red adenine system (Roman, 1956) provides a unique opportunity to obtain large numbers of spontaneous and mutagen-induced adenine auxotrophs. Mutations in either the *ade1* or *ade2* gene result in the accumulation of large amounts of a bright red pigment in such amounts that strains containing these mutations form colonies with a characteristic red color. On certain media the accumulation of this red pigment leads to retarded growth. Mutations that block the production of the pigment have a selective advantage and give rise to fast-growing white colonies. Mutations in three genes (*ade5, 7, ade 3,* and *ade8*) lead to blocks in the adenine pathway earlier than *ade1* or *ade2* and give white colonies. None of the several hundred mutations isolated in the *ade5, 7* locus (Dorfman, 1964) appeared to be a deletion. The tests used to study the mutations would not have revealed a small deletion within the locus. Twenty-two spontaneous mutations at the *ade8* locus were examined by reversion and recombination. Two mutations failed to give Ade8$^+$ revertants and one failed to recombine with two recombining point mutations. In a comprehensive study of *ade3* mutations, Jones found that 2 out of 26 spontaneous mutations failed to recombine with 2 or more mutations at the locus (Jones, 1972). Thus, in the red-white adenine system deletions are rare even among independently isolated spontaneous mutations.

cyc1

An elegant system that permits the screening of both forward and reverse mutations is the *cyc1*-iso-1-cytochrome c system studied by Sherman and his collaborators (Sherman et al., 1970, 1974). Mutants deficient in cytochrome c grow on chlorolactate medium, whereas wild-type strains containing a functional iso-1-cytochrome c fail to form colonies. Medium containing lactate as the carbon source fails to support the growth of Cyc1$^-$ mutants and is used to select for Cyc1$^+$ revertants or Cyc1$^+$ recombinants. Of 353 spontaneous and induced Cyc1$^-$ mutants selected from the same parent strain, none contained extended deletions or gross chromosomal aberrations of the *cyc1* locus (Sherman et al., 1974).

The paucity of deletions in standard strains stands in direct contrast to the high frequency of Cyc1$^-$ deletions found in two special circumstances. One of these cases involves a mutator gene that causes a high frequency of deletions of the *cyc1* locus (Liebman et al., 1979). These deletions appear to include the adjacent gene *rad7*, which controls UV sensitivity, and *osm1*, which controls osmotic sensitivity. Genetic analysis indicates that the mutable property is due to a single gene called *DEL1*, which maps adjacent to the *cyc1* locus. The remarkable fact is that *DEL1* appears to be both cis- and trans-dominant, although it fails to affect genes other than *cyc1*.

The second instance of high frequency deletion formation involves crosses between two different Cyc1$^-$ mutants each carrying a different deletion (Sherman et al., 1975). Each deletion removed a functionally unimportant segment of the *cyc1* gene and, therefore, was Cyc1$^+$ and chlorolactate-sensitive. The cross was sporulated, and the meiotic products were plated on chlorolactate medium on which only Cyc1⁻ strains could grow. Among the chlorolactate-resistant colonies resulting from this cross, Sherman observed many new deletions. From some crosses over one-fourth of all the chlorolactate-resistant mutants contained deletions of various lengths. Remarkably, these deletions often encompassed a region different from those defined by either of the parents. This result was unexpected since one mechanism proposed for the generation of meiotic deletions in these crosses involves heteroallelic mispairing between the altered segments in the parental strains. This mechanism appears to be unlikely since more than 40% of the deletions did not encompass the segments altered in the parental strains. Speculation on the mechanism by which these deletions are produced should take into account the possibility that *DEL1* or an element like it could be present in these strains.

can1

Mutations in the gene for the arginine permease (*can1*) can be selected in both directions. Wild-type strains fail to grow on medium containing the arginine analogue canavinine, whereas *can1* mutants grow. If the *can1* mutant is selected in a strain carrying an *arg6* or *arg8* mutation, the *can1 arg6* or *can1 arg8* strains will fail to grow on medium containing arginine since *can1* mutants are defective in arginine transport. These strains can be propagated on media containing ornithine as a source of arginine, and revertants to Can1⁺ can be selected by growth on medium containing arginine as a supplement. A recent study of the *can1* locus showed that only 5 of 111 independently isolated Can1⁻ mutations behaved as deletions (Whelan et al., 1979). These 5 mutations appeared to delete the entire gene and in addition to material outside of the *can1* locus.

sup4

A variation of the CAN1 selection has been used to study mutations within suppressor genes of yeast. The SUP4 gene of yeast codes for a tyrosine-inserting suppressor tRNA. Mutations at this locus can be obtained easily by selection of co-revertants from a multiply-auxotrophic strain containing ochre (UAA) mutations in a number of genes. These revertants have a high probability of containing a suppressor mutation (Mortimer and Gilmore, 1968; Sherman et al., 1973; Fink and Styles, 1974). If the strain also contains a *can1-100* mutation (a UAA mutation in the *can1* gene), then the strain can be analyzed further for recombination and reversion of the suppressor. Strains carrying *can1-100* and the suppressor have a canavinine-sensitive phenotype, permitting selection for loss of suppressor by growth on medium containing canavinine. Using this system the reversion of SUP4 (UAA) was studied. Loss of suppression in 16 of 66 spontaneous revertants tested occurred by deletion of the entire SUP4 locus (Rothstein, 1979). Deletions at SUP4 were verified by hybridization to Southern blots of total DNA from the mutant strains. The unusual suspectibility of the SUP4 gene to deletions contrasts with most other yeast loci studied and is comparable only to the DEL1 situation described for iso-1-cytochrome c.

lys2 and met15

Two systems that have been studied recently show a special promise for deletion isolation and characterization. Wild-type strains of yeast fail to grow on α-aminoadipic acid, whereas *lys2* auxotrophs can grow. Thus, α-aminoadipic acid medium provides a positive selection for *lys2* auxotrophs (Chattoo et al., 1979a, b). A study of 53 independently isolated spontaneous *lys2* auxotrophs obtained from the α-aminoadipate selection failed to uncover a single deletion as defined by reversion and recombination tests (Fink, unpublished). A study of methylmercury-resistant mutants by Singh and Sherman (1975) revealed a high proportion of *met15* auxotrophs among the resistant colonies. They examined 133 spontaneous and induced *met15* auxotrophs isolated as methylmercury-resistant colonies. Only 30 of these were examined by both recombination and reversion analysis. Two X ray-induced methylmercury-resistant *met15* auxotrophs appeared to be deletions by these criteria. Thus, even in systems like *lys2* and *met15*, where large numbers of spontaneous mutations are easily obtained and analyzed, deletions are rare.

Mating Type

Perhaps the most illustrative case for the mechanism of deletion production is one that originally seemed to have baroque qualities. In a study of homosexual, alpha by alpha, crosses, Hawthorne (1963) found a deletion that extended from the mating type locus to *MAL2* (see fig. 1). This deletion, known as the "Hawthorne deletion," created a recessive lethal as a result of the loss of genetic information, and also conferred mating type *a* function. Subsequent studies in a number of laboratories have shown that the frequency of this deletion is approximately 10^{-5}. Although a great deal remains to be learned about the mechanism by which Hawthorne deletions are generated, the cassette hypothesis proposed by Herskowitz and his colleagues (Hicks and Strathern, 1977) offers a reasonable explanation for their occurrence. According to the cassette hypothesis,

CHROMOSOME III

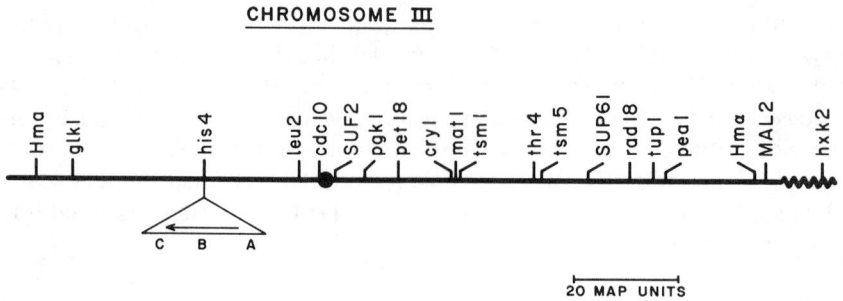

Fig. 1.   A linkage map of chromosome III of yeast.

there are multiple gene copies of the mating type locus (MAT1) distributed on chromosome III. In particular, standard strains contain a silent copy of MAT1-*a*(HMα) on the same arm of chromosome III as the expressed copy, but mapping farther out on the chromosome near *MAL2*. A reasonable explanation for the high frequency of deletions is that homology between the expressed copy at *MAT1* and the silent copy of *a* information at HMα permits an intrachromosomal recombination event between the homologous segments. The recombination event places the HMα gene at MAT and deletes the sequences between MAT and HMα. The Hawthorne deletion is the only product of such a recombination event that would contain a centromere and thus be perpetuated mitotically. These data suggest that deletions may occur at high frequencies in yeast by a recombination event between duplicated DNA segments on the same chromosome.

The Search for Deletions in the his4 System

More than 400 EMS, UV, ICR170, and nitrous acid-induced mutations of *his4* have been studied by reversion, recombination, and suppression analysis (Shaffer et al., 1970; Fink and Styles, 1974; Culbertson et al., 1977). Missense mutations, nonsense mutations (of all three types), and frameshift mutations are abundant among these mutagen-induced *his4* auxotrophs. However, none of the mutations appeared to be an extensive deletion.

We have also examined 61 spontaneous mutations of the *his4* locus (Fink and Chaleff, unpublished). These *his4* mutations were isolated by the inositol starvation procedure (Henry et al., 1975). The principle of the procedure is that an inositol-requiring mutant will die if deprived of inositol unless it has acquired a second mutation. The second mutation will prevent growth in the absence of inositol, effectively blocking inositolless death. (The conclusion that auxotrophs obtained by inositolless death are spontaneous is based upon the untested assumption that starvation for inositol is not itself mutagenic.) None of the 60 *his4* auxotrophs obtained by the inositol procedure behaves like a deletion in recombination and mutational studies. Similar results were obtained in an examination of 44 X ray-induced mutations at *his4*.

To investigate the frequency of deletions in more detail, we devised a positive selection procedure for viable haploid strains carrying deletions of the *his4* region. The strategy employed in the selection of deletions depends upon several unique features of the *his4* region. The *his4* gene specifies a single protein (Keesey et al., 1979) that catalyzes three distinct enzymatic reactions. Each of these reactions is specified by a distinct subregion of the *his4* locus called A, B, and C (see fig. 2). The A, B, and C segments control respectively the third, the second, and the last steps in the pathway of histidine biosynthesis. Each segment has functional autonomy as shown by the fact that missense mutations in one of the segments fails to affect the enzyme activity specified by the other two. The *his4* region is transcribed and translated from A → C so that nonsense mutations in his4A are polar into his4B and his4C (i.e., $A^-B^-C^-$). Nonsense mutations in his4A abolish his4C function (and histidinol dehydrogenase activity) and, therefore, are

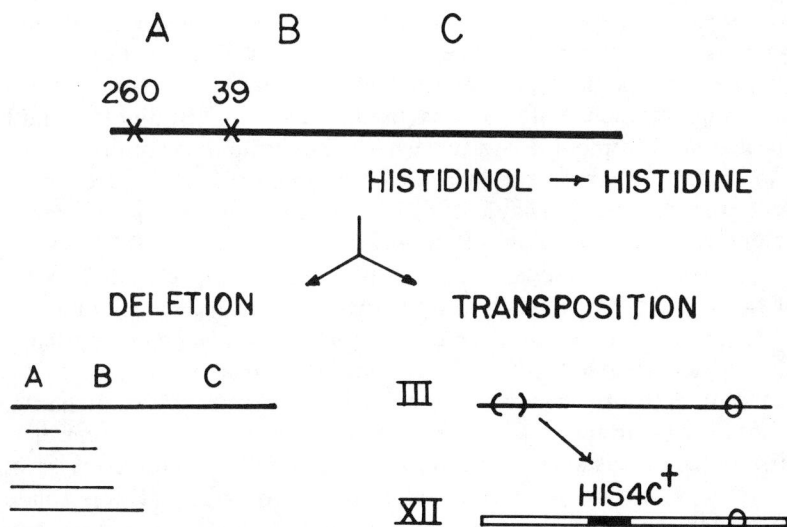

Fig. 2. Low frequency reversion of a double polar mutation. The basis for the selection is that polar his4A mutations (260, 39) prevent growth on histidinol. Strains carrying these mutations are plated on histidinol, and revertants capable of growth are analyzed. Some turn out to be deletions and others transpositions. All have his4C function as a result of the aberration and are, therefore, capable of growth on histidinol.

incapable of growth on histidinol. We have constructed a strain with 2 nonsense mutations, his4-260 (UGA) and his4-39 (UAA) within the his4A region (fig. 2). Deletions were obtained in this strain by selecting for growth on histidinol (Fink and Styles, 1974). The efficiency of this selection system is high since only revertants carrying deletions form colonies on histidinol medium. The coupling of a UGA and UAA mutation within the same gene excludes suppressor revertants because no suppressor in yeast suppresses both mutations. Thus, the only colonies that should appear on the histidinol plates carry deletions that remove the polar mutations. More than 50 deletions have been uncovered in the his4A and his4B region by this selective technique. The remarkable fact is that these deletions occur very infrequently; even after irradiation with ultraviolet light only 1 in 10^{10} cells plated can grow on histidinol. Strains carrying these deletions are viable in the haploid state and grow at rates equivalent to wild type if supplemented with either histidinol or histidine.

Deletions of *his4* may occur at a frequency greater than 10^{-10} but may often give rise to recessive lethals. For example, simultaneous deletion of *his4* and an adjacent gene encoding a vital function, would lead to inviability. According to this model most deletions encompass a region larger than *his4* and include the vital gene. To investigate this possibility, revertants capable of growth on histidinol were selected from diploids homozygous for the *his4* double polar mutations. If the vast majority of deletions in the haploid go undetected because they create recessive lethals, then a high frequency of deletions giving growth on histidinol should be expected in the diploid. However, when diploids homozygous for the two polar mutations were examined, deletions occurred at low frequencies similar to those found in haploid strains. The only other events found in the diploid were transpositions of his4C$^+$ to chromosome XII (Greer and Fink, 1979). These results with the diploid strains make it unlikely that deletions are rare because they are associated with recessive lethals.

Another explanation for the paucity of deletions at *his4* could be that the deletions are frequent but so large that they cause dominant lethality. Dominant lethality would exclude their recovery either from haploid strains or from diploids. A chromosome aberration isolated in some of our earlier studies casts serious doubt on this interpretation. A strain containing a deletion of the entire left arm of chromosome III was isolated from a population of cells that had been heavily irradiated with X rays. This deletion, which had removed not only *his4* but all the other known markers on the left arm of chromosome III (see fig. 1), was lethal as a haploid, but was completely viable in diploids containing one normal chromosome III. The viability of this extensive deletion in the diploid rules out the existance of genes in the vicinity of *his4* whose deletion would cause a dominant lethal. Although classical mutational studies have revealed a great deal about the structure of the *his4* region, they have failed to provide an insight into the mechanism by which deletions are formed.

TRANSFORMATION STUDIES REVEAL HIGH FREQUENCIES OF DELETIONS
AMONG STRAINS CARRYING DUPLICATIONS

The development of transformation (Hinnen et al., 1978) has permitted a more detailed investigation of the mechanism by which deletions are formed in yeast. Two types of transformation have been uncovered in yeast. High-frequency transformation occurs when plasmids capable of autonomous replication in yeast are used as a source of transforming DNA (Struhl et al., 1979; Hicks et al., 1979; Beggs, 1979). In high-frequency transformation the transforming DNA remains as an autonomously replicating, covalently closed, circular DNA molecule. A replication origin either from the yeast plasmid or from a yeast chromosome is capable of conferring the property of autonomous replication. In low-frequency transformation, transforming DNA in circular form integrates into the linear chromosomal DNA by a Campbell-like recombination event. This crossover results from pairing between homologous sequences on the circular, transforming DNA and the chromosome and results in insertion of the sequences on the transforming DNA into the chromosome. This recombination event creates a duplication of the homologous yeast sequences. If bacterial (nonhomologous) DNA is also on the plasmid, then the transformation will create an integrated bacterial sequence flanked by direct repeats of the yeast sequence from the transforming plasmid (see fig. 3).

The duplications created by transformation are unstable and frequently segregate strains that have lost the duplication. Loss of the duplication occurs in *his4* transformants in about 1% of the population. This frequency is related to the size of the duplication; large duplications are considerably more unstable than smaller ones. Loss of the duplication is easily verified both genetically and physically. If the two components of the duplication are genetically different, then segregants that have lost the duplication can be identified. For example, if the structure of the duplication is *His4⁺/his4⁻*, segregants that are His4⁻ can be detected on minimal medium. If the structure is His4-a/his4-b where allele his4-a and his4-b complement, the appearance of *His4⁻* segregants signals loss of the duplication. Loss of the duplication can be verified by colony hybridization since the bacterial plasmid sequences surrounded by the duplication represent a

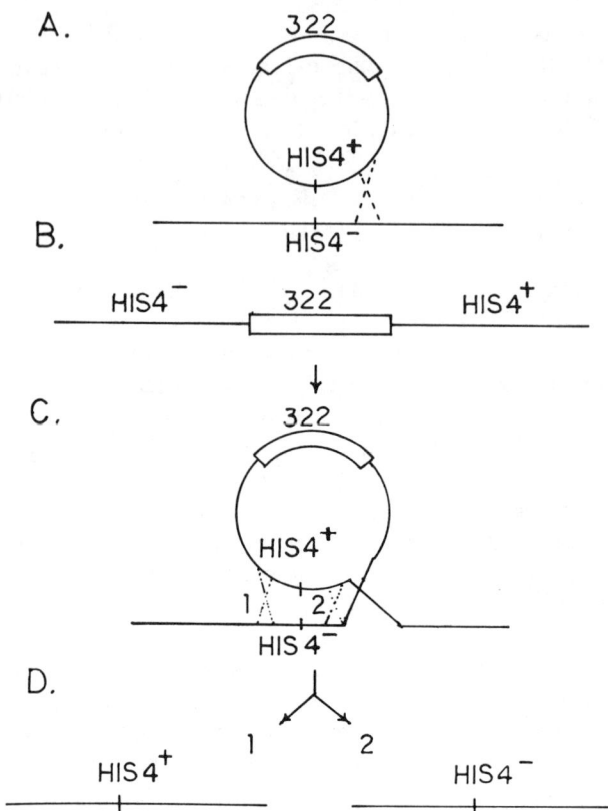

Fig. 3.   Duplication and deletion formation. Duplications are formed by a transformation event. Recombination within the duplication leads to loss of the duplicated element and all information between the elements of the duplication as well. If recombination occurs at (1), the recombinant will be His⁺, whereas if it occurs at (2), the recombinant will be His⁻.

unique sequence for hybridization. Colonies that have lost the duplication fail to hybridize to radioactively labeled plasmid DNA, a sign that these sequences were lost in concert with a portion of the duplication.

DISCUSSION

Work on prokaryotes has indicated that deletions occur frequently in regions of repeated DNA sequences. In an analysis of 140 independently occurring spontaneous mutations in the *lacI* gene of *E. coli*, Farabaugh and coworkers (1978) found that small frameshift and deletions constitute 80% of the *lacI* mutations. The majority of these events occurred in the tandemly repeated sequence C-T-G-G-C-T-G-G-C-T-G-G. Mutations having an additional C-T-G-G sequence appear at a frequency of about 2×10^{-6} and those having a deletion of a C-T-G-G sequence appear at a frequency of about 0.5×10^{-6}. Repeated sequences are also involved in the formation of the larger deletions at *lacI*. In 7 out of 12 cases, repeats of five or eight bases are found at each end in the wild-type sequence. Some of the deletions at *lacI* do not occur in repeated sequences.

The exact mechanism by which deletions are formed is not known in *E. coli*. In the *lacI* study of Farabaugh and colleagues (1978), the reversion of a four base insertion (C-T-G-G) to LacI$^+$ occurs at similar frequencies in both *recA*$^-$ and *1recA*$^+$ background. This result suggests that *recA*$^-$-mediated recombination is not involved in this type of deletion. Furthermore, deletions in the tonB-trp region have been shown to arise at approximately the same rate in *recA*$^-$ as in *recA*$^+$ (Franklin, 1967). By contrast, studies on the stability of duplications in the histidine operon of *S. typhimurium* have shown that deletion of repeated sequences is dramatically reduced in the *recA*$^-$ background. The conflicting results concerning the involvement of the *recA* gene product in forming deletions may be due to the existence of several different mechanisms for deletion formation. Some regions may be adjacent to an insertion sequence (IS element). IS sequences have been reported to be "hot-spots" for *recA*$^-$ independent deletions (Rief and Saedler, 1975). The enzyme systems involved in IS-mediated deletion do not seem to require repetitive DNA. In the case of large repetitive sequences, *recA* is certainly involved. Small repetitive sequences may not be recognized by *recA*$^+$ for homologous recombination but rather (as in the case of some *lacI* duplications) by some other enzyme system. So far there is no evidence concerning the extent of homology required for *recA* mediated events.

In yeast, deletions have been analyzed by conventional genetic analysis and more recently by DNA-DNA hybridization techniques. In some regions deletions are infrequent in large samples of spontaneous and mutagen-induced mutations. Nevertheless, at a number of these loci, rare deletions have been detected by genetic analysis. The mechanism of formation of these deletions is obscure. In *cyc1* (where the DNA sequence is known), these rare deletions occur in regions devoid of repetitive DNA.

Other regions like mating type, SUP4, and *cyc1* (in a DEL1 background) appear to give rise to deletions at high frequencies. One explanation for the high deletion frequency of these genes is that they are located between the elements of a duplication or are themselves one of the elements of the duplication. The study of duplications formed by transformation shows clearly that wild-type yeast cells have the enzymes required to catalyze intrachromosomal recombination events between the homologous segments of a duplication. These recombination events lead to a loss of one element of the duplication. In accord with these observations, preliminary studies have identified reiterated segments in or near the "deletion prone" yeast genes. For example, mating type appears to be a duplicated segment of DNA. Homologous DNA sequences exist at the silent copies of mating type (Hma and Hmα in fig. 1). In addition, in certain strains highly duplicated DNA segments encompass the region around SUP4 and *cyc1* (Cameron et al., 1979). It is not known whether the formation of deletions in these deletion-prone regions is under the control of a *recA* like protein. Studies on genes affecting homologous recombination could provide important insights into the mechanism(s) by which deletions are formed in yeast.

REFERENCES

Anderson, R. P., and J. R. Roth. 1977. Tandem genetic duplications in phage and bacteria. Ann. Rev. Microbiol. 31:473–505.

Beggs, J. D. 1978. Transformation of yeast by a replicating hybrid plasmid. Nature 275:104–9.

Cameron, J., E. Loh, and R. W. Davis. 1979. Evidence for transposition of dispersed repetitive DNA families in yeast. Cell 16:739–51.

Chattoo, B. B., E. Palmer, B. Ono, and F. Sherman. 1979. Patterns of genetic and phenotype suppression of *lys2* mutations in the yeast *Saccharomyces cerevisiae*. Genetics (in press).

Chattoo, B. B., F. Sherman, D. A. Azubalis, T. A. Fjellstedt, D. Mehnert, and M. Ogur. 1979. Selection of *lys2* mutants of the yeast *Saccharomyces cerevisiae* by the utilization of α-aminoadipate. Genetics (submitted).

Culbertson, M., L. Charnas, T. Johnson, and G. R. Fink. 1977. Frameshifts and frameshift suppressor in *S. cerevisiae*. Genetics 86:745–64.

Davies, J. E., and R. Rownd. 1972. Transmissible multiple drug resistance in enterobacteriaceae. Science 176:758–68.

Dorfman, B. Z. 1964. Allelic complementation at the ade5, 7 locus in yeast. Genetics 50:1231–43.

Esposito, M. S. 1968. X-ray and meiotic fine-structure mapping of the adenine 8 locus in *Saccharomyces cerevisiae*. Genetics 58:507–27.

Farabaugh, P. J., U. Schmeissner, M. Hofer, and J. Miller. 1978. Genetic studies of the *lac* repression VII on the molecular nature of spontaneous hot spots in the *lacI* gene of *E. coli*. J. Mol. Biol. 126:847–63.

Franklin, N. C. 1967. Extraordinary recombinational events in *Escherichia coli*: their dependence on the rec[+] function. Genetics 55:699–707.

Fink, G. R. 1966. A cluster of genes controlling three enzymes in histidine biosynthesis in yeast. Genetics 53:445–59.

Fink, G. R., and C. Styles. 1974. Gene conversion of deletions in the *his4* region of yeast. Genetics 77:231–44.

Greer, H., and G. R. Fink. 1979. Unstable transpositions of *his4* in yeast. Proc. Nat. Acad. Sci. USA 76:4006–10.

Hashimoto, H., and R. H. Rownd. 1975. Transition of the R factor NR1 in *Proteus mirabilis*: level of drug resistance of non-transitioned and transitioned cells. J. Bacteriol., 123:56–68.

Hawthorne, D. C. 1963. A deletion in yeast and its bearing on the structure of the mating type locus. Genetics 48:17–27.

Henry, S., T. Donahue, and M. Culbertson. 1975. Selection of spontaneous mutants by inositol starvation. Mol. and Gen. Genet. 143:5–11.

Hicks, J., and J. Strathern. 1977. Interconversion of mating type in *S. cerevisiae*. Brookhaven Symp. 29:233–42.

Hicks, J., A. Hinnen, and G. R. Fink. 1979. Properties of yeast transformation. Cold Spring Harbor Symp. Quant. Biol. 14:1305–11.

Hinnen, A., J. Hicks, and G. R. Fink. 1978. Transformation of yeast. Proc. Nat. Acad. Sci. USA 75:1929–33.

Jones, E. W. 1972. Fine-structure analysis of the ade3 locus in *Saccharomyces cerevisiae*. Genetics 70:233–50.

Keesey, J. K., R. Bigelis, and G. R. Fink. 1979. The product of the *his4* gene cluster in *Saccharomyces cerevisiae*, a trifunctional polypeptide. J. Biol. Chem. 254:7427–33.

Liebman, S. W., A. Singh, and F. Sherman. 1979. A mutator affecting the iso-1-cytochrome *c* gene in yeast. Genetics (submitted).

Mortimer, R., and R. Gilmore. 1968. Suppressors and suppressible mutations in yeast. Adv. Biol. Med. Phys. 12:319–31.

Reif, H. J., and H. Saedler. 1975. IS1 is involved in deletion formation in the gal region of *E. coli* K12. Mol. Gen. Genet. 137:17–28.

Roman, H. 1956. A system selective for mutations affecting the synthesis of adenine in yeast. Compt. Rend. Trav. Lab. Carlsberg Ser. Physiol. 26:299–314.

Rothstein, R. 1979. Deletions of a tyrosine tRNA gene in *S. cerevisiae*. Cell 17:185–90.

Shaffer, B., J. Rytka, and G. R. Fink. 1969. Nonsense mutations affecting the *his4* enzyme complex of yeast. Proc. Nat. Acad. Sci. USA 63:1198–1205.

Sherman, F., J. W. Stewart, J. H. Parker, G. J. Putterman, B. L. Agrawal, and E. Margoliash. 1970. The relationship of gene structure and protein structure of iso-1-cytochrome c from yeast. Symp. Soc. Exptl. Biol. 24:85–107.

Sherman, F., S. Liebman, J. W. Stewart, and M. Jackson. 1973. Tyrosine substitutions resulting from suppression of amber mutants of iso-1-cytochrome c in yeast. J. Mol. Biol. 78:157–68.

Sherman, F., J. W. Stewart, M. Jackson, R. A. Gilmore, and J. H. Parker. 1974. Mutants of yeast defective in iso-1-cytochrome c. Genetics 77:255–84.

Sherman, F., M. Jackson, S. Liebman, A. M. Schweingruber, and J. W. Stewart. 1975. A deletion map of *cyc1* mutants and its correspondence to mutationally altered iso-1-cytochromes c of yeast. Genetics 81:51–83.

Singh, A., and F. Sherman. 1975. Genetic and physiological characterization of *met15* mutants of *Saccharomyces cerevisiae*: a selective system for forward and reverse mutations. Genetics 81:75–97.

Struhl, K., D. Stinchcomb, S. Scherer, and R. W. Davis. 1979. High frequency transformation of yeast: autonomous replication of hybrid DNA molecules. Proc. Nat. Acad. Sci. USA 76:1035–39.

Tartoff, K. 1975. Redundant genes. Ann. Rev. Genetics, pp. 355–85.

Whelan, W., E. Gocke, and T. R. Manney. 1979. The *CAN1* locus of *Saccharomyces cerevisiae*: fine structure analysis and forward mutation rates. Genetics 91:35–51.

A. M. CHAKRABARTY

Practical Applications of
Genetic Engineering Techniques

7

There has been tremendous progress in recent times in the development of technology that allows deliberate modification of the genome of prokaryotes and eukaryotes for specific purposes. This is usually referred to as genetic engineering. Until recently, the techniques available for genetic improvement have relied on *in vivo* alteration of the structure or composition of the genetic material, i.e., mutations, recombinations, or introduction of plasmids from one species, or in some cases from one genus, to another. Such manipulation used to be based on processes that occur in nature, i.e., transformation, conjugation, and so on. In the early seventies the powerful techniques of recombinant DNA, which is the major theme of this volume, were introduced, and since then it has been possible to introduce completely novel genetic entities into both prokaryotes and eukaryotes. As a result, there is great excitement in the construction of new microbial strains capable of producing such unique compounds as human growth hormone, chicken ovalbumin, or rat insulin. Numerous genetic improvements that have hitherto been thought to be difficult or impossible to accomplish are now rendered possible because of the advent of the recombinant DNA technology. Numerous articles and books have been written regarding the benefits and biohazards of this technology (Curtiss, 1976; Chakrabarty, 1979; Beers and Bassett, 1977; National Academy of Sciences, 1977; Chakrabarty, 1978). Since many of the potential applications have already been discussed (Curtiss, 1976; Chakrabarty, 1979), I will confine the potential applications of the genetic engineering

Department of Microbiology and Immunology, University of Illinois Medical Center, Chicago, Illinois 60612 techniques in those areas, with which I am somewhat familiar and where I think the application of these techniques will be highly rewarding in an accelerated understanding of the molecular basis of bacterial virulence, in the development of alternate sources of energy or in specific construction of strains capable of biodegrading and removing toxic, persistent pollutants from the environment.

IN VITRO GENETIC ENGINEERING (RECOMBINANT DNA)

The *in vitro* recombinant DNA (molecular cloning) techniques allow the insertion of foreign DNA, either prokaryotic or eukaryotic, into *E. coli* or other bacteria as part of a plasmid or phage. A similar system, using genomes of defective animal viruses as vectors, has also been developed for insertion of foreign DNA into mammalian cells. The technology therefore allows a barrier-free flow of genetic materials from prokaryotes and eukaryotes to various bacterial and mammalian cells. Depending upon the nature of vectors used, the foreign gene segments can be amplified greatly within the host cell to yield large quantities of the genetic materials or their gene products. Depending upon the nature of the products, the application of the recombinant DNA techniques may be classified under five broad categories.

Biomedical

This is an area where recombinant DNA technology is considered to have wide applicability. The challenge of producing animal proteins of pharmacological and medical importance such as interferons, immunoglobulins, or hormones in bacterial cells has already met with some success, and is dealt with by Dr. Villa-Komaroff in this volume. The development of cloning foreign DNA into the penicillinase gene, leading to the formation of hybrid DNA molecules that can be functionally expressed to fused proteins carrying the antigenic determinant of the foreign gene product, has been described (Villa-Komaroff et al., 1978). Since penicillinase is a secretable protein, such fused proteins are also secreted into the periplasmic space, thus enhancing the yield and ease of isolation. This technique, where a protein such as penicillinase acts as a carrier leading to the expression and secretion of eukaryotic proteins cloned as cDNA copies, may be an extremely useful tool as a general technique for the bacterial production of important animal proteins of biological value. The expression of antigenic determinants of the eukaryotic proteins further points out the usefulness of this technique in cloning viral DNA segments encoding the surface antigens in the manufacture of vaccines against a large number of viral diseases. Cloning of chicken ovalbumin gene as part of *E. coli lac z* gene leads to the production of hybrid, β-galactosidase-ovalbumin protein that can be detected by radioimmuno assays in *E. coli* extracts (Mercereau-Puijalon et al., 1978). Recent observations by Fraser and Bruce (1978) indicate that the *E. coli* secretory apparatus can recognize ovalbumin, which is normally synthesized and secreted in the chicken oviduct as a secretable protein. The microbially synthesized ovalbumin has been found to secrete through *E. coli* cell membrane into the periplasmic space, where it can be easily isolated in large quantities. Similar cloning of the human growth hormone gene linked in phase to a fragment of the *trp D* gene of *E. coli* in a plasmid vehicle has been described (Martial et al., 1979). The human growth hormone gene is expressed at a very high level on derepression of the *trp* operon, and the fusion protein reacts specifically to antibodies to human growth hormone.

Apart from production of proteins of biological value, the recombinant DNA technology may be usefully employed in the manufacture of vaccines or in the development of drugs for prevention of bacterial infections. The application of recombinant DNA techniques in the manufacture of vaccines has already been discussed (Chakrabarty, 1979), and will not be repeated here. As an example of the potential applicability of the recombinant DNA techniques in an understanding of the mode of bacterial infections in specific human diseases, I will describe some experiments conducted in our laboratory during the past several months. The problem involves the infection of cystic fibrosis patients with *Pseudomonas aeruginosa*. Unlike most other strains of *P. aeruginosa* isolated from burn, eye, or urinary tract infections, the *P. aeruginosa* strains isolated from the sputum of cystic fibrosis patients are highly mucoid. Although initially the nonmucoid forms predominate, the proportion of mucoid cells increases drastically with the severity of the infections. The recovery of mucoid *P. aeruginosa* from the sputum is thus almost a diagnostic test for cystic fibrosis, and a bad prognosis for the patient. The bacterial mucus is an exopolysaccharide (alginic acid) composed of mannuronic and guluronic acids in varying proportions (Evans and Linker, 1973). The significance of alginic acid production by the mucoid cells in order to establish themselves in the respiratory tract of cystic fibrosis patients is unknown.

We have recently demonstrated that a number of mucoid *P. aeruginosa* strains, isolated from the sputum of cystic fibrosis patients in various hospitals in the United States, produce alginic acid primarily when grown with sugars or benzoate as growth substrates. When grown in an environment containing patient sputum as the only source of carbon, such strains produce very little alginic acid, and large quantities of other exopolysaccharides lacking uronic acids. Since host sputum is composed of proteins and glycoproteins, and since the carbohydrate moiety of the host glycoproteins consists of mannose, fucose, and N-acetylneuraminic acid, none of which can be used as a sole source of carbon by the mucoid *Pseudomonas*, it is likely that the mucoid strains proliferate in the respiratory tract by utilizing the protein components of the sputum. It therefore seems that the mucoid strains also produce alginic acid when grown with proteinaceous materials. This view is confirmed by the fact that when several mucoid strains are grown with L broth and the exopolysaccharides are isolated, they demonstrate the presence of alginic acid, and large quantities of a second type of polysaccharide that has little uronic acid in it.

In order to explain why mucoid *P. aeruginosa* strains are so prevalent in the respiratory tract of cystic fibrosis patients, we have presented a model (Hamada, Medenis, and Chakrabarty, in preparation) where the exopolysaccharides are believed to help the mucoid strains colonize the respiratory tract of the cystic fibrosis patients through specific attachment with the bronchial epithelial lining cells. There is growing evidence that specific attachment to the target tissues is an important factor in bacterial virulence as well as for symbiotic relationships with eukaryotic hosts (Gibbons, 1977; Dazzo and Brill, 1979). In some cases as in *Rhizobium*-clover symbiosis, the selective adherence may be initiated by a specific cross-bridging of antigenically related polysaccharide determinants on the surface of both the bacterium and the cell wall of the clover root hair by a multivalent, plant host-coded protein (lectin) called trifoliin. If such surface polysaccharides are involved in the specific attachment and colonization of the host lung tissues by the mucoid strains, then it is important to know whether it is the alginic acid or the other polysaccharide (or both) that may facilitate this attachment. This can be readily tested by the application of recombinant DNA techniques. Thus, it might be possible to clone in a nonmucoid strain of *P. aeruginosa* the genes that code for alginic acid formation. Similarly, it should be possible to clone the genes that specify formation of the other polysaccharides in a different strain. Such strains could then be checked for specific attachment to cystic fibrosis patient bronchial epithelial cells to determine which polysaccharide, if any, may help in the colonization process.

How can the determination of the nature of the exopolysaccharide promoting specific attachment to host tissue help in eliminating the bacterial infection? We have found that mannose, a sugar commonly present in the carbohydrate moiety of the host glycoproteins, strongly inhibits alginic acid formation from most carbon sources (excepting fructose). If alginic acid is found to be responsible for specific attachment of the mucoid cells with the host lung tissues, then it should be possible to introduce mannose, perhaps as an aerosol through the nasal passages, into the lung tissues of cystic fibrosis patients so as to inhibit *in situ* alginic acid formation by the strains. A lack of alginic acid synthesis by the mucoid cells may lead to ultimate elimination of the infection.

Industrial

The major application of recombinant DNA technology in industry can be envisaged in the manufacture of bulk chemicals, enzymes, solvents, vitamins, and so on. The usefulness of this technology, as well as some of the problems, in the manufacture of industrial chemicals, fuels, and pharmaceutical products, have been discussed before (Chakrabarty, 1979). The development of *Bacillus subtilis* as a host for recombinant DNA work, as described by Dr. Dean in this volume, is extremely important in this regard, since *B. subtilis* is known to produce a large number of extracellular enzymes and antibiotics that are industrially useful. This is also the type of bacteria that has been consumed for decades in countries such as Japan. In this respect *B. subtilis* does not appear to pose any hazard as a host for recombinant DNA experiments. A similar situation exists with yeasts, which have been used in the industrial production of ethanol. Yeasts are also used widely in food industries, so that the development of the recombinant DNA technology with baker's yeasts, as described by Dr. Fink in this volume, is a welcome development of this technology. As a result of these developments, it may be possible to improve the quality of the cell mass so that cells of better taste or nutritional value may be used as hosts. Cloning of foreign genes with such hosts may lead to production of biomass that could be utilized as cattle feed after the fermentation broth has been separated for extraction of desired products. The development of the recombinant DNA technology with *Actinomyces* may similarly lead to production of hybrid antibiotics of wider spectrum but less toxicity as well as enhanced production of antibiotics, steroids, enzymes, amino acids, and other industrially useful chemicals.

Agricultural

The potential of the recombinant DNA technology in enhanced agricultural productivity as well as in crop improvement has been repeatedly emphasized by various workers (Curtiss, 1976; Chakrabarty, 1979; Setlow and Hollaender, 1979). Appropriate vectors for introducing foreign genes into plants are presently under development. The impact of introducing bacterial nitrogen fixation genes into plants to make them fix their own nitrogen (and thereby eliminate the use of petroleum-based fertilizers), or the introduction of genes affording resistance against pests has been emphasized. The improvement in photosynthetic efficiency leading to enhanced crop yield appears similarly to be achievable through use of the recombinant DNA technology.

Genetic Therapy

A discussion of the potential application of recombinant DNA technology to human genetic therapy is beyond the scope of this review. Because of the social, moral, and ethical issues involved, the actual application of the recombinant DNA technology must await the resolution of such issues, even when the technical procedures for genetic therapy might have been perfected. Roblin (1978) has summarized many of the techniques used for cloning human genes, inherent problems, and technical feasibility of developing this technology for solving the problems of human genetic diseases.

Energy

Although the potential of the recombinant DNA technology in biomedical as well as in agricultural areas has been repeatedly emphasized, virtually nothing has been said about any future role of this technology in energy generation, particularly from biomass or cellulosic wastes. I feel this is an area where the fruits of the recombinant DNA technology can be realized sooner than we anticipate. Given the current shortage of energy, and no impending solution to the energy problem, any newer way of converting ligno-cellulosic wastes to fuels, or ways to recover secondary oil from abandoned oil wells will be an extremely rewarding scientific and economic venture. The application of the recombinant DNA technology may allow a tangible solution to both these problems.

Production of ethanol or methane from ligno-cellulosic wastes. Lignin and cellulose together comprise the most abundant naturally occurring carbon compounds. The association of lignin with cellulose, however, greatly retards the biodegradation of the latter since the ligno-cellulosic complexes are seldom amenable to microbial enzymatic attack. Lignin, which is a complex polymer, is known to be slowly biodegraded by fungi (Kirk and Chang, 1975), and on rare occasions by bacteria (Haider et al., 1978). In general, fungi such as white rot fungi are believed to depolymerize lignin to monomers, but are incapable of utilizing the monomers. Many bacteria are known to be capable of utilizing the monomers as sole sources of carbon, but are incapable of depolymerizing lignin itself. In nature, successive attacks by fungi and bacteria therefore lead to a slow degradation of lignin, and a lack of genetic exchange between fungi and bacteria may explain the absence of any natural microbial strain capable of rapid lignin degradation. Since bacteria such as lignobacter are known to oxidize a large number of lignin monomers such as α-conidendrin, veratric, ferulic, vanillic, and isovanillic acids (Salkinoja-Salonen et al., 1979), it might be possible to clone the lignin depolymerase gene(s) from the white rot fungi to the lignobacter. This may allow the lignobacter to convert lignins to simple metabolites at a reasonably rapid rate. The cloning of cellulase genes from *Trichoderma viride* to such a bacterium may greatly enhance the ability of this bacterium to decompose ligno-cellulosic wastes. Since this bacterium has also been shown to fix nitrogen aerobically (Salkinoja-Salonen et al., 1979), construction of such strains may provide a simple way to convert ligno-cellulosic wastes to products that may be converted to methane by a second stage anaerobic fermentation with methane bacteria.

Another way to produce fuels such as ethanol from cheap substrates such as agricultural wastes might be to clone in yeasts the genes specifying cellulolytic and lignolytic activities. Gasohol, which is finding increasing use, is composed of about 10% ethanol in gasoline. Much of the ethanol in the midwestern states of this country is made from corn and other agricultural products. With the perfection of the cloning techniques in yeasts, as described by Dr. Fink in this volume, it might be possible to clone the lignolytic and cellulolytic genes from various fungi and bacteria in yeasts. Such yeast strains may then be used for rapid one-step conversion of ligno-cellulosic wastes to ethanol.

Recovery of secondary oil from abandoned wells. Another very potent application of the recombinant DNA technology in energy generation will be the construction of strains that may be used for recovery of crude oil from abandoned wells. It should be noted that most abandoned wells still contain more than 60 to 70% of the original oil; however, such oil is usually very viscous and difficult to extract because of poor rock permeability. Attempts to introduce microorganisms with appropriate nutrients so as to generate surfactants and gas to mobilize the oil have been sporadic but occasionally successful (Bubela, 1978). It should be emphasized that in order to effect free flow of oil, the microorganisms should have certain desirable properties. They should be able to consume sulfur-containing or highly viscous components of the oil, produce surfactants and biopolymers to help solubilize and mobilize the oil, and should produce acids to improve the permeability and increase the porosity of rocks. Many of these properties can be incorporated into a single strain by use of the recombinant DNA technology. For example, the release and migration of the oil can be greatly facilitated by the bacterial production of detergent-like substances (surfactants). These substances reduce the surface tension between the oil, the water, and the rock and improve the release of the oil present in the pores of the rock. Similarly, many polymers are added to the water used for flooding to increase its viscosity. Such viscosifiers may be synthetic polymers or materials produced by biological activity, and are often polysaccharide in nature. I have previously mentioned the occurrence of *Pseudomonas aeruginosa* strains capable of producing large amounts of a viscous polysaccharide called alginic acid. Evidence is also accumulating that dissimilation of sulfur-containing hydrocarbons such as dibenzothiophene or complete biodegradation of chlorinated aromatics such as chlorobenzoates with the liberation of HCl may be coded by plasmids that are transmissible to *P. aeruginosa* (W. R. Finnerty, personal communications; Chakrabarty, 1980). By transferring such plasmids to alginate-producing strains and by supplying chlorobenzoates as nutrients, it should be possible to generate large quantities of HCl and alginate *in situ* at the expense of sulfur-containing hydrocarbons and chlorobenzoates. Plasmids specifying dissimilation of other naphthenic or highly viscous components may also be introduced into such strains to accelerate their growth with hydrocarbons, reduce the viscosity of the oil, increase rock porosity through acid production, and enhance oil migration by decreasing the surface tension of the rock-oil-water system and increasing the viscosity of flooding water through production of exopolysaccharides.

IN VIVO GENETIC ENGINEERING

Improvement of the yield or quality of the products has long been achieved by *in vivo* genetic manipulations such as mutations, recombinations, and so on. This has resulted in the isolation of strains capable of producing antibiotics, vitamins, amino acids, or enzymes in high yield. Another widely used technique employs plasmid transfer between different bacterial species or genera. Thus, entirely new genetic functions can be transferred from the chromosome of one bacterial genus to different genera in the form of plasmids. The role of phage Mu in promoting transposition of blocks of genes from the chromosome onto plasmids, including such broad host range plasmids as RP4, has been intensively studied, and new strains having a variety of useful functions have been constructed (Faelen et al., 1977). In addition to plasmid-mediated transfer of chromosomal genes from one bacterium to another, plasmids may themselves specify functions that can be utilized for construction of novel strains. The role of the chlorobenzoate or dibenzothiophene degradative plasmids, or other hydrocarbon degradative plasmids, has been mentioned before (Chakrabarty, 1979). Evidence is also accumulating that suggests that the biosynthesis of some antibiotics in Actinomyces may be specified by plasmid-borne genes (Hopwood, 1978). It may thus be possible to manipulate the plasmid copy number or their transfer to new hosts to greatly enhance the yield of antibiotics or even produce hybrid antibiotics having some desirable characteristics. Potential agent for transferring bacterial DNA to plants is the common segment (T-DNA) of the *Agrobacterium* Ti plasmids (Nester and Montoya, 1979), which is transferred to the plant. Insertion of a transposon near the T-DNA segment allows the subsequent transfer of the transposon into the plant cells. If this can be developed as a general technique for introduction of foreign DNA into plants or their protoplasts, from which plants can be regenerated, it would certainly open up a new field of engineering plant species having novel characteristics. It should be stressed, however, that there is no definitive evidence at the present time that prokaryotic genes, including the T-DNA segment, are expressed functionally within the plants to form prokaryotic protein products. With increasing knowledge about the structure and functions of T-DNA or other plant virus genomes, it might be possible to fuse desirable foreign DNA segments with nonvirulent forms of such vehicles, so that the foreign inserts are expressed functionally within the plants.

GENETIC ENGINEERING AND ENVIRONMENTAL POLLUTION

The problems of environmental pollution have become acute due to manufacture and release of large quantities of chlorinated hydrocarbons such as PCBs, Kepone, and DDT (Carson, 1962), which usually accumulate in nature because of the inability of the soil and aquatic microflora to biodegrade them to any significant extent. To a large measure, hydrocarbons are utilized by a limited number of aerobic microorganisms; it is not clear why the vast majority of facultative anaerobes have not acquired the ability to utilize hydrocarbons. It is, however, known that when the hydrocarbon-degradative plasmids are transferred from hydrocarbon-utilizers such as *Pseudomonas* species to nonutilizers such as *E. coli, Klebsiella,* or *Serratia* species, they are rarely expressed by the latter group of microorganisms (Nakazawa et al., 1978; Jacoby et al., 1978; Chakrabarty et al., 1978). It is also known that the introduction of the chlorine substituents into the organic molecules makes such compounds inaccessible to microbial attack. Evidence is, however, accumulating that suggests that various bacteria are evolving new genetic competence, often associated with plasmids, that allows them to convert toxic chlorinated compounds to lesser toxic forms. During such conversion chlorine atoms may or may not be released from the molecule (Chakrabarty, 1980). A major finding in such studies is that dehalogenation occurs only at a late stage of the biodegradative pathway, so that in order to dissimilate a chlorinated compound, a bacterium must evolve a completely new set of degradative enzymes. It appears that there are two basic problems in the microbial biodegradation of halogenated compounds: (1) lack of ability of many aerobic and facultatively anaerobic bacteria to express hydrocarbon degradative genes and (2) a slow rate of evolution of dehalogenation functions in nature.

Given the two major constraints mentioned above, one may wonder if genetic engineering could play a role in an accelerated evolution of the microbial biodegradative competence against synthetic environmental pollutants. The large quantities of PCBs accumulated in the bottom sediments of the Hudson River for the last few decades prompted us to look for facultatively anaerobic cultures that may have evolved the ability to dissimilate PCBs in the semi-anaerobic environment of the sediments. We have recently reported the occurrence of a p-chlorobiphenyl (pCB) degradative plasmid in *Klebsiella pneumoniae* and probably in other facultative anaerobes such as *Serratia marcescens* (Kamp and Chakrabarty, 1979). Such strains, unlike laboratory strains of *Klebsiella* or *Serratia*, can functionally express hydrocarbon degradative genes transferred from *Pseudomonas* (Kamp and Chakrabarty, 1979). Thus, such strains have evolved the capability to functionally express hydrocarbon degradative genes and grow at the expense of hydrocarbons. In order to determine if such expression genes evolved on the chromosome, or as part of the pCB plasmid (pAC21), we have transferred RP4-TOL to the pAC21-negative *K. pneumoniae* cells. While pAC21-positive cells could utilize xylene and toluene, pAC21-negative cells could not, even though they harbored the intact RP4-TOL plasmid. The expression genes therefore appear to have evolved as part of the pAC21 plasmid (Farrell and Chakrabarty, 1979).

Since pAC21 is a transmissible PCB-degradative plasmid, it might be possible to transfer this plasmid to a large number of facultatively anaerobic bacteria, thereby enabling them to express hydrocarbon degradative genes. Indeed, preliminary evidence indicates that it is possible to transfer this plasmid to RP4-TOL[+] *E. coli*, thereby allowing such cells to functionally express toluene-degradative genes and grow at the expense of toluene. Naturally occurring plasmids carrying dechlorination genes have also been described (Chakrabarty, 1980). It may thus be possible to introduce genes encoding dechlorination steps into a transposable element. Although the dechlorination enzymes are substrate-specific, a rapid spread of the genes specifying such enzymes among various microorganisms capable of functionally expressing hydrocarbon degradative genes may allow a change in substrate specificity by mutation, so that various chlorinated substrates may become amenable to such enzymatic attack. An understanding of the nature of genes allowing functional expression of hydrocarbon-degradative genes or deliberate introduction of dehalogenation genes on transposons, may greatly accelerate the evolution of the degradative competence among the soil and aquatic microbiota against synthetic environmental pollutants, and help remove such pollutants from the environment.

CONCLUDING REMARKS

In this article I have tried to describe a few problematic areas where techniques of genetic engineering, including those of recombinant DNA, may be usefully applied, leading to practical solutions of such problems. Needless to say, the ultimate solution will depend upon the magnitude of the problem and the ingenuity of the workers. In many cases the technical solutions will be achieved only after a thorough discussion of the relevant social, moral, or political issues. I have deliberately avoided recapitulating on those areas of application, which are more profound but either described before or difficult to attain. Instead, I have based my discussion on those areas that I feel are amenable to genetic studies. I would like to emphasize that the potential application areas of genetic engineering studies are essentially as vast as the ingenuity of the workers themselves.

REFERENCES

Beers, R. F., Jr., and E. G. Bassett (eds.). 1977. Recombinant molecules: impact on science and society. Raven Press, New York.

Bubela, R. 1978. Role of geomicrobiology in enhanced recovery of oil: status quo. APEA Journal, 1978, pp. 161–66.

Carson, R. 1962. Silent spring. Houghton Mifflin Co., Boston, Massachusetts.

Chakrabarty, A. M., D. A. Friello, and L. H. Bopp. 1978. Transposition of plasmid DNA segments specifying hydrocarbon degradation and their expression in various organisms. Proc. Nat. Acad. Sci. USA 75:3109–12.

Chakrabarty, A. M. (ed.). 1978. Genetic engineering. CRC Press, West Palm Beach, Florida.

Chakrabarty, A. M. 1979. Recombinant DNA: areas of potential applications. *In* D. A.

Jackson, and S. P. Stich (eds.), Recombinant DNA debate, pp. 56–66. Prentice-Hall, Englewood Cliffs, New Jersey.

Chakrabarty, A. M. 1980. Plasmids and dissimilation of synthetic environmental pollutants. *In* C. Stuttard and K. R. Rozee (eds.), Plasmids and transposons: environmental effects and maintenance mechanisms, pp. 21–30. Academic Press, New York.

Curtiss, R. III. 1976. Genetic manipulation of microorganisms: potential benefits and biohazards. Ann. Rev. Microbiol. 30:507–33.

Dazzo, F. B., and W. J. Brill. 1979. Bacterial polysaccharide which binds *Rhizobium trifolii* to clover root hairs. J. Bacteriol. 137:1362–71.

Evans, L. R., and A. Linker. 1973. Production and charactierization of the slime polysaccharide of *Pseudomonas aeruginosa*. J. Bacteriol. 116:915–24.

Faelen, M., A. Toussaint, M. Van Montagu, S. Van den Elsacker, G. Engler, and J. Schell. 1977. In vivo genetic engineering: the Mu-mediated transposition of chromosomal DNA segments onto transmissable plasmids *In* A. I. Bukhari, J. A. Shapiro, and S. Adhya (eds.), DNA Insertion Elements, Plasmids, and Episomes, pp. 521 35. Cold Spring Harbor Laboratory, Cold Spring Harbor, New York.

Farrell, R. and A. M. Chakrabarty. 1979. Degregative plasmids: molecular nature and mode of evolution. *In* K. Timmis and A. Puhler (eds.), Plasmids of medical, environmental, and commercial importance, pp. 97–109. Elsevier/North-Holland Biomedical Press, Amsterdam, Netherlands.

Fraser, T. H., and B. J. Bruce. 1978. Chicken ovalbumin is synthesized and secreted by *Escherichia coli*. Proc. Nat. Acad. Sci. USA 75:5936–40.

Gibbons. R. J. 1977. Adherence of bacteria to host tissue. *In* D. Schlessinger (ed.), Microbiology 1977, pp. 395–406. American Society for Microbiology, Washington, D.C.

Haider, K., J. Trojanowshi, and V. Sundman. 1978. Screening for lignin degrading bacteria by means of H^{14}C labelled lignins. Arch. Microbiol. 119:103–6.

Hopwood, D. A. 1978. Extrachromosomally determined antibiotic production. Ann. Rev. Microbiol. 32:373–92.

Jaboby, G. A., J. E. Rogers, A. E. Jacob, and R. W. Hedges. 1978. Transposition of *Pseudomonas* toluene-degrading genes and expression in *Escherichia coli*. Nature 274:179–80.

Kamp, P. F., and A. M. Chakrabarty, 1979. Plasmids specifing ρ-chlorobiphenyl degradation in enteric bacteria. *In* K. Timmis and A. Puhler (eds.), Plasmids of medical, environmental, and commercial importance, pp. 275–85. Elsevier/North-Holland Biomedical Press, Amsterdam, Netherlands.

Kirk, T. K., and H. Chang. 1975. The composition of lignin by white-rot fungi. Holztorschung 29:56–64.

Martial, J. A., R. A., Hallewell, J. D. Baxter, and H. M. Goodman. 1979. Human growth hormone complementary DNA cloning and expression in bacteria. Science 205:602–6.

Mercereau-Puijalon, O., A. Royal, B. Cami, A. Garapin, A. Krust, F. Gannon, and P. Kourilsky. 1978. Synthesis of an ovalalbumin-like protein by *Eshcerichia coli* K12 harbouring a recombinant plasmid. Nature 275:505–10.

Nakazawa, T., E. Hayaishi, T. Yokota, Y. Ebina, and A. Nakazawa. 1978. Isolation of TOL and RP4 recombinants by integrative suppression. J. Bacteriol. 134:270–77.

National Academy of Sciences. 1977. Research with recombinant DNA. Washington, D.C.

Nester, E. W., and A. Montoya. 1979. Crown gall: a natural case of genetic engineering. ASM News 45:283–87.

Roblin, R. O. 1978. Potential application of molecular cloning to genetic therapy. *In* A. M. Chakrabarty (ed.), Genetic engineering, pp. 159–64. CRC Press, West Palm Beach, Fla.

Salkinoja-Salonen, M. S., E. Vaisanen, and A. Paterson. 1979. Involvement of plasmids in the bacterial degradation of lignin-derived compounds. *In* K. Timmis and A. Puhler (eds.), Plasmids of medical, environmental, and commercial importance, pp. 301–14. Elsevier/North-Holland Biomedical Press, Amsterdam, Netherlands.

Setlow, J. K., and A. Hollaender (eds.). 1979. Genetic engineering: principles and methods. Plenum Press, New York.

Villa-Komaroff, L., A. Efstratiadis, S. Broome, P. Lomedico, R. Tizard, S. P. Nater, W. L. Chick, and W. Gilbert. 1978. A bacterial clone synthesizing proinsulin. Proc. Nat. Acad. Sci. USA 75:3727–31.

M. PIATAK, M. OLIVE, K. N. SUBRAMANIAN,
P. K. GHOSH, V. B. REDDY, P. LEBOWITZ,
AND S. M. WEISSMAN

The Structure and Expression of the Late Region of SV40 DNA

8

INTRODUCTION

Simian Virus 40 (SV40) is a small DNA virus capable of lytic infection of monkey cells and of *in vitro* transformation of cells from a wide variety of species. The virus consists of a protein capsid and a nucleoprotein core (Fareed and Davoli, 1977; Fried and Griffith, 1977; Kelly and Nathans, 1977; Lebowitz and Weissman, 1979). This capsid is made up principally, if not solely, of three virus-coded structural proteins: VP1, the major structural protein; VP3, a minor structural protein; and VP2, present in even smaller quantity than VP3. The nucleoprotein core is composed principally of host histones complexed with viral DNA; in addition, variable amounts of viral structural proteins may also be associated with the core. The DNA of SV40 is a double-stranded, supercoiled circular molecule. The number of base pairs in SV40 DNA varies with the particular isolate. DNAs from the two plaque-purified strains we have worked with were originally assigned lengths of 5,226 and 5,262 base pairs (Fiers et al., 1978; Reddy et al., 1978). However, an additional stretch of 17 base pairs has recently been demonstrated in the shorter DNA (Van Heuverswyn and Fiers, 1979, in press), and the correct chain lengths of the DNAs in these two viral strains are 5,243 and 5,262 base pairs. It appears likely that the shorter genome arose by deletion of a segment of a tandemly repeated sequence in the longer DNA.

M. Piatak, Department of Human Genetics, Yale University School of Medicine, New Haven, Connecticut 06510

Expression of the viral genome is temporally regulated during the course of lytic infection (Fareed and Davoli, 1977; Fried and Griffith, 1977; Kelly and Nathans, 1977; Lebowitz and Weissman, 1979). Starting shortly after infection, but continuing throughout the lytic cycle, messenger RNA (mRNA) encoded in one contiguous half of the minus, or early, strand of the viral genome is expressed. This so-called early mRNA encodes the virus tumor antigens including small t and large T antigens. Early mRNA and the tumor antigens are also expressed in transformed cells. Large T antigen binds preferentially to SV40 DNA in the origin of replication (Reed et al., 1975; Jessel et al., 1976; Tjian, 1978), is involved in the stimulation of both viral (Tegtmeyer, 1972; Chou et al., 1974) and host (Chou and Martin, 1975) DNA synthesis in lytic infection and is required for the initiation of transformation by SV40 (Abrahams and van der Eb, 1975) and its maintenance as well in certain cell lines (Brugge and Butel, 1975; Martin and Chou, 1975; Osborn and Weber, 1975; Tegtmeyer, 1975; Rassoulzadegan et al., 1978). Small t antigen has no known role in lytic infection; its role in cellular transformation is unclear.

Shortly after the initiation of viral DNA replication, usually 15 to 18 hours after infection at 37°, a second set of mRNAs, designated collectively as late mRNA, appears in abundance in the cytoplasm of infected cells (Fareed and Davoli, 1977; Fried and Griffith, 1977; Kelly and Nathans, 1977; Lebowitz and Weissman, 1979). This mRNA is transcribed from the opposite strand (plus or late strand) and the opposite half of the viral genome from that coding for early mRNA. Analyses of the 5' termini of early and late mRNAs have clearly shown that transcription of the early and late gene regions diverges from a common genomic region lying within the origin of DNA replication (Dhar et al., 1977; Ghosh et al., 1978; Reddy et al., 1979). Furthermore, the regions coding for the 3' termini of early and late mRNAs also converge and actually overlap for about 70 bases (Reddy et al., 1978a; Reddy et al., 1979). Late mRNA encodes and can be translated *in vitro* into the capsid proteins VP1, VP2, and VP3 (Fareed and Davoli, 1977; Fried and Griffith, 1977; Kelly and Nathans, 1977; Lebowitz and Weissman, 1979). Following the onset of viral DNA replication in normal lytic infection, late mRNA is about fiftyfold more abundant than early mRNA and may constitute as much as 5 to 10% of total cellular polyadenylated RNA. Late mRNA is not synthesized in significant quantity in transformed cells. If cells are infected with SV40 in the presence of inhibitors of DNA synthesis, or if cells are infected at the nonpermissive temperature with a mutant virus whose early proteins are temperature-sensitive, then little or no late cytoplasmic mRNA is formed. However, once DNA replication has begun, late mRNA production continues for some time, even in the presence of inhibitors of DNA synthesis (Brandner and Mueller, 1974) or at elevated temperatures in infections by viruses with temperature-sensitive early proteins (Cowan et al., 1973). Therefore, under normal circumstances, initiation of DNA replication seems necessary for the establishment, but not for the maintenance, of late mRNA production. However, under unusual circumstances, late mRNA synthesis may occur in the absence of any early gene expression or DNA replication. For example, when cells are injected with large amounts of SV40 DNA containing single-stranded regions, they may produce late SV40 proteins, in the absence of normal early protein synthesis and any viral DNA replication (Graessmann et al., 1977). Also, oocytes injected with SV40 DNA produce late proteins in the absence of viral DNA synthesis (J. Mertz, personal communication); in this system it is not clear whether early viral proteins are synthesized. These exceptions have not been completely investigated, and their relationship to the normal process of late mRNA production is not clear.

The mechanism of switching from early to late mRNA production and the coupling between initiation of DNA synthesis and late mRNA production in SV40 are of interest not only from the point of view of regulation of viral macromolecular syntheses, but also because of their possible relevance for understanding how eukaryotic cells in general might couple expression of different proteins to DNA replication and cell division and to different states of differentiation. For these reasons we have been investigating the detailed structure of the 5′ termini of SV40 early and late mRNA, their genomic templates, and the structural relationships of their templates to the origin of DNA replication. As our investigations have proceeded, it has become apparent that SV40 late mRNA is composed of not one or a few mRNA species but of a heterogeneous array of mRNA molecules (Ghosh et al., 1978; Reddy et al., 1978a), some with complex structures. Moreover, both early and late mRNAs include species with 5′ termini lying within the origin of replication (Dhar et al., 1977; Ghosh et al., 1978; Reddy et al., 1979). Unfortunately, however, our understanding of the biogenesis of these multiple mRNA species has not kept pace with our understanding of their structure. Furthermore, the relationship between the structure of the SV40 late mRNAs and the proteins they encode show several features that have not yet been detected in cellular mRNAs, and in some cases, not even in the mRNAs of other viruses. One might hope that although some of these features may be specialized for SV40, others may represent more general features of gene organization and expression in animal cells that will become appreciated as other systems are investigated in detail.

Recently, we reviewed the organization and expression of the SV40 early gene region (Ghosh et al., 1979a, in press). In this paper we review the organization of the SV40 late gene region and the complexity of its expression and discuss alterations in its expression in one class of SV40 mutants containing deletions in the so-called leader segment of the late gene region.

THE STRUCTURE OF VIRAL DNA IN THE LATE GENE REGION

As presently formulated, the late gene region of SV40 extends from a genomic site within the origin of DNA replication coding for the most distal established 5′ terminus of the late mRNAs (designated the 5′ terminus of the late gene region) halfway around the SV40 circle to a site coding for the 3′ terminus of the late mRNAs (3′ terminus of the late gene region) (fig. 1). Extending from DNA residues 5189-2592, the late gene region covers over 2,600 nucleotides. From both structural and functional points of view, the late gene region can be subdivided into a number of separate regions. We discuss the structure and function of each of these, proceeding in a 5′ → 3′ direction from the origin of replication through the region coding for the viral capsid proteins.

The structure of the origin of viral DNA replication has been discussed

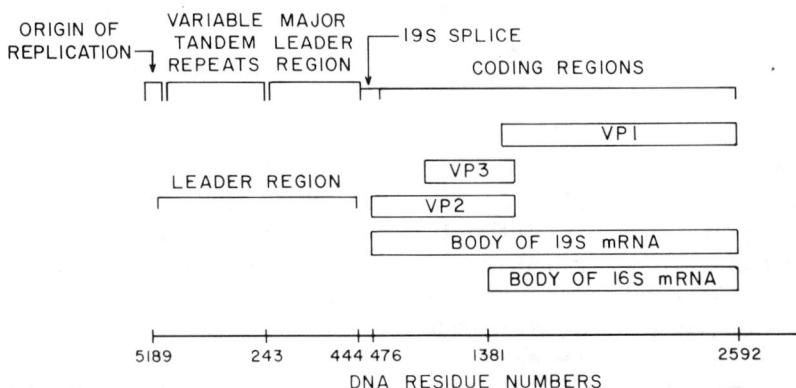

Fig. 1. Schematic diagram of the SV40 late gene region depicting in brackets major subdivisions of this region and in boxes the regions coding for the bodies of 16S and 19S mRNAs and VP1, 2, and 3 capsid proteins. Transcription of this region is from left to right. Numbering of DNA residues is taken from Reddy et al., 1978b); because of circularity of SV40 genome, numbering terminates at residue 5243 and restarts with residue 1. Residue 5189 marks the 5′ end of the late gene region (see text); 243, the 5′ end of the region coding for the major SV40 leaders; 444, the 3′ end of the leader region; 476, the 5′ end of the template for the body of 19S mRNA; 1381, the 5′ end of the template for the body of 16S mRNA; and 2592, the 3′ end of the late gene region.

extensively elsewhere (Subramanian et al., 1977; Lebowitz and Weissman, 1979) and will not be reviewed here, except to point out that it contains a rich array of symmetrical and repeated deoxyadenylate and deoxythymidylate-rich sequences and that a protein closely related to SV40 T antigen binds preferentially and in graded fashion to three sites within and overlapping it (Tjian, 1978).

Just beyond the AT-rich region of the origin of replication lies a region of viral DNA that is quite heterogeneous from strain to strain and even from isolate to isolate of virus. This heterogeneity has caused some confusion in the literature concerning DNA sequences in this region and has created difficulty in comparing DNA and RNA sequences of mutants prepared from different virus stocks. This region of viral DNA is remarkable in that it contains a number of tandem repeated sequences of variable length (fig. 1) (Fiers et al., 1978; Reddy et al., 1978). The number of these tandem repeats and their chain lengths also vary from one viral isolate to another. In one strain of SV40, this region contains a perfect tandem duplication of 91 nucleotides, constituting approximately 1.8% of the viral genome. It is curious that a virus that in all other respects seems to utilize the utmost economy in the organization of its genome maintains long tandem duplications of genetic sequence in this region. The function of these tandem repeats is not known. It is apparent that they are not necessary for viral replication, since mutants with deletions in this area, including one mutant containing only one copy of the longest tandem repeat in the wild-type virus, are fully viable (Mertz and Berg, 1974; Shenk et al., 1976; Subramanian, 1979). Furthermore, the related papovavirus, polyoma, does not contain tandem repeating sequences in this region (B. Griffin, personal communication). Tandem repeats may not be reflections of "accidental" duplication events in the history of SV40, however, since the DNA of another related papovavirus, BK, also contains tandem repeated sequences in this region (Yang and Wu, 1979). In this virus the organization of the tandem repeats also varies from strain to strain; however, the repeats of BK virus do not share any obvious sequence homology with the tandem repeats of SV40. Despite our lack of knowledge of the function of the tandem repeats, it has been suggested that they may provide a model for repeating sequences found in abundance in eukaryotic cellular DNA. As discussed in detail below, this region of tandemly repeated and variable DNA sequences serves as the template for the 5' ends of long leaders bound to the 19S late mRNAs.

Proceeding farther into the late gene region from the region of the tandem repeated sequences (fig. 1), there is an adjacent region of DNA approximately 200 nucleotides long whose transcript is abundantly represented in cytoplasmic late mRNA (Dhar et al., 1975; Ghosh et al., 1978; Reddy et al., 1978a). It is now clear that transcripts of this genomic region form the smaller leader sequences and constitute the 3' portion of the longer leaders of the late mRNAs (Ghosh et al., 1978; Reddy et al., 1978a). This genomic region is curious in that it contains a translational initiation signal, followed by 60 triplets coding for amino acids and then a translational termination signal (Fiers et al., 1978; Ghosh et al., 1978; Reddy et al., 1978b). There is strong homology between this DNA region in SV40 and BK virus, and this region in BK also contains information for a very similar peptide of about 60 amino acids (Yang and Wu, 1979). However, no peptide corresponding to the codons in this region of SV40 or BK DNA has yet been identified in infected cells. This DNA region contains not only coding information for a potential peptide but also translation termination triplets in all three reading frames toward its 3' end; thus, it could not code for the amino-terminal portions of any of the late viral capsid proteins.

Just beyond the DNA encoding the late mRNA leaders lies a sequence of 31 bases (fig. 1) whose transcript was found in early studies to be missing from late cytoplasmic mRNA whereas transcripts of the regions downstream as well as upstream of this segent were present in abundance (Dhar et al., 1975). It is now appreciated that this sequence is an "intervening" sequence, the transcript of which is spliced out of most of the mature cytoplasmic late mRNAs in SV40 infected cells (Ghosh et al., 1978; Reddy et al., 1978a). However, as emphasized below, certain late mRNAs remain unspliced and do contain transcripts of this area (Ghosh et al., 1978). The transcript of the late strand of this short DNA segment is notably uridylic acid-rich.

Beyond this 31 nucleotide segment lies a long expanse of DNA, the full transcript of which constitutes the body of the 19S mRNAs, while a transcript of the distal two-thirds forms the body of the 16S mRNAs (fig. 1). The former mRNAs serve as templates for all three capsid proteins, and the latter serve as templates for only VP1 in *in vitro* translational systems (Fareed and Davoli, 1977; Fried and Griffith, 1977; Kelly and Nathans, 1977; Lebowitz and Weissman, 1979, in press). Five nucleotides into this DNA, there is a translation initiation codon that is followed by 351 sense codons and then a translation termination triplet. This coding region (residues 480–1535, or .766–.969 map units $= m\mu$) is of appropriate size to serve as the template for VP2 synthesis and has been proposed as the VP2 template (Fiers et al., 1978; Reddy et al., 1978b; Lebowitz and Weissman, 1979) (fig. 1). Unfortunately, the amino acid sequence ofVP2 has not yet been determined, and so it has not been possible to compare the amino acid sequence predicted from these 351 codons with the VP2 amino acid sequence. However, deletion mutants in this genomic region code for altered VP2 proteins (Cole et al., 1977), suggesting the validity of the assignment of this region as the template for this protein.

Comparison of tryptic peptide maps of VP2 and VP3 of polyoma (Fey and Hirt, 1974; Gibson, 1974; Hewick, 1975), and later of SV40 (Prives and Shure, 1979), and studies of the VP2 and VP3 proteins synthesized in cells 6nfected with deletion mutants of SV40 (Cole et al., 1977) suggested that the genomic template for VP3 was included within the 3' terminal portion of the template for VP2. Examination of the sequence of the 19S mRNAs showed the first internal AUG sequence in the presumptive VP2 template to be situated about one-third of the way into this region. If this internal methionine codon were to serve as a translation initiation site, then the remaining portion of the VP2 coding region would be of the appropriate size to code for VP3. Given all this evidence, the stretch of DNA from the triplet coding for first internal methionine residue of VP2 through the codons serving as template for the carboxy-terminus of VP2 (residues 834–1535, .834–969 mμ) has been proposed as the template for VP3 (Fiers et al., 1978; Reddy et al., 1978b; Lebowitz and Weissman, 1979). However, comparison of the amino acid sequence of VP3 with the codons in this region will be required to confirm this template assignment. Attempts to chase radioactivity from VP2 into VP3 *in vivo* have failed. Furthermore, both VP2 and VP3 incorporate labels from formylmethionine (A. Smith, personal communication). These two lines of evidence establish that VP2 and VP3 are the products of separate translation initiation events and that VP3 is not derived by cleavage from VP2. It therefore appears that VP2 and VP3 are encoded in the same reading frame of a segment of late DNA with initiation of the latter taking place at an internal methionine codon in the template for the former. Analogous examples of gene organization and expression have been demonstrated in the small DNA bacteriophages (Sanger et al., 1977).

The body of the 16S mRNAs, coding for VP1, extends on the SV40 map from residues 1381–2592 (.939–.170 mμ) (see below). Within this expanse of DNA, there is a long open reading frame starting with a translation initiation triplet at residues 1423–1425 (.947 mμ), continuing with 361 successive amino acid coding triplets and concluding with a termination signal at residues 2509–2511 (.155 mμ) (Fiers et al., 1978; Reddy et al., 1978b). The size of the protein predicted by this sequence, approximately 46,300 daltons, is in good agreement with experimental estimates of the molecular weight of VP1; the amino-terminal amino acid sequence and the carboxy-terminal amino acid predicted by this sequence agree with the analytically determined amino-terminal sequence (Lazarides et al., 1974) and the carboxy-terminal amino acid, glutamine (W. Kempe, W. Beattie, S. Weissman, and W. Konigsberg, unpublished observations) of VP1; and deletion mutants that direct the synthesis of altered VP1 proteins fall within this genomic region (Lai and Nathans, 1976; Cole et al., 1977). Thus, it is virtually certain that the genomic template for VP1 extends from residues 1423–2508 on the SV40 map (fig. 1). It is seen that the codons for the amino-terminal region of VP1 overlap codons for the carboxy-terminus of VP2 and VP3 for a relatively short distance. This overlap is out of phase with respect to translational frame. The functional significance of this overlap is not clear, but it has been suggested that it may ensure that the initiation AUG triplet for VP1 is blocked by ribosomes on mRNA molecules on which VP2 and VP3 are being translated.

STRUCTURE OF LATE mRNAs

Although early nucleic acid hybridization and sequencing studies provided the general outlines of the structures of, and templates for, the SV40 late 16S and 19S mRNAs, more recent studies involving electron microscopic examination of SV40 DNA-late mRNA hybrids (Hsu and Ford, 1977; Bratosin et al., 1978), analysis of single-stranded radioactive SV40 DNA segments protected from S1 nuclease digestion by hybridization to late mRNAs (Lai et al., 1978) and especially sequence analysis of radio-labeled cDNAs reverse transcribed on late mRNA (Ghosh et al., 1978; Reddy et al., 1978a; Bina-Stein et al., 1979) have provided very detailed information on the structure of the late mRNAs. The latter approach has been especially fruitful in establishing the precise splicing patterns and pointing out the multiplicity of 5′ termini of both 16S and 19S mRNAs. It has been especially advantageous in permitting structural analysis of mRNA species present in cells in relatively small quantity. This methodology has been described in detail previously (Ghosh et al., 1978; Ghosh et al., 1979b, in press) and is discussed briefly in the last section of this paper. cDNA analysis and direct analysis of capped oligonucleotides derived from *in vivo* labeled late messages (Haegeman and Fiers, 1978b) have in some cases permitted the precise genomic localizations of 5′ termini of late mRNAs. Utilizing the results of all these studies, one can now be quite certain of the structure of a number of species of late mRNA. The structure of the SV40 late mRNAs and the relationship of their structure to function have been reviewed in detail elsewhere (Lebowitz and Weissman, 1979) and are presented in summary fashion here.

The 16S late mRNAs all contain a body extending from residues

1381–2592 (.939–.170 mμ) covalently linked to leaders that have a 3′ terminus at residue 444 (.760 mμ) (figs. 1, 2) (Reddy et al., 1978a). Thus, 936 nucleotides are spliced out of a precursor RNA in the biogenesis of the various 16S mRNAs. The 16S mRNAs can be subcategorized into three classes on the basis of the structure of their leaders (fig. 2). The first class contains leaders that are colinear with SV40 DNA. The principal 16S mRNA, accounting for 85–90% of all 16S mRNAs, falls into this class. Its leader is 202 nucleotides in length extending from residues 243–444 (.721–.760 mμ). The second class is characterized by leaders with an internal splice that fuses residues 211–352 (0.714–.741 mμ). This class contains three known species, two with 5′ termini at residues 180–182 (.709 mμ) and one with a 5′ terminus upstream from residue 180. These are the only known SV40 mRNAs with m4re than one splice, but multiply spliced mRNAs are also specified by adenoviruses (Chow et al., 1977; Klessig, 1977). The third class of 16S mRNAs contain leaders with a duplication of 93 nucleotides at their 3′ termini. As shown in figure 2, the termini of the duplicated segment serve as splice sites, suggesting that this duplication arose by splicing within a transcript containing tandem copies of the SV40

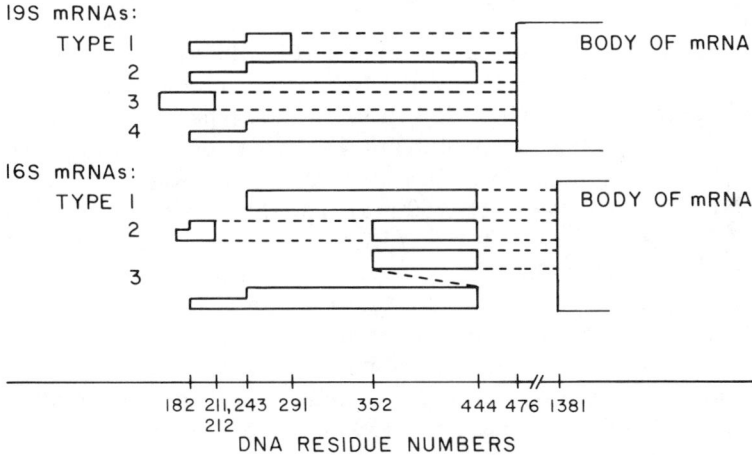

Fig. 2.   Schematic diagram depicting the principal types of leaders and splices of the late 16S and 19S mRNAs. Transcription proceeds from left to right; thus, the bodies of the 16S and 19S mRNAs are depicted on the right side of the figure. Expanses outlined by dotted lines indicate segments of RNA spliced out of the mature cytoplasmic mRNAs, and boxes outlined by solid lines indicate the spans of the principal mRNA leaders. RNA residue numbers indicate genomic localizations of the 5′ termini of principal leaders and nucleotides that undergo fusion in the formation of splices. See text for discussion of additional leaders with 5′ termini upstream from (to the left of) residue 182. Numbering of RNA residues corresponds to numbering of SV40 DNA residues by Reddy et al. (1978b).

genome. Such a transcript could arise by either transcription of an oligomeric SV40 DNA or by transcription about the monomeric SV40 circle more than once. Although the quantity of greater than unit genomic length nuclear transcripts in SV40 infected cells is small, this explanation is rendered likely by the relative abundance of both 16S mRNAs with leaders with similar sequence duplications (Legon et al., 1978) and transcripts of greater than unit genomic length in cells infected with polyoma (Acheson, 1978). One SV40 mRNA of this type has a 5′ terminus at residue 243 (.720 mμ), and at least one additional species has a 5′ terminus farther upstream. The significance of these minor 16S mRNA species is not known.

To date 19S late RNAs of four classes have been identified in cells infected by SV40 (fig. 2). All contain a body extending from residues 476–2592 (.765–.170 mμ) (Ghosh et al., 1978). RNAs of the first three classes are spliced, whereas RNAs of the fourth class are unspliced in their 5′ terminal region. All are present on polysomes of $>$80S as well as in whole cytoplasm; it is thus likely that they all serve as mRNAs. The first class of 19S mRNAs is characterized by a splice that joins the 19S body to leaders with a 3′ terminus at residue 291 (.730 mμ), and the second class consists of species joining the body to leaders with a 3′ terminus at residue 444 (.760 mμ). Within each of these classes are species with leaders having 5′ termini at residues 243 (.721 mμ) and 182 (.709 mμ) as well as more upstream sites. mRNAs of the first two classes, but especially Class 1, constitute the most abundant 19S species in wild type virus-infected cells. Of note, only one 19S and one 16S mRNA share leaders; a second class of 19S species and the principal 16S mRNA share the leader that extends from residues 243–444 (.721–.760 mμ). The third class of 19S mRNAs consists of only one member, a relatively nonabundant species of mRNA, containing a leader with a 5′ terminus at, or upstream from, residue 90 (.692 mμ) and a 3′ terminus at residue 212 (.714 mμ). Two Class 4 (unspliced) 19S mRNAs have their 5′ termini at the aforementioned residues 243 and 182; the former mRNA is a relatively abundant species. In addition, one or more unspliced species exist with 5′ termini at or upstream from residue 110 (.695 mμ). By cDNA sequencing a 19S mRNA 5′ terminus has also been identified at residue 5189 (.667 mμ); however, it is not presently clear which classes of 19S species include members with this terminus. To date tandem duplications of DNA sequences have not been detected in the leaders of the SV40 19S mRNAs. Although it is conceivable that they do not occur, it appears equally likely, given the small quantities of 19S mRNAs relative to 16S mRNAs, that they occur in quantities too small to be detected even by cDNA sequencing technology.

It has been known for a considerable period of time that the principal capped 5' terminus of the late SV40 mRNAs is m^7GpppAmUp(Up) (Lavi and Shatkin, 1975; Groner et al., 1977; Haegeman and Fiers, 1978b). In agreement, the terminal transcribed sequence of mRNAs with 5' termini at residues 243, 182, and 5189 is ApUpUp.

In addition to the aforementioned late mRNAs, reverse transcription studies have suggested the existence of a small quantity of an RNA in infected cells that is a contiguous transcript of the residues 548–2592 (.779–.170 mμ) of the SV40 genome (Ghosh et al., 1978). Its 5' terminus also appears to be ApUpUp. This RNA lacks the VP2 initiation codon; however, it appears to be present in insufficient quantity to account for the total synthesis of VP3. It thus appears likely that 19S species containing both VP2 and VP3 initiation codons code for the latter protein. It has been noted that 19S mRNAs of the second class contain leader sequences that can base-pair with sequences overlapping the VP2 initiation site; in contrast, Class 1 19S mRNAs lack such sequences (Ghosh et al., 1978). Thus, it has been hypothesized that perhaps Class 1 19S mRNAs serve as templates for VP2 and Class 2 species code for VP3.

It is noteworthy that the VP1 initiation codon lies 244 nucleotides downstream from the 5' end of the principal 16S mRNA. Furthermore, the VP3 initiation site lies as many as 400–500 nucleotides from the 5' end of certain 19S mRNA species. Since at least the former is translated effectively, leader sequences of about 250 nucleotides, and very likely longer stretches, do not appear to present any significant obstacle to mRNA translation.

THE PROBLEM OF THE 5' ENDS OF THE LATE MRNAS

Most eukaryotic cell messages described to date have unique 5' ends. The principal forms of adenovirus late mRNA also are said to have unique 5' ends. In contrast, SV40 early mRNAs appear to have two closely spaced 5' ends and, as noted, the 5' ends of both 16S and 19S late mRNAs are remarkably heterogeneous. From the direct mRNA sequencing and cDNA sequencing studies, a minimum of three and probably five or six sites on SV40 DNA give rise to late mRNAs with the 5' terminal ApUpUp sequence. However, analysis of the capped ends of the SV40 late mRNAs suggests a number of additional 5' termini with other sequences (Canaani et al., 1979). Furthermore, with an abundant 5' terminus at .721 mμ and the possibility that one or more 5' termini take origin from sequences upstream from the origin of replication, as many as 300–400 nucleotides may separate sites on SV40 DNA serving as templates for the 5' termini of the late mRNAs.

A question basic to understanding the biogenesis of the SV40 late mRNAs is whether the capped ends of individual mRNAs represent transcription initiation sites, or RNA processing sites. Conceivably, certain termini could arise from transcription initiation, and others may be the products of processing events. Unfortunately, the information available on this important point is limited. On the one hand, the only host cell capping enzymes discovered to date require RNAs with 5′ di- or triphosphate ends as receptors (Wei and Moss, 1977). Furthermore, although an enzyme that will convert 5′ terminal mononucleotides on RNA chains to diphosphates has been identified in vaccinia-infected cells (Spencer et al., 1978), no such enzyme has been found to date in any noninfected cell type. Since transcription is initiated with nucleoside 5′ triphosphates, and since cells apparently lack the means for phosphorylating 5′ terminal monophosphates of RNAs that might arise by internal cleavage of RNA chains, it seems reasonable to conclude that at least some, and perhaps most, capped termini may mark transcription initiation sites. On the other hand, eukaryotic transcription was thought to be initiated solely with purine nucleoside triphosphates, but capped termini have been identified adjacent to terminal uridylic acid residues as well as purine nucleotides (Schibler and Perry, 1976). However, it has been shown that transcription in eukaryotic systems may begin with cytidine triphosphate (Kates and Beeson, 1970), suggesting that caps adjacent to uridylate residues may also represent transcription initiation sites. Still, the possibility cannot be entirely discarded that some 5′ capped termini may represent internal RNA processing sites. Recently, Plotch and colleagues (1978) have observed that influenza virus RNAs synthesized *in vitro* may have at their 5′ ends short capped sequences transferred from globin or other eukaryotic mRNAs. Although a similar cap transfer mechanism operating *in vivo* could generate the heterogeneous capped structures observed in SV40 mRNA, it could not account very well for the variable and long lengths of SV40 encoded leaders. Furthermore, the 5′ terminal leader sequences of the late SV40 mRNAs, as far as can be determined, match SV40 DNA sequences all the way through to their capped nucleotides. Thus, this mechanism does not seem to be operative in SV40 mRNA biogenesis.

Goldberg and Hogness (personal communication) have pointed out that the $5' \rightarrow 3'$ sequence, $T^{AAAAA}_{TTTT}A^{A}_{T}$, lies on the noncoding DNA strand approximately 25 to 30 nucleotides from templates coding for the 5′ ends of a number of eukaryotic mRNAs, e.g., *Drosophila* and sea urchin histone, and rabbit and mouse globin mRNAs. The presumption is that this sequence, also called the Hogness-Goldberg box, in some way is analogous to the Pribnow box (Pribnow, 1975) or other sequence features that are partially conserved in bacterial RNA polymerase promoters. The arrangement of potential Hogness boxes near the 5' ends of SV40 early and late mRNAs is of interest. There is an adequate Hogness-Goldberg box 21 and 27 nucleotides upstream from the 5' ends of SV40 early mRNAs, within the AT-rich region of the origin of DNA replication. The localization of this box raises the possibility that a promoter for early transcription lies within the origin of DNA replication. On the other hand, there is no Hogness-Goldberg box immediately preceding the major 5' ends of the late SV40 mRNAs at residues 182 and 243. However, the sequence complementary to the early Hogness-Goldberg box is itself a potential Hogness-Goldberg box, and we have noted previously that there is at least one species of SV40 late 19S mRNA with a 5' terminus at residue 5189, within the origin of replication and downstream from this potential Hogness-Goldberg sequence. Thus, a fraction of late transcription may involve an initial interaction of RNA polymerase molecules with a late promoter within the origin of replication followed by initiation of transcription at a downstream site(s). Of note, this model of complementary early and late promoter sites would not allow for simultaneous interaction of more than one RNA polymerase molecule with these presumptive promoters on a given molecule of DNA. Furthermore, it would appear that a factor or factors acting in the vicinity of this site could influence the interaction of RNA polymerase molecules with either the early or late promoter and in a sense provide a switch between early and late transcription. By binding to SV40 DNA within the origin of replication and the Hogness-Goldberg box, large T antigen may serve as such a factor, playing a direct role in the initiation of at least a portion of late transcription.

THE EFFECT OF DELETIONS IN THE LEADER REGION ON THE STRUCTURE AND PRODUCTION OF THE SV40 LATE mRNAs

A number of deletion mutants of SV40 lacking segments of DNA that code for late mRNA leader sequences have become available for investigation in the past few years. Certain of these mutants lack segments of DNA encoding the principal 5' ends of the 19S and 16S leaders, and others lack sequences coding for internal leader regions. Despite their deletions, these mutants grow in the absence of helper viruses. Thus, an intact late leader region is not required for synthesis of VP1 and at least one of the minor capsid proteins and for complete viral replication.

Despite the fact that sequences coding for the late leaders are not essential for viral replication, the possibility must be considered, because of their crucial location, that they play some role in the regulation of late transcription. Thus, the effects of deletion mutations in the late leader region on late mRNA structure and the means by which mutants compensate for these deletions have become topics of great interest. Our knowledge in these two areas, derived from two recent studies (Villareal et al., 1979; Piatak et al., 1979), may be summarized as follows. Cells infected with deletion mutants synthesize both 16S and 19S mRNAs, the former always containing the wild-type 16S splice joining nucleotides 444–1381 and the latter including both species with one or more of the three wild-type 19S splices and species lacking splices upstream from the 19S body. However, the leaders of these mRNAs lack sequences complementary to the deleted DNA segments and the distribution of 5′ termini of mutant mRNAs differs from the distribution of 5′ termini seen in wild-type infection. Specifically, the 5′ termini of mutant mRNAs tend to be transcribed mainly from sequences upstream from sequences encoding the principal 5′ termini for wild-type mRNAs; however, at least some of the upstream 5′ termini of mutant mRNAs do correspond to minor 5′ termini of wild-type mRNAs.

In order to investigate both late mRNA splicing patterns and the redistribution of 5′ termini in late leader deletion mutants more extensively, we have now studied the structure of the late mRNAs synthesized in cells infected with three newly isolated deletion mutants lacking nonoverlapping segments of DNA encoding the major late leaders (Subramanian, 1979). All three grow to titers of at least 10^7 in the absence of helper virus. One mutant, dl 1659, lacks a stretch of DNA from residue 217–265 (.718–.727mμ) (fig. 3). Reference to figure 2 shows that the deleted segment in this mutant includes sequences encoding the 5′ ends of the major 16S leader and a number of 19S leaders; however, it does not include sequences contributing to 19S splice sites or the characteristic 16S splice site. The second mutant, dl 1626, lacks DNA residues 267–420 (.727–.755 mμ). The deleted segment in this mutant includes most of the DNA encoding the internal sequences of the principal 16S mRNA leader and Type 2 19S leaders; it also includes the residue 291 splice site involved in Type 1 19S mRNA leaders. However, this mutant maintains intact the residue 444 splice site for Type 2 19S leaders and all 16S leaders and also sequences coding for the 5′ termini of all the major 16S and 19S mRNA leaders. The third mutant, dl 1613, lacks a small segment of DNA from residue 420–440 (.755–.759 mμ); this DNA segment includes sequences at the 3′ end of the 16S and Type 2 19S leaders; furthermore, the deleted segment comes within six nucleotides of the residue 444 splice site, but does not actually delete any splice sites.

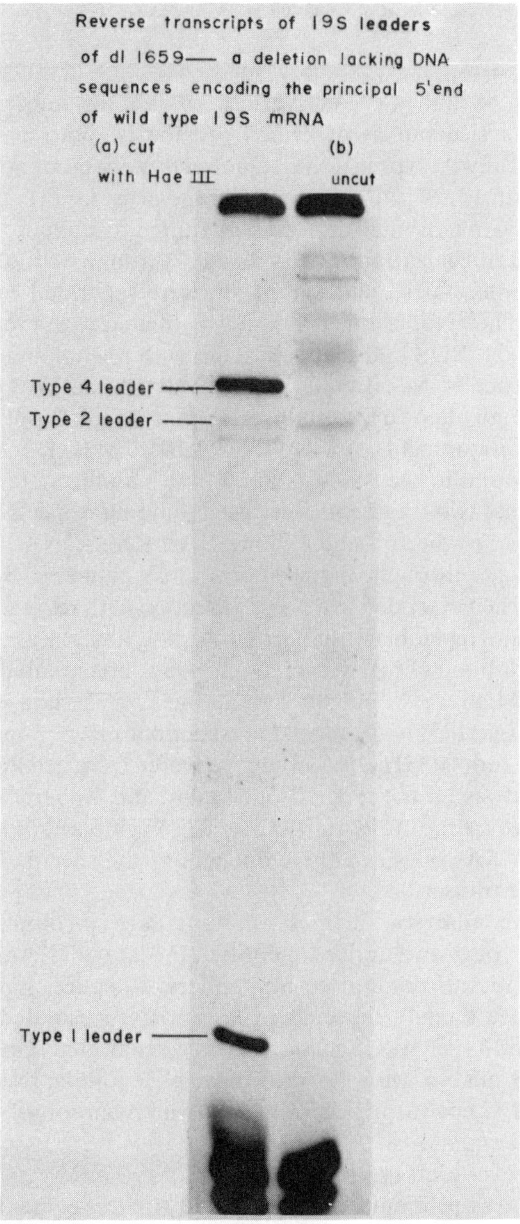

Fig. 3. 7M urea-8% polyacrylamide gel electrophoretic separation of cDNAs reverse transcribed on polyadenylated late RNA from cells infected with dl 1659 using as a primer the SV40 restriction fragment extending from DNA residues 478–509. This fragment binds exclusively to the 5' terminus of the body of the 19S late mRNAs. In the right channel cDNAs have been electrophoresed without prior treatment; in the left channel cDNAs were digested with *Hae*III restriction enzyme prior to electrophoresis. The cleaved cDNAs derived from unspliced 19S mRNAs and mRNAs with Type 1 and 2 splices and leaders are indicated.

Our analyses of the mRNAs synthesized in cells infected by these mutants were carried out as described previously (Ghosh et al., 1978, 1979b, in press) for wild type mRNAs. Confluent Vero monkey kidney cells were infected with 10–20 pfu/cell of the mutant viruses. Forty to forty-four hrs. later, cells were harvested and then disrupted in a hypotonic medium by Dounce homogenization and passage through a no. 26 gauge hypodermic needle. Nuclei and cytoplasm were separated by low speed centrifugation. The cytoplasmic fraction was treated with proteinase K in the presence of .5% SDS and then extracted with phenol and chloroform also in the presence of .5% SDS. Cytoplasmic RNA was then precipitated and passed through oligo-(dT)-cellulose columns to isolate polyadenylated RNA. Restriction fragments of SV40 DNA labeled at their 5′ termini with ^{32}P were then bound to the RNA in an 80% formamide system at 50°. In these experiments, two fragments were used, binding respectively to the 5′ ends of the bodies of the 16S and 19S mRNAs. RNA-DNA hybrids were isolated by passage through oligo-(dT)-cellulose columns. Bound DNA fragments were then extended in a 3′ → 5′ direction with respect to template RNA using avian myeloblastosis virus reverse transcriptase. Following alkaline degradation of RNA, cDNAs were precipitated and then separated on 7M urea-8% polyacrylamide gels. In certain experiments cDNAs were digested with *Hae*III restriction enzyme prior to gel electrophoresis. Individual cDNAs were recovered from gels and subjected to sequence analysis by the method of Maxam and Gilbert (1977). Since reverse transcriptase faithfully transcribes RNA templates, the sequences of individual cDNAs provided the sequences of the specific RNAs from which they were transcribed.

Although our analyses of these mutants are incomplete, several conclusions may be drawn at this time. First, the late mRNAs specified by these deletion mutants contain no new splices; all splices have the same structure as those already detected in late mRNAs isolated from cells infected with wild-type virus. Second, all three deletion mutants direct the synthesis of 16S mRNA with the wild-type splice joining residues 444 to 1381; this result is not surprising in view of the retention of these splice points in all three mutants.

Third, all three mutants synthesize a variety of 19S mRNAs, both spliced and lacking a splice upstream from the 19S body (in this context designated "unspliced"). In our initial nucleotide sequencing studies of cDNAs transcribed from the 5′ ends of the mutant 19S mRNAs, it was our impression that the ratio of spliced to unspliced 19S species synthesized by the three mutants varied greatly and differed substantially from the ratio of spliced to unspliced 19S species synthesized by wild-type virus. We therefore developed a relatively simple assay for estimating the relative quantities of 19S mRNAs that were unspliced and that contained Type 1 and 2 splices and leaders. Specifically, we digested the unfractionated cDNAs transcribed from the 5' ends of the 19S mRNAs with *Hae*III restriction enzyme prior to their separation on polyacrylamide gels. Since each of the three deletion mutants retains at least one *Hae*III cleavage site within its late leader templates, this procedure reduces reverse transcripts of 19S species having a specific type of leader but different 5' termini to one characteristic length. Figure 3 shows the results of *Hae*III cleavage of the reverse transcripts of the leaders of 19S species for dl 1659. From the location of the *Hae*III cleavage sites in this mutant, knowledge of the precise nucleotides deleted in this mutant and the approximate sizes of the cDNAs present after enzymatic digestion, we were able to deduce that the longest cleaved cDNA was copied from unspliced 19S species, that the next longest was copied from species with a Type 2 splice and leader, and that the shortest was copied from species with a Type 1 splice and leader. These conclusions were confirmed by sequence analysis of the three cDNAs. A cDNA corresponding to the reverse transcript from a Type 3 leader was not identifiable, and a very faint additional cDNA migrating slightly faster than the Type 2 leader was found by sequencing to be a reverse transcript of an RNA with a Type 2 leader but with a 5' terminus several nucleotides short of the *Hae*III cleavage site. It is apparent from the relative intensities of these *Hae*III digestion products that dl 1659 directs the synthesis of about equal quantities of unspliced 19S mRNAs and mRNAs with the Type 1 splice and relatively little mRNA with the Type 2 splice. With respect to the quantities of 19S species of these three types found in cells infected with wild-type virus, the relative amount of unspliced mRNAs synthesized in dl 1659-infected cells is greatly increased, the amount of mRNAs with the Type 2 splice is significantly decreased and the amount of mRNAs with the Type 1 splice remains about the same. This striking alteration in the distribution of spliced and of unspliced mRNAs is of great interest, since the deletion in mutant dl 1659 lies about 180 nucleotides upstream from the Type 2 splice site. This result thus points out that a deletion in a region some distance upstream from a particular splice site may alter the frequency with which splicing occurs at that site.

Figure 4 shows the relative abundance of unspliced and Types 1 and 2 19S mRNAs in cells infected with the dl 1613 and dl 1626 mutants, as well as dl 1659. Whereas dl 1613 produces these three types of mRNAs in approximately the same relative quantities as does wild-type virus, dl 1626 (which cannot form the Type 1 splice because of deletion of the residue 291 splice site) is remarkable for its synthesis of predominantly unspliced 19S

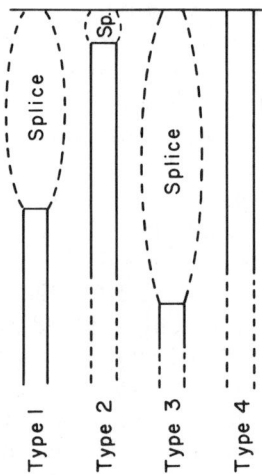

ABUNDANCE OF INDICATED WILD TYPE
SPLICES IN MUTANT-INFECTED CELLS

	dl 1659	dl 1626	dl 1613
Type 1 (Splice)	Abundant	Absent	Similar to WT
Type 2 (Sp.)	Less than in WT	Less than in WT	Similar to WT
Type 3 (Splice)	N.A.	Present	N.A.
Type 4	More than in WT	More than in WT; Predominant speices	Similar to WT

MUTANT	EXTENT OF DELETION
dl 1659	
dl 1626	
dl 1613	

PRINCIPAL 19S SPLICES AND LEADERS

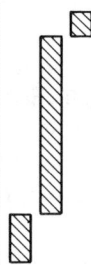

Type 1

Type 2

Type 3

Type 4

211, 217 265 291
212
420 476
440 444

DNA/RNA RESIDUE NUMBER

Fig. 4.   Schematic diagram depicting the deletions in three mutants of SV40 and the splices and leaders of the wild type SV40 19S mRNAs, DNA and RNA residues (Reddy et al., 1978b) are given for the termini of the deletions and splices. Shown in tabular form is the abundance of wild-type splices and leaders in 19S mRNAs from cells infected by the three deletion mutants. WT indicates wild-type virus; NA indicates "not ascertained."

mRNAs. The abundance in the cytoplasm of cells infected with dl 1659 and 1626 of 19S mRNA species lacking a splice upstream from the VP2 and VP3 coding regions is significant in view of the recent findings of Hamer and colleagues (1979) and Lai and Khoury (1979) that viral mutants containing coding sequences for specific proteins but lacking upstream intervening sequences and splice junctions fail to accumulate stable late cytoplasmic mRNAs for these proteins. Since the genes for these proteins were transcribed adequately, one interpretation of these results is that RNAs lacking splices in this location cannot be exported from nucleus to cytoplasm. Our finding that 19S late mRNAs lacking a splice upstream from the VP2 and VP3 coding regions can reach the cytoplasm in abundance disputes this interpretation. We will shortly attempt to determine if such 19S species may contain a splice at some internal or 3′ terminal location that could facilitate their entry into the cytoplasm. We have not yet examined polysomal RNA from cells infected with our SV40 mutants for the presence of unspliced RNA species; however, unspliced 19S RNAs are present in polysomal RNA in cells infected with wild-type virus (Ghosh, personal communication). It thus seems likely that the 19S RNAs lacking a splice upstream from the body of this RNA function as messengers.

The fourth and final conclusion from our studies is that the relative abundance of 5′ termini of late mRNAs in mutant-infected cells differs from the relative abundance of 5′ termini of late mRNAs specified by wild-type virus. For example, we found that mutant dl 1626 makes relatively little mRNA with a 5′ end at residue 243, the site of the principal 5′ end of wild-type messages. Rather, the most abundant mRNA species, both 16S and 19S, have 5′ termini lying 50-60 nucleotides upstream from residue 243. Furthermore, 19S mRNA species with a 5′ terminus mapping within the origin of replication are considerably more abundant in dl 1626 mutant-infected cells than in cells infected with wild-type virus. These changes in the distribution of 5′ termini in dl 1626-infected cells occur even though the 5′ end of the deletion in this mutant is located at residue 265, 22 nucleotides downstream from residue 243. A similar shift in the distribution of late mRNA 5′ termini to upstream positions has also been noted for dl 1613 and 1659. Although it is hypothetically possible that the utilization of upstream 5′ termini in mutant mRNAs is a consequence of destabilization of mRNAs with downstream 5′ termini, it seems more likely that downstream, as well as upstream, mutations may alter regulation of the processes—transcription initiation and/or RNA processing—that determine the relative abundance of the multiple 5′ termini of the late mRNAs. The sensitivity of 5′ termini of the 19S late mRNAs to downstream deletions further suggests that generation of 5′ termini may be governed by a measuring mechanism that preferentially selects 5′ termini lying some distance upstream from the intervening sequences of the 19S mRNAs. One possible analogy to this type of control is the recent remarkable observation of D. Brown and his colleagues (personal communication) that internal sequences within the *Xenopus* 5S RNA gene appear to be the determinants for the site of initiation of transcription of 5S RNA; provided that these sequences are retained, transcription is initiated at an appropriate distance upstream from them.

ACKNOWLEDGMENTS

This research was supported by Grant No. CA-16038, awarded by the National Cancer Institute, DHEW, and grants from the Leonard Eckstein Living Fund for Leukemia Research and the Swebilius Trust of Yale University.

REFERENCES

Abrahams, P. J., and A. J. van der Eb. 1975. In vitro transformation of rat and mouse cells by DNA from Simian Virus 40. J. Virol. 16:206–9.

Acheson, N. H. 1978. Polyoma virus giant RNAs contain tandem repeats of the nucleotide sequence of the entire viral genome. Proc. Nat. Acad. Sci. USA 75:4754–58.

Bina-Stein, M., M. Thoren, N. Salzman, and J. A. Thompson. 1979. Rapid sequence determination of late Simian Virus 40 16S mRNA leader by using inhibitors of reverse transcriptase. Proc. Nat. Acad. Sci. USA 76:731–35.

Brandner, G., and N. Mueller. 1974. Cytosine arabinoside and interferon-mediated control of polyoma and SV40 genome expression. Cold Spring Harbor Symp. Quant. Biol. 39:305–8.

Bratosin, S., M. Horowitz, O. Laub, and Y. Aloni. 1978. Electron microscopic evidence for splicing of SV40 late mRNAs. Cell 13:783–90.

Brugge, J. S., and J. S. Butel. 1975. Role of Simian Virus 40 gene A function in maintenance of transformation. J. Virol. 15:619–35.

Canaani, D., C. Kahana, A. Mukamel, and Y. Groner. 1979. Sequence heterogeneity at the 5′-termini of late Simian Virus 40 19S and 16S mRNAs. Proc. Nat. Acad. Sci. USA 76:3078–82.

Chou, J. Y., J. Avila, and R. G. Martin. 1974. Viral DNA synthesis in cells infected by temperature-sensitive mutants of Simian Virus 40. J. Virol. 14:116–24.

Chou, J. Y., and R. G. Martin. 1975. DNA infectivity and the induction of host DNA synthesis with temperature-sensitive mutants of Simian Virus 40. J. Virol. 15:145–51.

Chow, L. T., R. E. Gelinas, T. R. Broker, and R. J. Roberts. 1977. An amazing sequence arrangement at the 5′ ends of adenovirus 2 messenger RNA. Cell 12:1–8.

Cole, C., T. Landers, S. Goff, S. Manteuil-Brutlag, and P. Berg. 1977. Physical and genetic characterization of deletion mutants of Simian Virus 40 constructed *in vitro*. J. Virol. 24:277–94.

Cowan, K., P. Tegtmeyer, and D. D. Anthony. 1973. Relationship of replication and transcription of Simian Virus 40 DNA. Proc. Nat. Acad. Sci. USA 70:1927–30.

Dhar, R., K. N. Subramanian, J. Pan, and S. M. Weissman. 1977. Nucleotide sequence of a fragment of SV40 DNA that contains the origin of DNA replication and specifies the 5′ ends of "early" and "late" viral RNA. J. Biol. Chem. 252:368–76.

Dhar, R., K. N. Subramanian, B. S. Zain, A. Levine, C. Patch, and S. M. Weissman. 1975. Sequences in SV40 DNA corresponding to the "ends" of cytoplasmic mRNA. INSERM 47:25–32.

Fareed, C. G., and D. Davoli. 1977. Molecular biology of papovaviruses. Ann. Rev. Biochem. 46:471–522.

Fey, G., and B. Hirt. 1974. Fingerprints of polyoma virus proteins and mouse histones. Cold Spring Harbor Symp. Quant. Biol. 39:235–42.

Fiers, W., R. Contreras, G. Haegeman, R. Rogiers, A. Van de Voorde, H. Van Heuverswyn, J. Van Heereweghe, G. Volchaerstand, and M. Ysebaer. 1978. Complete nucleotide sequence of SV40 DNA. Nature 273:113–20.

Fried, M., and B. E. Griffith. 1977. Organization of the genomes of polyoma virus and SV40. Adv. Cancer Res. 24:67–113.

Ghosh, P. K., V. B. Reddy, J. Swinscoe, P. Lebowitz, and S. M. Weissman. 1978. Heterogeneity and 5′ terminal structures of the late RNAs of Simian Virus 40. J. Mol. Biol. 126:813–46.

Ghosh, P. K., M. Piatak, V. B. Reddy, J. Swinscoe, P. Lebowitz, and S. M. Weissman. 1980a. Transcription of the Simian Virus 40 genome in virus transformed cells and early lytic infection. Cold Spring Harbor Symp. Quant. Biol. (in press).

Ghosh, P. K., V. B. Reddy, M. Piatak, P. Lebowitz, and S. M. Weissman. 1980b. Determination of messenger RNA sequences by primer directed synthesis and sequencing of their cDNA trancripts. *In* L. Grossman and K. Moldave (eds.), Methods in enzymology, 65: 580–95. Academic Press, New York.

Gibson, W. 1974. Polyoma virus proteins: a description of the structural proteins of the virion based on polyacrylamide gel electrophoresis and peptide analysis. Virology 62:319–36.

Graessmann, A., M. Graessmann, and C. Mueller. 1977. Regulatory function of Simian Virus 40 DNA replication for late viral expression. Proc. Nat. Acad. Sci. USA 74:4831–34.

Groner, Y., P. Carmi, and Y. Aloni. 1977. Capping structures of Simian Virus 40 19S and 16S mRNAs. Nucl. Acids Res. 4:3050–60.

Haegeman, G., and W. Fiers. 1978a. Evidence for splicing of SV40 16S mRNA. Nature 273:70–73.

Haegeman, G., and W. Fiers. 1978b. Characterization of the 5′ capped structures of late Simian Virus 40-specific mRNA. J. Virol. 25:824–30.

Hamer, D. H., K. D. Smith, S. H. Boyer, and P. Leder. 1979. SV40 recombinants carrying rabbit β-globin gene coding sequences. Cell 17:725–35.

Hewick, R. M., M. Fried, and M. D. Waterfield. 1975. Nonhistone virion proteins of polyoma: characterization of the particle proteins by tryptic peptide analysis by use of ion exchange columns. Virology 66:408–19.

Hsu, M. T., and J. Ford. 1977. Sequence arrangement of the 5′ ends of Simian Virus 40 16S and 19S mRNAs. Proc. Nat. Acad. Sci. USA 74:4982–85.

Jessel, D., T. Landau, J. Hudson, T. Lalor, D. Tenen, and D. M. Livingston. 1976. Identification of regions of the SV40 genome which contain preferred SV40 T-antigen binding sites. Cell 8:535–45.

Kates, J., and J. Beeson. 1970. Ribonucleic acid synthesis in *Vaccinia* virus. I. The mechanism of synthesis and release of RNA in *Vaccinia* cores. J. Mol. Biol. 50:1–18.

Kelly, T. J., Jr., and D. Nathans. 1977. The genome of Simian Virus 40. Adv. Vir. Res. 21:85–173.

Klessig, D. F. 1977. Two Adenovirus mRNAs have a common 5' terminal leader sequence encoded at least 10Kb upstream from their main coding regions. Cell 12:9–21.

Lai, C. J., and D. Nathans. 1976. The B/C gene of Simian Virus 40. Virology 75:335–45.

Lai, C. J., R. Dhar, and G. Khoury. 1978. Mapping the spliced and unspliced late lytic SV40 RNAs. Cell 14:971–82.

Lai, C. J., and G. Khoury. 1979. Deletion mutants of Simian Virus 40 defective in biosynthesis of late viral mRNA. Proc. Nat. Acad. Sci. USA 76:71–75.

Lavi, S., and A. J. Shatkin, 1975. Methylated Simian Virus 40-specific RNA from nuclei and cytoplasm of infected BSC-1 cells. Proc. Nat. Acad. Sci. USA 72:2012–16.

Lazarides, E., J. Filer, and K. Weber. 1974. Simian Virus 40 structural proteins: amino terminal sequence of the major capsid protein. Virology 60:584–87.

Lebowitz, P., and S. M. Weissman. 1979. Organization and expression of the Simian Virus 40 genome. Current Topics in Microbiology and Immunology 87:43–172.

Legon, S., A. Flavell, A. Cowie, and R. Kamen. 1978. Amplification in the leader sequence of "late" polyoma virus mRNAs. Cell 16:373–88.

Martin, R. G., and J. Y. Chou. 1975. Simian Virus 40 functions required for the establishment and maintenance of malignant transformation. J. Virol. 15:599–612.

Mertz, J. E., and P. Berg. 1974. Viable deletion mutants of Simian Virus 40: selective isolation by means of a restriction endonuclease from *Hemophilus parainfluenzae*. Proc. Nat. Acad. Sci. USA 71:4879–83.

Osborn, M., and K. Weber. 1975. Simian Virus 40 gene A function and maintenance of transformation. J. Virol. 15:636–44.

Piatak, M., P. K. Ghosh, P. Lebowitz, V. B. Reddy, and S. M. Weissman. 1979. Complex structures and new surprises in SV40 mRNA. *In* Extrachromosomal DNA, pp. 199–215. ICN-UCLA Symp. Academic Press, New York.

Plotch, S. J., M. Bouloy, and R. M. Krug. 1979. Transfer of 5'-terminal cap of globin mRNA to influenzae viral complementary RNA during transcription *in vitro*. Proc. Nat. Acad. Sci. USA 76:1618–22.

Pribnow, D. 1975. Bacteriophage T7 early promoters: nucleotide sequences of two RNA polymerase binding sites. J. Mol. Biol. 99:419.

Prives, C. L., and H. Shure. 1979. Cell-free translation of SV40 16S and 19S L-strand specific mRNA classes to SV40 major VP-1 and minor VP-2 and VP-3 capsid proteins. J. Virol. 29:1204–12.

Rassoulzadegan, M., B. Perbal, and F. Cuzin. 1978. Growth control in Simian Virus 40 transformed rat cells: temperature-independent expression of the transformed phenotype in tsA transformants derived by agar selection. J. Virol. 28:1–5.

Reddy, V. B., P. K. Ghosh, P. Lebowitz, and S. M. Weissman. 1978a. Gaps and duplicated sequences in the leaders of SV40 16S RNA. Nucl. Acids Res. 5:4195–4214.

Reddy, V. B., B. Thimmappaya, R. Dhar, K. N. Subramanian, B. S. Zain, J. Pan, M. L. Celma, P. K. Ghosh, and S. M. Weissman. 1978b. The genome of Simian Virus 40. Science 200:494–502.

Reddy, V. B., P. K. Ghosh, P. Lebowitz, M. Piatak, and S. M. Weissman. 1979. Simian Virus 40 early mRNAs. I. Genomic localization of 3' and 5' termini and two major splices in mRNA from transformed and lytically infected cells. J. Virol. 30:279–96.

Reed, S. I., J. Ferguson, R. W. Davis, and G. R. Stark. 1975. T antigen binds to Simian Virus 40 DNA at the origin of replication. Proc. Nat. Acad. Sci. USA 72:1605–9.

Sanger, F., G. M. Air, B. G. Barrell, N. L. Brown, A. R. Coulson, J. C. Fiddes, C. A. Hutchinson, III, P. M. Slocombe, and M. Smith. 1978. Nucleotide sequence of bacteriophage ϕX174 DNA. Nature 265:687–95.

Schibler, U., and R. Perry. 1976. Characterization of the 5' termini of hnRNA in mouse L cells: implications for processing and cap formation. Cell 9:121–30.

Shenk, T. E., J. Carbon, and P. Berg. 1976. Construction and analysis of viable deletion mutants of Simian Virus 40. J. Virol. 18:664–71.

Spencer, E., D. Loring, J. Hurwitz, and G. Monroy. 1978. Enzymatic conversion of 5'-phosphate-terminated RNAs to di- and triphosphate-terminated RNA. Proc. Nat. Acad. Sci. USA 75:4793–97.

Subramanian, K. N., R. Dhar, and S. M. Weissman. 1977. Nucleotide sequence of a fragment of SV40 DNA that contains the origin of DNA replication and specifies the 5' ends of "early" and "late" viral RNA. III. Construction of the total sequence of the *Eco*RII-G fragment of SV40 DNA. J. Biol. Chem. 252:355–67.

Subramanian, K. N. 1979. Segments of Simian Virus 40 DNA spanning most of the leader sequence of the major late viral messenger RNA are dispensable. Proc. Nat. Acad. Sci. USA 76:2556–60.

Tegtmeyer, P. 1972. Simian Virus 40 deoxyribonucleic acid synthesis: The viral replicon. J. Virol. 10:591–98.

Tegtmeyer, P. 1975. Function of Simian Virus 40 gene A in transforming infection. J. Virol. 15:613–18.

Tjian, R. 1978. The binding site on SV40 DNA for a T antigen related protein. Cell 13:165–80.

Van Heuverswyn, H., and W. Fiers. 1979. Nucleotide sequence of the *Hind* C fragment of Simian Virus 40 DNA: comparison of the 5'-untranslated region of wild-type virus and of some deletion mutants. Eur. J. Biochem. (in press).

Villarreal, L. P., R. T. White, and P. Berg. 1979. Mutational alterations within the Simian Virus 40 leader segment generate altered 16S and 19S mRNAs. J. Virol. 29:209–19.

Wei, C. M., and B. Moss. 1977. 5'-terminal capping of RNA by guanylyltransferase from HeLa cell nuclei. Proc. Nat. Acad. Sci. USA 74:3758–61.

Yang, R. C. A., and R. Wu. 1979. BK virus DNA sequence: extent of homology with Simian Virus 40 DNA. Proc. Nat. Acad. Sci. USA 76:1179–83.

LOUISE T. CHOW AND THOMAS R. BROKER

The Elucidation of
RNA Splicing in the Adenoviral System

9

INTRODUCTION

The discovery of RNA splicing in human adenoviral late RNA transcripts
in the spring of 1977 and its subsequent detection in a wide range of
eukaryotic transcripts greatly altered the scientific understanding of
eukaryotic gene structure and regulation. The revelation of this
phenomenon depended upon assimilating a series of seemingly unrelated,
anomalous observations and on designing and executing several critical
experiments to test emerging ideas. The research involved a large number
of people at Cold Spring Harbor Laboratory and at other institutions
working on adenoviruses and other eukaryotic systems.

   This report is a personal perspective on our own electron microscopic
investigations of adenoviral transcription and how they correlated with
other studies. We emphasize the reasons for performing certain
experiments, the significance ascribed to the results at the time, and the
different approaches contributing to the unprecedented concepts.
Following the basic characterization of RNA splicing, new questions arose
as to how and why splicing occurs. A number of laboratories have
addressed these problems with many different approaches. At the present
time knowledge of RNA splicing patterns, DNA and RNA nucleotide
sequences, and protein sizes and compositions have made it possible to
draw a number of conclusions about the effects of RNA splicing on genetic
expression and versatility and about some of the patterns or "rules" of
splicing.

   Human adenoviruses cause relatively mild infections of the upper and

Cold Spring Harbor Laboratory, Cold Spring Harbor, New York 11724 lower respiratory tract and conjunctiva. The purified virus particles infect human cell cultures in a lytic fashion that takes two to three days. In rat or hamster cells, however, the infection is abortive. Nonetheless, some of the rodent cells become transformed, as characterized by a loss of contact inhibition, acquisition of cell immortality, and the expression of viral-specific tumor antigens. Hence, adenoviruses are considered bona fide tumor viruses.

The adenoviral chromosome is a linear double-stranded DNA molecule of about 36,500 nucleotide pairs. Efforts to define the viral transcription and translation maps have proceeded over many years with the application of a variety of techniques. Using saturation hybridization, Green and colleagues (1970) reported that about 30% of the genome is transcribed into early RNAs during the first 6 to 8 hr of productive infection. Thereafter until cell death, the rest of the genome is transcribed and expressed as late proteins, most of which are involved in viral morphogenesis. DNA replication begins about 6 to 8 hours after infection and reaches a peak rate 10 to 12 hours later. In the absence of replication, much, but not all, of the late transcription is blocked. When restriction endonucleases became available and their cleavage sites in adenoviral DNA were determined, the early and late transcription blocks were located by RNA hybridization to restriction fragments (Sharp et al., 1974; Pettersson et al., 1976). Four widely separated early regions were found, two transcribed from each of the DNA strands. For the most part, late transcripts came from regions not expressed at early times, and the early and late regions were often separated by a change in the template strand and direction of transcription. Based upon mobility coefficients determined by sucrose gradient centrifugation or polyacrylamide gel electrophoresis, the RNA transcripts were shown to be reasonably reproducible in length (Anderson et al., 1974; Tal et al., 1974; Craig and Raskas, 1976; Büttner et al., 1976).

The proteins encoded by the early and late viral RNAs were identified in several ways: mature virion particles were disrupted and fractionated by sodium dodecyl sulphate-polyacrylamide gel electrophoresis. Capsid components were assigned Roman numerals to designate their positions on the gels. These and several other proteins were detected in infected cells at late times after infection. Studies of individual capsid proteins by electron microscopic and immunologic techniques allowed correlations of many of the morphological components with protein bands observed on gels (Valentine and Pereira, 1965; Maizel et al., 1968; Russell and Skehel, 1972; Anderson et al., 1973; Everitt et al., 1973). Early proteins were studied with considerably more difficulty because they are not present in substantial quantities in infected cells. Nevertheless, comparisons of cytoplasmic and nuclear proteins in uninfected and infected cells (van der Vliet and Levine, 1973; Bablanian and Russell, 1974; Walter and Maizel, 1974; Saborio and Öberg, 1976; Neuwald et al., 1977) and the enhancement of certain proteins by the inclusion of drugs during infection (Harter et al., 1976) allowed the correlation of a number of early viral antigens and other early proteins with bands present on SDS-polyacrylamide gels, most of which are designated according to their apparent molecular weight, expressed in thousand daltons (K).

The genetic regions encoding the various early and late proteins were mapped by three basic methods. Transformed rodent cells have only part of the viral genome integrated into the cell DNA (Gallimore et al., 1974; Graham et al., 1974) and therefore can express only those regions as messenger RNAs; they are a subset of the early transcripts seen in lytically infected cells (Flint et al., 1975; Bachenheimer and Darnell, 1976; Chinnadurai et al., 1976; Flint, 1977). Similarly antisera prepared against the proteins of various transformed cells (i.e., T-antigens) precipitate subsets of the early proteins in lytically infected cells (Gilead et al., 1976; Levinson and Levine, 1977; Johansson et al., 1978; Green et al., 1979; van der Eb et al., 1979). Therefore, the coding regions of the T-antigens are defined. Second, temperature-sensitive mutations in some of the proteins were isolated and the mutations mapped by classical recombinational crossover analysis (Williams et al., 1974) as well as by physical placement with respect to restriction sites that differed between the parental viruses (Grodzicker et al., 1974). More recently, marker rescue methods have been used with isolated DNA restriction fragments (Frost et al., 1978; Galos et al., 1979). Most accurately, the proteins have been mapped by hybridization-selection of messenger RNAs with specific DNA restriction fragments followed by translation in cell-free systems (Lewis et al., 1975, 1976, 1977, 1979; Spector et al., 1979; van der Eb et al., 1979). Most of the proteins synthesized *in vitro* are also detected in infected cells. This allowed reasonably close correlations between the protein bands formed by gel electrophoresis and the regions from which the mRNAs were derived. A recent variation of this technique, the hybridization-arrested translation method (Paterson et al., 1977), has resulted in further refinement of the map coordinates of the late mRNAs and their coding regions (Miller et al., in press).

R-LOOP MAPPING OF THE ADENOVIRAL TRANSCRIPTS

In 1976 we commenced an electron microscopic project to map individual transcripts of adenovirus serotype 3 (Ad2) by the newly developed R-loop technique of Thomas and colleagues (1976) and White and Hogness (1977). In the presence of high concentrations of formamide, RNA:DNA heteroduplexes are more stable than the corresponding DNA:DNA homoduplexes (Birnstiel et al., 1972; Thomas et al., 1976; White and Hogness, 1977; Casey and Davidson, 1977). When DNA is heat-denatured, the strands separate over a rather broad transition range, usually 8° to 15°C, depending upon the variability in average AT/GC composition in different regions of the DNA. By combining these two facts, it is possible to hold DNA in the transitional state, with the AT-rich regions preferentially denatured and the GC-rich regions generally paired. When RNA molecules complementary to the DNA are present during the incubation, they can pair with transiently dissociated DNA strands. Once hybridized, the higher stability of the RNA:DNA heteroduplexes favors the retention of the RNA, and the DNA gradually incorporates complementary RNA. Upon dilution of the formamide and/or cooling of the mixture, DNA strands renature wherever possible. But opposite a hybridized RNA, the homologous DNA strand remains displaced as an "R-loop" and is a landmark easily observed in the electron microscope for determining the map coordinates of the RNA. At the time we began these studies, it was not known whether the 5' and 3' ends of a sufficiently high percentage of any one individual RNA would be reproducible or intact so that discrete RNA species could be assigned map coordinates and ulti-mately ascribed to particular proteins, although some of the cytoplasmic adenoviral RNAs seemed to have reasonably homogeneous S-values and were translatable, suggesting a certain degree of RNA stability.

The R-loops were formed by incubation of total early or late cytoplasmic RNA with Ad2 DNA for about 16 hours at the melting temperature of the DNA. Samples were prepared for electron microscopy in the presence of 40% or 45% formamide, which allows single-stranded DNA segments to remain extended. R-loops were found in many regions of the genome. Many thousands of heteroduplexes were scanned with an electron microscope and about a thousand selectively photographed and traced. The end-point coordinates and lengths of each of the R-loops were recorded. Two problems were handled without great difficulty: we found that, under the conditions used, the RNA:DNA heteroduplexes had nearly the same length per base pair as DNA:DNA duplexes and that the left-right orientation of the molecules could be determined by forming R-loops in defined restriction endonuclease fragments that represented either the left half or the right half of the Ad2 chromosome. The DNA molecules were aligned for best fit of the R-loops, and it became clear that R-loops in a number of different regions of the chromosome had reproducible lengths and coordinates (Chow et al., 1977b). The early regions and the late regions abutted but did not overlap, and, when early and late RNAs were mixed during the same annealing reaction, convergent R-loops were formed in which two RNA molecules that were transcribed from the l- or r-strand respectively annealed to adjacent segments of opposite DNA strands. Three transcription strand switch points were defined in this way. The R-loops fell into two basic categories. Some had reproducible 5' and 3' ends, while others shared the same 3' end point, but had alternative 5' ends or vice versa. Many of the map coordinates of cytoplasmic RNAs were assigned with a standard deviation of better than 0.5% of the chromosome length. The key advantages of this technique are the ability to use electron microscopy for the mapping of unfractionated RNA transcripts, to work with microgram amounts of RNA and DNA, and to map strand switches. By combining the R-loop map with the *in vitro* translation map of Lewis and coworkers (1975, 1976), we were able to assign many specific RNAs to individual proteins and to deduce the relative order of the pVI and hexon genes and of the 100K and pVIII genes. In addition, we showed that a short, late message near coordinate 10 is transcribed from the r-strand and could be assigned to virion peptide IX. Independent work on peptide IX RNA by Pettersson and Mathews (1977) arrived at the same conclusion. We further demonstrated that early region 1 is subdivided and gives rise to nonoverlapping transcripts from its left and right portions, in agreement with the hybridization data of Craig and Raskas (1976) and of Büttner et al. (1976). R-loop investigations of early and late adenoviral transcripts were also carried out at the same time by Neuwald and colleagues (1977), Meyer and colleagues (1977), and Westphal and Lai (1977). Their studies evaluated the relative abundances of RNAs from different regions of the genome at different times after infection and located the blocks of early and late regions and some of the strand switch points between them.

UNUSUAL HYBRIDIZATION BEHAVIOR OF ADENOVIRAL RNAS

Different aspects of adenoviral transcription have been studied by many of our colleagues at Cold Spring Harbor Laboratory. In 1976 several experiments indicated that adenoviral RNAs had anomalous behavior during RNA:DNA hybridization. In retrospect, these results collectively pointed to the ultimate interpretation, but, at the time, each could be rationalized by more conventional explanations. The key observations contributing to the discovery of RNA splicing have been described in detail by Broker and colleagues (1977) and are summarized briefly:

1. *There is only one capped 5'-terminal RNAase T1 digestion product from late RNAs transcribed from the r-strand* (Gelinas and Roberts,

1977). This finding was unexpected because Anderson and colleagues (1974), Lewis and colleagues (1975), and we (1977b) had shown that at least a dozen late r-strand transcripts mapped at different positions along the r-strand. One explanation considered at the time was that the RNA samples had a universal, low molecular weight contaminant that gave rise to the common capped undecanucleotide. This was largely ruled out by RNA fractionation in the presence of glyoxal, a potent denaturant of nucleic acid secondary structures. Therefore, they concluded that either only one of the late RNAs was capped or all the late RNAs had one and the same capped T1 undecanucleotide. They further postulated that, if the latter was true, different r-strand mRNAs might be primed by the same short RNA species according to a model proposed by Dixon and Robertson (1976).

2. *Two late r-strand RNAs have the same 5′-capped undecanucleotide.* To eliminate the possibilities that a universal RNA contaminant was the source of the common undecanucleotide or that it came from only a single RNA species, Klessig (1977b) purified the mRNAs for the fiber and the 100K proteins to 95% and 70% homogeneity, respectively, and found that both had the same 5′ capped RNAase T1 undecanucleotide as found in bulk RNA by Gelinas and Roberts (1977).

3. *The 5′-terminal capped undecanucleotide is RNAase sensitive when RNA is hybridized to the coding DNA.* When heteroduplexes between total late RNA and DNA or between purified fiber or 100K RNAs and DNA restriction fragments encoding these messages were treated with ribonuclease T1, the capped undecanucleotide was removed, indicating that it was not protected in an RNA/DNA heteroduplex structure (Gelinas and Roberts, 1977; Klessig, 1977b). Two explanations seemed plausible: either that the 5′ end of the RNA was breathing during ribonuclease treatment and became sensitive to cleavage or that the 5′ end simply did not pair with the DNA in the immediate vicinity of the main message coding region. A control experiment in which the 5′-capped oligonucleotide of an early region RNA remained insensitive to ribonuclease T1 was compatible with the latter and supported the hypothesis of common RNA primer for many late transcripts.

4. *The expected pyrimidine tract is absent from the 5′ end of the fiber gene.* The capped undecanucleotide includes a C_5U_4 pyrimidine tract followed by a G residue. Zain (cf. Broker et al., 1977) failed to find it

in the DNA sequence containing the 5' portion of the fiber gene, the location of which had been determined by hybridization data (Zain and Roberts, 1979) and by R-loop mapping studies (Chow et al., 1977b). This negative observation, together with the sensitivity of the capped sequence to RNAase T1 when mRNAs were hybridized to their respective genes, strengthened the possibility that the 5' portion of the fiber message and, by inference, those of many other late r-strand messages, were encoded by DNA sequences other than those at the 5' ends of their respective genes.

5. *Many late RNAs show second site hybridization.* If the 5' ends of the late RNAs were not encoded adjacent to the main body of the messenger RNAs, then from where were they derived? Were they of host or viral origin? Three independent experiments provided a tentative answer. Klessig (1977b) hybridized purified 100K mRNA or fiber mRNA to various DNA restriction fragments representing the entire adenovirus chromosome. A weak hybridization to the *Hin*dIII B restriction fragment was noted. The *Hin*dIII B fragment extends between map coordinates 17.0 and 31.5, a region to the left of all of the late r-strand mRNAs. Ribonuclease T1 treatment of heteroduplexes between these mRNAs and the *Hin*dIII B restriction fragment resulted in the degradation of the major portion of the 100K and fiber RNAs but the protection of identical sets of oligo-nucleotides totaling about 150 nucleotides. This was the first strong evidence that the common sequence present at the 5' end of late RNAs might be substantially longer than the capped RNAase T1 undecanucleotide. Concurrently, Dunn and Hassell (1977), working on several adeno-SV40 hybrid viruses, developed a very sensitive hybridization assay, the "sandwich" blotting technique. They were able to detect adenoviral sequences covalently linked to SV40 sequences in hybrid transcripts that crossed the adenovirus-SV40 recombination junction. In addition to the expected annealing to the Ad2 sequences immediately upstream from the SV40 sequences, they noticed a low-level hybridization to several other regions of the genome, including the *Hin*dIII B restriction fragment. These observations of second site hybridization were reinforced by experiments of Lewis and colleagues (1977), who mapped mRNAs and the proteins encoded using hybridization selection-*in vitro* translation analysis. They repeatedly noticed a low level hybridiza-tion of many late r-strand mRNAs to the *Hin*dIII B restriction fragment as well as the expected strong hybridization to the different segments of the genome that encode the various messages.

RNA SPLICING

Because of the multitude of suggestive evidence, our colleagues and we at Cold Spring Harbor Laboratory considered the possibility that the 5′ segments of many late r-strand mRNAs consisted of primer sequences encoded in the *Hin*dIII B restriction fragment. The most appealing candidates for such sequences were the viral-associated (VA) RNA molecules that have a length of about 155 nucleotides and are encoded near map coordinates 29-30, within the *Hin*dIII B restriction fragment on the r-strand (Mathews, 1975; Söderlund et al., 1976). In collaboration with Gelinas and Roberts, we carried out the following experiment, as reported by Chow and coworkers (1977a): R-loops between total late RNA and double-stranded Ad2 DNA were formed and then mixed with the fast or the slow strand (as prepared by agarose gel electrophoresis) of the *Hin*dIII B restriction fragment in separate hybridization mixtures. The incubation temperature was gradually lowered from 52° C to 0° C, with the expectation that the mixture would pass through the optimal hybridization temperature for the short RNA "primer"/DNA heteroduplexes. These mixtures were prepared for electron microscopy using the formamide spreading technique. Upon examination it was immediately obvious that the slow strand, but not the fast strand, of the *Hin*dIII B restriction fragment was specifically associated with the 5′ end of each one of several different R-loops. Furthermore, each of the DNA strands almost always adopted a configuration of a 2,400 nucleotide-long loop situated about 900 nucleotides from one end of the strand and 1,800 nucleotides from the other end. The *Hin*dIII B slow and fast strands formed no such looped structures by themselves. This suggested that sequences near the 5′ ends of many different late r-strand messages hybridized with two separate segments of the *Hin*dIII B restriction fragment.

In order to assign the slow *Hin*dIII B strand to the 1- or r-strand of Ad2 DNA and to orient the two sites of hybridization with respect to the left and right ends of the physical map, it was necessary to repeat the same R-loop experiment with other DNA restriction fragments that extended through this region. Furthermore, to show that these sequences were not reiterated at the beginning of each gene, it was necessary to use fragments from other regions of the genome. Therefore, the separated strands of various DNA restriction fragments encompassing the entire Ad2 chromosome were prepared and individually added to R-loop hybridization mixtures, as just described. To everyone's surprise, a third site of hybridization was found, and the three sites caused the slow strand of one of the DNA fragments to form a 1,000-base loop, that of another to form both the 1,000-base and 2,400-base loops, and two others to form heteroduplexes with one point of contact without any loop formation (fig. 1). Alignment of the locations of hybridization with respect to the ends of the various DNA fragments resulted in a consistent and unique interpretation that RNA sequences present at the 5′ ends of fiber, 100K, hexon, and at least five other late r-strand mRNAs paired with three remote sequences on the r-strand of the adenoviral DNA located at coordinates 16.5-16.6, 19.5-19.7, and 26.5-26.8, in that order. None of these coincided with the positions of VA RNAs.

At this time in our project, we received a preprint of studies by S. Berget, C. Moore, and P. Sharp at the Massachusetts Institute of Technology. They reported that the 5′ end of the hexon RNA remained unpaired in heteroduplexes with short single-stranded adenoviral DNA restriction fragments from the vicinity of the hexon gene.

The results of our electron microscopic experiments provided an explanation for the failure of the *Hin*dIII B restriction fragment (17.0-31.5) to protect the 5′ capped undecanucleotide. Klessig (1977b) was then able to protect the capped undecanucleotide from RNAase T1 by hybridizing the RNAs to the *Hin*dIII C (coordinates 7.5 to 17.0) or *Bal*E (coordinates 14.7 to 21.5) restriction fragments, localizing the cap between coordinates 14.7 and 17.0, consistent with the electron microscopic conclusion.

The evidence demonstrating the multisegmented nature of the late Ad2 RNAs was announced at a symposium on adenoviruses at Rockefeller University in April of 1977. It was presented in more detail in May at the Brookhaven Symposium in Biology (Gelinas et al., 1977), during which we first learned of additional studies by Berget and Sharp (1977), in which they identified the 5′ end sequences of hexon RNA. They reported that, when annealed to long single-stranded DNA restriction fragments that extended as far as the leader region, the hexon mRNA paired with the same main coding region and three upstream sites as we had mapped (Berget et al., 1977).

The sequences joined to the 5′ ends of the late RNAs were termed "leader segments," and their discovery quickly accounted for all of the anomalies noted in earlier biochemical studies, specifically, the presence of a common 5′-terminal sequence in many of the late RNAs, the failure to detect it in the DNA at the beginning of the fiber gene, the failure of the 5′ ends of RNAs to hybridize to DNA adjacent to their main coding sequences, and the weak hybridization of all these RNAs to sequences between coordinates 17.0 and 31.5.

Several models seemed plausible for the generation of the 5′ leader sequences: (1) post-transcriptional ligation of a leader; (2) initiation of transcription by a common primer; (3) a stuttering RNA polymerase capable of alternating periods of transcription and sliding; (4) transcription

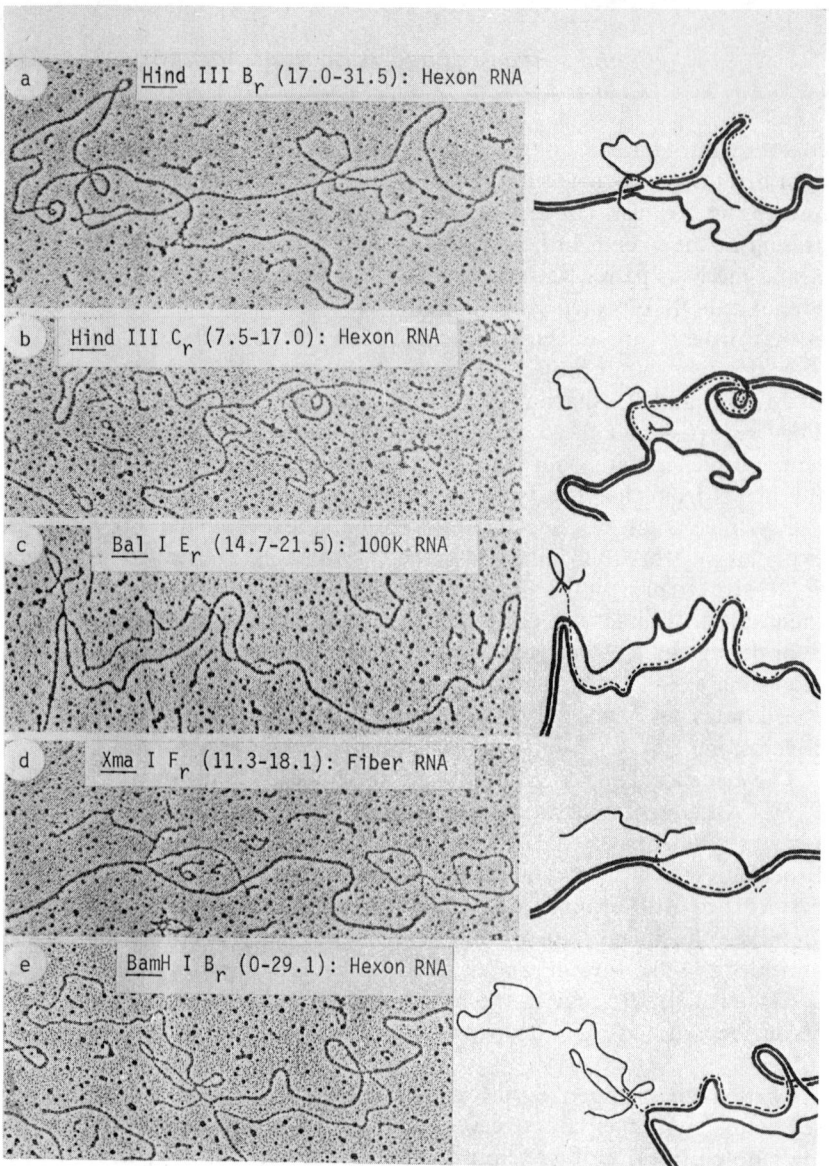

Fig. 1. Hybridization of r-strands of restriction fragments to the common 5′ leader sequences of late Ad2 mRNAs in R-loops. (a) *Hind*III B_r annealed to mRNA for hexon; (b) *Hind*III Cr annealed to mRNA for hexon; (c) *Bal*I E_r annealed to mRNA for the 100K protein; (d) *Xma*I F_r annealed to mRNA for fiber; (e) *Bam*HI B_r annealed to mRNA for hexon. The map coordinates covered by each restriction fragment are given in parentheses. Illustrative tracings are provided. (▬▬) Ad2 DNA; (────) restriction fragments; (_____) mRNA. In (c) and (d), most of the "bridge" between the R-loop and the restriction fragment is due to branch migration of the mRNA. The remaining portion is due to the unhybridized second and third leader segments.

across the base of stem-loop structures in single-stranded DNA; (5) transcription of a subset of DNA templates that had undergone prior deletions to bring appropriate segments together; and (6) complete synthesis of a long primary transcript followed by intramolecular deletion and religation, that is, RNA splicing.

As unlikely as it seemed to have RNA:RNA recombination and recombination within single-stranded polynucleotides, the last possibility, as discussed by Klessig (1977b) and by Berget and colleagues (1977), received the weight of the evidence in the ensuing several months' research. The multisegmented nature of the cytoplasmic messages ruled heavily against intermolecular recombination or priming by independent small RNAs. Examination of virion (Garon et al., 1973) or intracellular (Lechner and Kelly, 1977; Horwitz et al., 1978) adenoviral DNA by electron microscopy and by restriction enzyme analysis (many investigations) has not revealed evidence for deleted DNA populations corresponding to each of the mRNAs. No stem-loop structures between pairs of sequences that are conserved in RNAs were seen in single-stranded adenoviral DNA (Wu et al., 1977; Peterlin et al., 1978).

In support of the concept of intramolecular RNA splicing, Philipson and colleagues (1971) have shown that adenoviral nuclear and cytoplasmic RNAs are polyadenylated at their 3' ends and that poly(A) is necessary for processing and transport of RNA to the cytoplasm. Puckett and Darnell (1977) calculated that there is almost quantitative transfer of poly(A) present on long heterogeneous nuclear RNA to the cytoplasmic mRNA in uninfected HeLa cells. The late and the early adenoviral cytoplasmic RNAs have a ^{7m}G-cap structure at their 5' ends (Moss and Koczot, 1976; Sommer et al., 1976; Gelinas and Roberts, 1977; Klessig, 1977b; Hashimoto and Green, 1979), analogous to caps found on many other eukaryotic RNAs (Shatkin, 1976). In mouse L-cells, for example, caps are conserved in the cytoplasmic RNAs (Perry and Kelley, 1976). There was, however, no direct demonstration that both ends of the same RNA from adenovirus or any other system were conserved during RNA processing. Adenoviral RNAs present in the nuclei of infected cells are long (Green et al., 1970; Wall et al., 1972) and are the precursors to the shorter cytoplasmic RNAs (Bachenheimer and Darnell, 1975; Nevins and Darnell, 1978b). Weber and colleagues (1977) and Goldberg and colleagues (1977) have presented evidence that late adenoviral transcription for most of the r-strand messages originates within the interval between map coordinates 15 and 30, suggesting the presence of a common late promoter. Subsequent to the discovery of the 3-part leader present on the cytoplasmic RNAs and the position of the first leader at coordinate 16.5-16.6, the site of initiation of most late messages was narrowed down to the vicinity of coordinate 16 (Evans et al., 1977; Ziff and Evans, 1977). Thus, the first conserved leader segment present on each of the late RNAs is promoter-proximal. The fact that nuclear RNAs are long is consistent with intramolecular RNA splicing instead of stuttered transcription. RNA splicing has received further support from three additional observations: (1) the detection of complex arrays of cytoplasmic (Chow and Broker, 1978; Kilpatrick et al., 1979; Klessig and Chow, 1980; L. Chow, T. Grodzicker, J. Sambrook, J. Lewis, and T. Broker, unpublished observations) and nuclear (Berget and Sharp, 1979) RNAs with additional conserved segments, suggesting that these molecules are intermediates in RNA splicing pathways; (2) biochemical pulse-chase experiments in which nuclear precursor molecules were shown to pass through partially deleted intermediates to molecules identical with those seen in the cytoplasm (Blanchard et al., 1978; Goldenberg and Raskas, 1979); (3) studies in other systems in which purified precursor transcripts are converted to spliced molecules using partially purified enzyme extracts (Knapp et al., 1978; Peebles et al., 1979; O'Farrell et al., 1978).

The questions that we and many others next asked centered on why splicing occurs, how it is achieved, and what parameters govern it. It was with these concerns in mind that a diversity of subsequent studies radiated since 1977. We set out to determine the map coordinates of each of the spliced RNAs present in the cytoplasm at early, intermediate, and late times after infection, to quantitate the relative abundance of each species in the different RNA samples, to search for RNA molecules that might be intermediates in splicing pathways, and to survey splicing in different host cell types, during different cell growth conditions, and in the presence of drugs that might alter the production of spliced RNAs. RNA samples were prepared from cells at 6 hr (early time), 13 and 16 hr (intermediate times), and 22 hr (late time) post-infection. RNAs were also extracted from cells infected in the presence of cycloheximide, an inhibitor of protein synthesis, or cytosine arabinoside, an inhibitor of DNA replication. The RNA species present were determined by analyzing heteroduplexes formed with single-stranded DNA. Spliced RNAs are easily detected because they draw the DNA into looped configurations. The relative abundances of RNAs were evaluated by quantitative scoring of the various heteroduplexes formed. The map positions of early and intermediate RNAs (Chow et al., 1979a, b) and of late RNAs (Chow and Broker, 1978) are summarized in figure 2 and the relative abundances are displayed in figures 3 and 4.

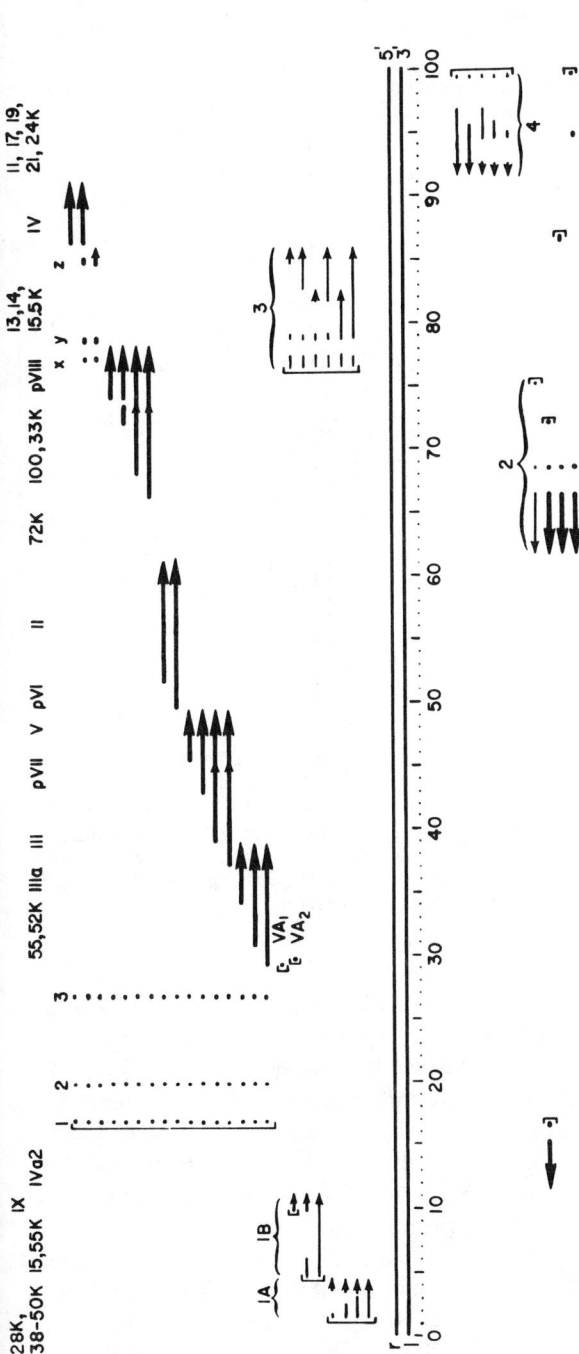

Fig. 2. The map of adenovirus-2 cytoplasmic RNA transcripts, determined by electron microscopy of RNA:DNA heteroduplexes. The 36,500 base pair chromosome is divided into 100 map units. Arrows show the direction of transcription along the r-strand or l-strand of DNA. The 5′ ends of the cytoplasmic RNAs correspond to the locations of promoters (indicated by vertical brackets). The conserved segments constituting early RNAs (Chow et al., 1979a) are depicted by thin arrows and those in late RNAs (Chow et al., 1977a; Chow and Broker, 1978; Broker and Chow, 1979) by thick arrows. Gaps in arrows represent intervening sequences removed from the cytoplasmic RNAs by splicing. Early regions 1A, 1B, 2, 3, and 4 are bracketed. At intermediate and late times, region 2 is expressed from several additional promoters and region 3 RNA can be made under the direction of the major r-strand late promoter. All derivatives of the late r-strand transcript have the same tripartite leader, the segments of which are labeled 1, 2, 3. Some of the RNAs for protein IV can also contain some combination of ancillary leader segments x, y, z. Viral-associated (VA) RNAs 1 and 2 were mapped by Mathews (1975) and Söderlund et al. (1976). The correlations of mRNAs with encoded proteins was based on cell-free translations of RNA selected by hybridization to DNA restriction fragments (Lewis et al., 1975, 1976, 1977, 1979; Harter and Lewis, 1978; Pettersson and Mathews, 1977; Spector et al., 1979; Miller et al., submitted). Alternatively spliced RNAs complementary to early regions 1A, 1B, 3 and 4 give rise to multiple proteins; some of them share common peptides. Proteins are designated by K. (1,000 daltons molecular weight) or by Roman numerals (virion components). Key to the late proteins: II (hexon); III (penton base); IIIa (peripentonal hexon-associated); IV (fiber); V (minor core); pVI (hexon-associated, precursor); pVII (major core, precursor); pVIII (hexon-associated, precursor); IX (hexon-associated).

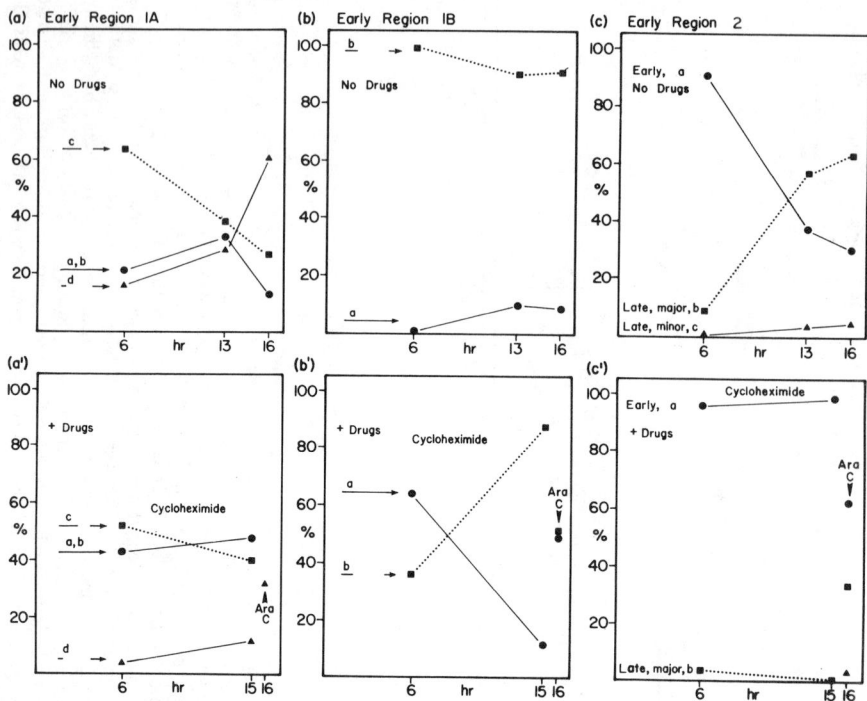

Fig. 3.   Relative abundances of RNAs from early regions 1A, 1B, and 2. The data are graphic summaries of table 5 in Chow et al. (1979a). All species are labeled on the figure. The value for each species is expressed as a percentage of the RNA from the same region at the time indicated. The lower panels a', b', and c' illustrate the effects of 25 μg/ml cycloheximide or 25 μg/ml cytosine arabinoside present from one hour post-infection.

EARLY RNAs

Each of the five early regions gives rise to families of cytoplasmic RNAs that share common 5' and/or 3' ends but have alternative internal splicing patterns. The relative abundances of the individual species within a region vary with time after infection and are influenced to different extents by cycloheximide and cytosine arabinoside. In general, the inclusion of cycloheximide to block protein synthesis results in a relative decrease in RNA species with the longer splices or multiple splices in steady-state populations. AraC causes similar, but less pronounced, effects.

The approximate map intervals within which independent transcriptions are initiated have been estimated by sensitivity of nuclear RNA synthesis to ultraviolet light or drugs (Berk and Sharp, 1977; Evans et al., 1977; Sehgal et al., 1979; Wilson et al., 1979a). These intervals include the coordinates of the 5' ends of the cytoplasmic RNAs that we and others have

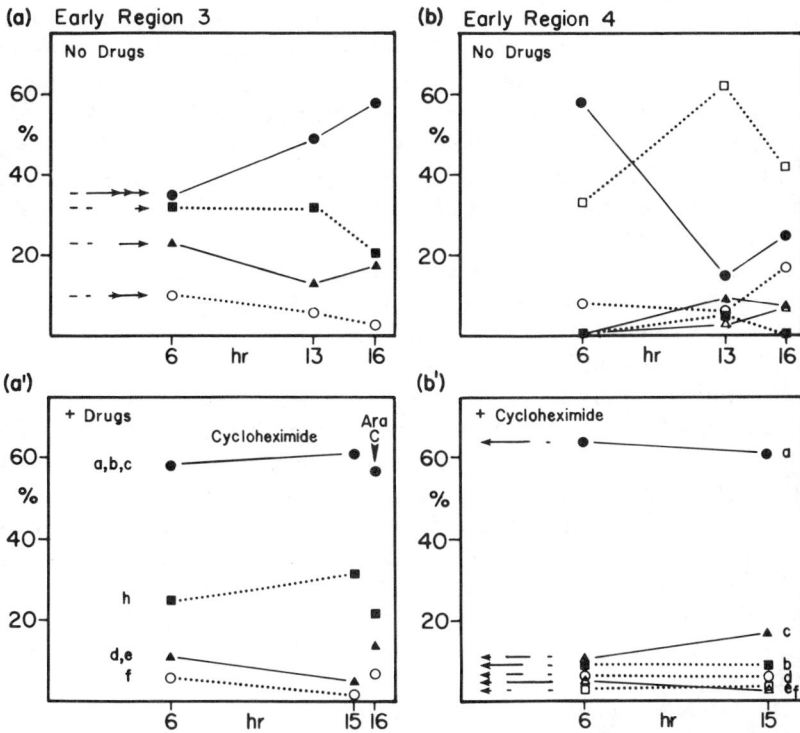

Fig. 4.   Relative abundances of RNAs from early regions 3 and 4 (see the legend to figure 2).

mapped, suggesting that the ends are conserved during post-transcriptional processing. Furthermore, the 5′ ends of the early and late RNAs are capped (Moss and Koczot, 1976; Sommer et al., 1976; Gelinas and Roberts, 1977; Klessig, 1977; Hashimoto and Green, 1979), a modification believed to take place at the 5′ ends of the primary transcripts (cf. Shatkin, 1976). Therefore, localization of the 5′ ends of the cytoplasmic RNAs provides refined mapping of the promoters. This expectation is substantiated by recent RNA and DNA sequence analyses that reveal promoter-like nucleotide sequences (D. Hogness, personal communication) just upstream from the 5′ ends of most mRNAs (Ziff and Evans, 1978; Baker and Ziff, 1979).

Early Regions 1A and 1B

Early regions 1A and 1B encode a group of proteins collectively known as T-antigens (Gilead et al., 1976; Levinson and Levin, 1977; Johansson et al., 1978; Green et al., 1979; van der Eb et al., 1979) that are necessary and sufficient for rodent cell transformation (Graham et al., 1974; Gallimore et al., 1975). Early region 1A is transcribed from the r-strand between coordinates 1.3 and 4.4. Three spliced RNAs are seen at all times examined; they have the same downstream splice site but different upstream sites (figs. 2 and 5a,b) (Berk and Sharp, 1978; Chow et al., 1979a). The species with the longest splice is in low abundance at 6 hr after infection but becomes the major product of region 1A by 16 hr (fig. 3) (Chow et al., 1979a, b; Spector et al., 1978).

Early region 1B is also transcribed from the r-strand and is separated from the region 1A transcription unit by 0.2-0.3%. Two forms of RNA are seen with a common 5' end at coordinate 4.6 and a common 3' end at coordinate 11.2. One of them has no deletion detectable by electron microscopy, but Berk and Sharp (1978) reported a very short splice near coordinate 9.8 (figs. 2 and 5a). The major product of the region has a single splice that removes sequences from coordinate 6.1 to 9.9 and is present in high abundance at intermediate and late times (figs. 3 and 5b) (Chow et al., 1977a, 1979a). Kitchingman and colleagues (1977) also reported these two species from region 1B. A third unspliced transcript seen at intermediate and, particularly, at late times after infection corresponds to the 3' proximal segment of the early region 1B spliced RNA (fig. 2) (Chow et al., 1977b; Pettersson and Mathews, 1977; Spector et al., 1978). It encodes virion peptide IX.

The DNA sequence of regions 1A and 1B has been determined for adenovirus-5 (van Ormondt et al., 1978; Maat and van Ormondt, 1979; van der Eb et al., 1979), a close relative of the Class C adenoviral serotype Ad2 which we studied. Two long nonoverlapping segments in early region 1A are in different reading frames, free of protein termination codons. Based on the map coordinates, the spliced RNAs delete segments of different lengths between the two open frames, in each case eliminating the termination codons. They were inferred to give rise to proteins that share common amino- and carboxy-termini. These expectations have been confirmed by direct analysis of spliced RNA sequences (Perricaudet et al., 1979) and of the proteins produced in infected cells or during cell-free translations of isolated early mRNAs (Harter and Lewis, 1978; Green et al., 1979; Spector et al., 1979; Lewis et al., 1979; van der Eb et al., 1979).

In early region 1B one long open reading frame extends from coordinates 4.6 to 9.6 and a second open frame extends from coordinates 9.8 to 11.2 (van der Eb, 1979). Region 1B gives rise to a 55,000 dalton protein, which is the product of the long RNA (Spector et al., 1979; Lewis et al., 1979), and a 15,000 dalton protein, which is the product of the short, spliced message (Lewis et al., 1976, 1979; Spector et al., 1979). Since there is

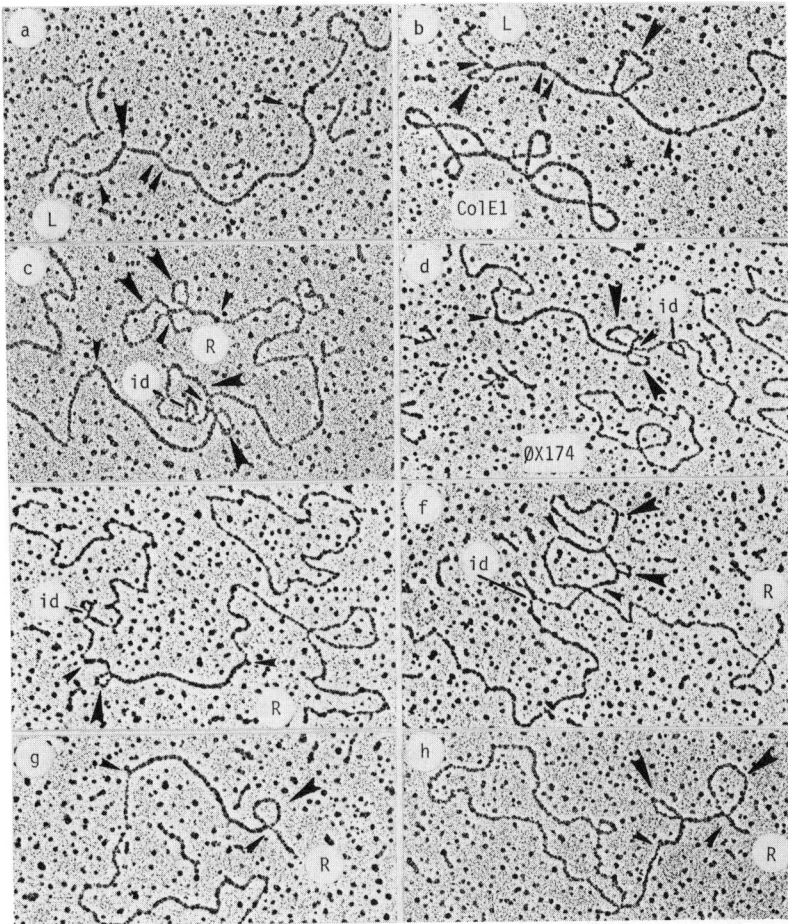

Fig. 5. Heteroduplexes of early cytoplasmic RNAs with single-stranded adenovirus-2DNA. Small arrowheads point to the 5′ and 3′ ends of the transcripts; large arrowheads indicate DNA deletion loops corresponding to intervening sequences. L denotes the left terminus and R the right terminus of the DNA strand. id represents the inverted duplication near coordinate 73.6 (Wu et al., 1978). (a) the species with the shortest splice from region 1A, and the seemingly unspliced species from region 1B; (b) the species with the longest splice from region 1A, and the spliced species from region 1B; (c) the early form of RNA from region 2, and the most extensively spliced species from region 4; (d) the major late form of RNA from region 2; (e) the species from region 3 with a single splice and a 3′ end at coordinate 82.7; (f) a species from region 3 with two splices and a 3′ end at coordinate 86.0; (g) a species from region 4 with a single short splice; (h) a species from region 4 with a longer first splice and the second splice.

no termination codon near coordinate 6.1 to account for the production of the 15,000 dalton protein, it can be inferred that the splice introduces a termination codon that results in the premature truncation of the 55,000 dalton protein to a 15,000 dalton polypeptide. The biological consequences of the RNA splicing in terms of genetic expression are summarized in table 1.

TABLE 1

BIOLOGICAL CONSEQUENCES OF ADENOVIRAL RNA SPLICING

1. Establishes alternative deletions within the coding sequences of the same primary transcripts to produce sets of related proteins. (Early regions 1A and 1B.)
2. Allows modulation of the relative abundances of the related messages. (Early regions 1A, 1B, 3, and 4.)
3. Bypasses protein termination codons to allow synthesis of longer polypeptides. (Early region 1A.)
4. Creates translation frameshifts to allow synthesis of longer polypeptides. (Early region 1A.)
5. Couples a protein initiation codon AUG to an open translation frame to allow its expression. (Early region 1A RNA species with the longest deletion.)
6. Couples a protein termination codon to an open translation frame to allow premature truncation of polypeptide synthesis. (Early region 1B.)
7. Allows utilization of a common promoter for coordinately expressed mRNAs. (Early regions 3, 4, and major late r-strand transcriptional unit.)
8. Provides translational access to interior messages in long primary transcripts. (Major late r-strand transcriptional unit.)
9.* Places at the 5′ end of RNAs a leader sequence that may be needed for ribosome binding. (Early region 4 and major late r-strand transcriptional unit.)
10. Allows utilization of different promoters for the same message at early and late stages of infection. (Early regions 2 and 3.)

Conclusions are based on correlations of RNA splicing patterns, DNA sequence, and protein composition determined in many laboratories, as cited in the text.

* Ziff and Evans (1978); Baker and Ziff (1979).

Early Region 2

Early region 2 is transcribed from the l-strand. Several different mRNAs have been identified (fig. 2). They encode a single protein, the DNA binding protein (DBP), which plays an important role in adenoviral DNA replication (van der Vliet and Levine, 1973; van der Vliet et al., 1975). Expression of region 2 is delayed with respect to the other early regions (Flint and Sharp, 1976; Westphal and Lai, 1977; Chow et al., 1979a; Wilson et al., 1979b), but the DBP transcript becomes the most abundant cytoplasmic mRNA at intermediate times after infection (Flint and Sharp, 1976; Chow et al., 1979b) when DNA replication is maximal (Green et al., 1970; Neuwald et al., 1977; Horwitz et al., 1978). At early times after infection, the 5' end of the cytoplasmic RNA is derived from sequences at coordinate 75 (fig. 5c) (Kitchingman et al., 1977; Berk and Sharp, 1978; Chow et al., 1979a). But at intermediate and late times after infection, the structure of the DBP message changes (Chow et al., 1979a). New species are seen with the 5' leader derived from coordinate 72 (fig. 5d) or from coordinates 86.7 to 86.2. In all cases these leaders are spliced to a common internal leader and the main body (66.5 to 61.6) of the mRNA. The internal leader is therefore a general receptor for 5' leaders originating from these three promoters and one other described below.

When cycloheximide, an inhibitor of protein synthesis, is present during the infection, all DBP messages isolated at intermediate times have the leader segment from coordinate 75, indicative of the early promoter, This strongly suggests that protein synthesis is necessary to turn on the late promoters. Cytosine arabinoside, which inhibits DNA replication, does not have this effect (fig. 3). Therefore, the activation of the late promoters is not tightly coupled to ongoing DNA replication (Chow et al., 1979a).

Early Region 3

Early region 3 is transcribed from the r-strand and gives rise to a multitude of different cytoplasmic transcripts. All have a common 5' segment (76.6-77.6) that is spliced to sequences beginning at coordinate 78.6. Some RNAs extend without further deletion to alternative 3' ends at 82.7 or 86.0 (fig. 5e) (Kitchingman et al., 1977; Berk and Sharp, 1978; Chow et al., 1979a). Others have a second deletion of intervening sequences of several different lengths (Chow et al., 1979a), as shown in figures 2 and 5f. At 6 hr post infection, substantially more cytoplasmic RNA is derived from early region 3 than any other region (Westphal and Lai, 1977; Eggerding and Raskas, 1978; Wilson et al., 1978; Chow et al., 1979a). Of all the early transcripts, the distribution of the region 3 primary transcript into the alternative cytoplasmic RNA species seems least affected by the inclusion of cycloheximide or araC during the infection (fig. 4) (Chow et al., 1979a). Early region 3 is embedded within the late r-strand transcription unit; the fiber gene is downstream, and all other late genes are upstream. For transcription of the fiber gene to occur, RNA synthesis initiated from the late promoter at coordinate 16.5 must pass through region 3. Polyadenylation can occur at the normal 3' ends of region 3 at late times, irrespective of the change in promoters. Subsequent splicing deletes the same internal intervening sequences, but the RNAs have the typical late tripartite leader (Chow et al., 1979a; Broker and Chow, 1979).

Region 3 encodes a protein of 15,500 daltons (Lewis et al., 1976) that is glycosylated *in vivo* (Persson et al., 1979b; Ross and Levine, 1979), and at least two other proteins of 13K and 14K (Harter and Lewis, 1978; Persson et al., 1978). Their functions have not been elucidated.

Early Region 4

Region 4 RNAs are transcribed from the l-strand and all have a short leader from coordinate 99.2 spliced to one of several alternative main body segments (figs. 2 and 5g) (Berk and Sharp, 1978; Chow et al., 1979a); some species have a second, common deletion (figs. 2, 5c and h) (Chow et al., 1979a; Kitchingman et al., 1977). All have their 3′ ends at coordinate 91.3. The relative abundances of the different species change with time after infection unless cycloheximide is present (fig. 4). The RNAs encode a group of small proteins (Harter and Lewis, 1978), but one-to-one correlations have not yet been possible.

At low frequency we have detected cytoplasmic transcripts in which sequences from early region 4 are spliced to sequences from region 2 (Chow et al., 1979a). We infer that transcription has initiated from the region 4 promoter at 99.2 and extended beyond the normal termination site to the region 2 polyadenylation site located at coordinate 61.6. As with other transcripts, the 5′ and 3′ ends of the primary RNA are conserved during processing such that the leader and one or two other segments from region 4 are spliced to the common, internal leader and main body segments of the region 2 transcripts. The usual 3′ proximal segment of region 4 RNAs is, however, deleted. This indicates that polyadenylate plays an important role in directing the conservation of sequences. When it is not at the usual 3′ end of region 4 transcripts, the normally 3′-proximal segment of the region 4 RNAs is not retained.

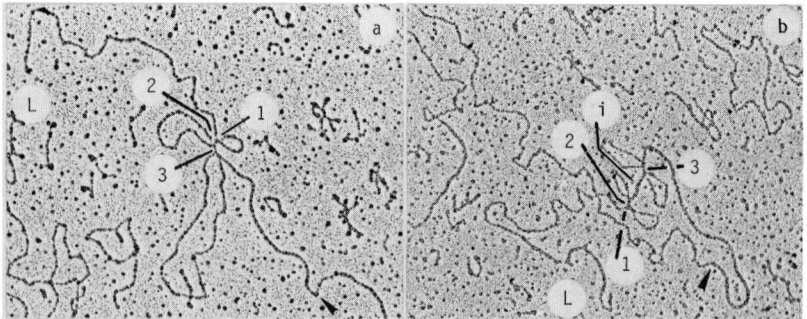

Fig. 6.   Heteroduplexes of late cytoplasmic RNAs with single-stranded adenovirus-2 DNA. Small arrowheads point to the 3′ ends of the transcripts. The leader segments are labeled. (a) A four-part leader, with segments 1, 2, i, and 3, on the 5′ end of the RNA for the 52K-55K protein (main body coordinates 30.5–39.0); (b) the usual three-part leader, on the RNA for pVII (coordinates 42.8–49.5.)

LATE RNAS

As described previously, many late RNAs derived from the primary r-strand transcription unit have a three-part leader at their 5' ends (Chow et al., 1977a). Subsequently, we have prepared RNA/single-strand DNA heteroduplexes and have shown that all late r-strand RNAs have this leader (figs. 6 and 7) (Chow and Broker, 1978; Chow et al., 1979a; Broker and Chow, 1979). Recent nucleotide sequence analyses demonstrated that the tripartite leaders of different late RNAs are identical (Akusjärvi and Pettersson, 1979; Zain et al., 1979). The 3' ends of these RNAs are also of

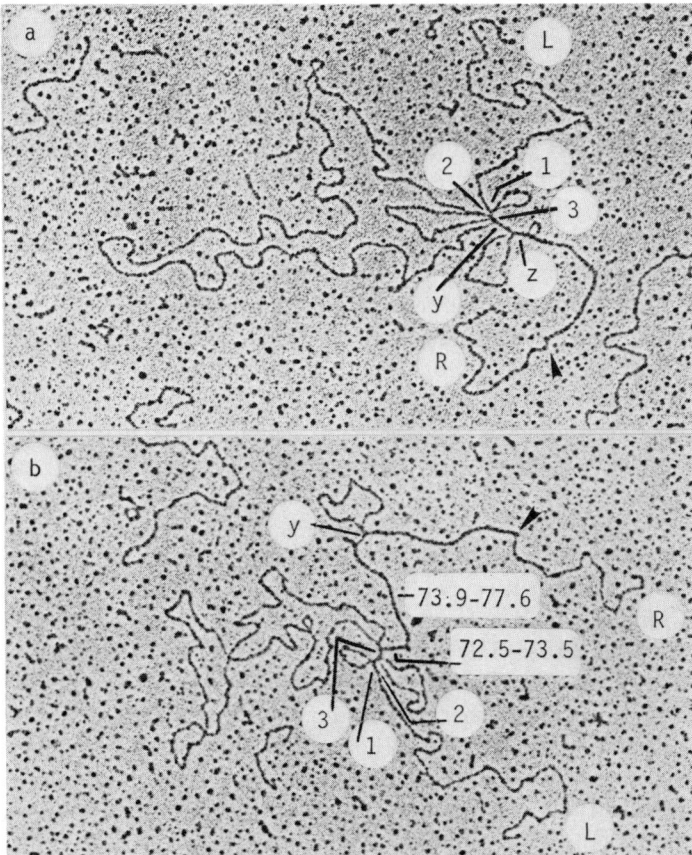

Fig. 7. Heteroduplexes of late fiber transcripts with single-stranded adenovirus-2 DNA. (a) Fiber RNA from monkey CV-1 cells with the consanguinous tripartite leader and the y (78.6–79.1) and z (84.7–85.1) ancillary leaders; (b) fiber RNA from monkey cells with the consanguinois tripartite leader and the extra intervening sequences 72.5–73.5/73.9–77.6/73.9–77.6/y.

considerable interest. We noted in our original R-loop characterization of the late RNAs (Chow et al., 1977b) that some of the RNAs such as the messages for pVI and hexon and for 100K and pVIII form 3′ co-terminal families. Further electron microscopic mapping (Chow and Broker, 1978) and biochemical experiments (McGrogan and Raskas, 1978; Nevins and Darnell, 1978a; Fraser and Ziff, 1978; Berget and Sharp, 1979) showed that there are five major families with 3′ ends located at coordinates 39.0, 49.5, 61.5, 78.3, and 91.3 map units. The longer members in any one family include the coding information for two or more proteins and therefore are physically polycistronic. However, they are functionally monocistronic (cf. Anderson et al., 1974) since only the 5′-proximal message is translated in eukaryotic systems (Kozak, 1978). Therefore, the alternative splicing within families of late messages effectively provides translational access to the interior messages within the families of RNAs.

The polyadenylation at the 3′ ends of the RNAs precedes the splicing of the tripartite leader to the main bodies of mRNAs (Nevins and Darnell, 1978b). The cytoplasmic RNAs do not usually contain sequences from two RNA families. The leaders are spliced to one of several alternative positions relatively near the poly(A). The only exception involves the fiber transcript, the promoter-distal message in the r-strand transcription unit. Fiber RNA is often seen with one or more ancillary leader segments, x, y, and z, interposed between the tripartite leader and the main body of the message (fig. 7a) (Chow and Broker, 1978; Dunn et al., 1978). The x, y, and z leaders coincide with segments that normally constitute early region 3 RNAs encoded immediately upstream. Their inclusion in fiber RNAs indicates that splicing at early and late times is very similar in terms of enzymes and splicing signals, even when the conserved segments are present in different primary transcripts.

Another example of incomplete segregation of families is observed when human adenovirus-2 infects cultures of monkey cells (Klessig and Chow, 1980). Monkey CV-1 cells are nonpermissive hosts for Ad2 infection (Rabson et al., 1964). The defect manifests itself as a reduction in the production of most late RNAs and their corresponding proteins (Klessig and Anderson, 1975; Farber and Baum, 1978), but the amount of fiber protein is anomalously low when compared with the level of fiber RNA sequence present (Klessig and Anderson, 1975). Electron microscopic and filter hybridization analysis of the infrequent fiber region transcripts present in the cytoplasm of monkey cells show that about half of this RNA population is improperly or incompletely spliced. Specifically, sequences from the upstream family of 100K-33K-pVIII transcripts remain between the tripartite leader and fiber-specific sequences (fig. 7b). These RNAs seem to be processing intermediates in which long intervening sequences between the tripartite leader and the fiber message have not been removed. Mutants of Ad2 and Ad5 capable of overcoming the block to growth in monkey cells have been isolated (Klessig, 1977a), and they map in the genetic region for the DNA binding protein (Klessig and Grodzicker, 1979). These mutants restore the production of fiber RNA and protein and do not accumulate incompletely spliced fiber RNAs (Klessig and Chow, 1980). Together these observations suggest that the DNA binding protein might play some role in RNA splicing. Examination of the array of such putative intermediates reveals the complexity of the splicing alternatives and suggests several patterns by which RNA splicing proceeds (table 2). In the presumptive splicing intermediates, the tripartite leader is attached to a bona fide splice site from an upstream main body. Several alternative internal splices within the 100K-33K-pVIII family and the early region 3 family can be noted in the fiber transcripts. They suggest that the long intervening sequences are removed by a series of short deletions. Often the tripartite leader plus short sequences at the 5′ end of an upstream gene are, together, spliced to the next downstream target, with the upstream 5′ end sequences acting as a carrier of the leader. It seems that the tripartite leader sequences are handed via the carrier sequences from splice site to splice site until they are associated with a stable site compatible with the position of polyadenylation. The carrier sequences are then deleted in a final step, attaching the tripartite leader directly to the downstream main body sequences. Many examples of these leader carrier structures involving RNAs other than the fiber transcript have also been seen in Ad2 RNA from HeLa cells (L. Chow, T. Grodzicker, J. Sambrook, J. Lewis, and T. Broker, unpublished observations). Thus, the existence of families of late transcripts and the apparent splicing intermediates suggest that splicing is multistep, that the intervening sequences may be removed in two or more successive stages to give rise to a single mature splice junction, and that splicing is not necessarily processive in a 5′ to 3′ direction but can delete some downstream intervening sequences prior to removing upstream sequences. Furthermore, the full array of apparent intermediates suggests that splicing can proceed by alternative pathways to arrive at the same final product (table 2).

These conclusions are reinforced by examination of cytoplasmic RNAs in which the consanguinous tripartite leader is not fully assembled (Chow et al., 1979a,b). The structures fall into three main groups. The most commonly observed has a fourth leader segment (the "i" leader) that is derived from coordinates 22.0 to 23.2 between the second and third leaders (fig. 6a). In the second type the normal third leader segment has the usual downstream splice site at coordinate 26.8 but extends upstream by variable but discrete distances. In the extreme case it merges with the i leader so

TABLE 2

GENERAL PATTERNS OF ADENOVIRAL RNA SPLICING

1. Almost all mature mRNAs are derived from two to four or five discontinuous segments of the genome.
2. Splices may remove a few nucleotides (e.g., 116 in early region 1A) to many (21,700 between the third consanguinous leader segment and the fiber message).
3.* The first two nucleotides of a deleted intervening sequence are GU and the last two are AG, as in virtually all eukaryotic mRNA splicing.
4. Splices are faithful to the nucleotide, since they can be within coding regions. (Early regions 1A, 1B.)
5. Conserved segments are always in the same order as in the genome, suggesting a directional, intrastrand event.
6. Most splicing occurs in the nucleus, but some cytoplasmic splicing has not been ruled out.
7. The capped 5' end and the polyadenylated 3' end of each primary transcript are conserved through all splicing events, and all deletions are therefore internal.
8.† The presence of the 5' cap nearby (within 50-100 bases) seems to potentiate certain splice sites; its absence might disqualify sites that can otherwise be used. (Early region 2.)
9. There are a number of early and late RNA families with common 3' ends. The sites of polyadenylation determine the sets of product messages derived from the precursor transcript such that the 5' end of the primary transcript is spliced to one of several sites relatively near the poly(A). (Early region 3; region 4-region 2 hybrid RNAs; late r-strand transcripts.)
10. A site that is a normal downstream junction site can be bypassed if poly(A) is not nearby. (Region 4-region 2 hybrid RNAs).
11. Deletions are not necessarily processive in the 5'→3' direction; certain downtown splices can occur prior to other upstream splices. (Early region 4; late r-strand leader segments; fiber transcripts.)
12. Some multiple deletions may occur in alternative orders. (Early region 4; late r-strand leader segments; fiber transcripts, and other late r-strand products.)
13. Splicing of certain segments does not necessarily occur in a single complete step, but can proceed by a series of partial deletions. (Early region 4; late r-strand leader segments, fiber transcripts, and other late r-strand products.)
14. Late tripartite leaders appear to undergo a gene-to-gene transfer during coupling to distal coding regions in polycistronic transcripts. The leaders are (apparently transiently) coupled to the 5' ends of upstream genes. (late r-strand transcripts.)
15. Leader sequences and main body sequences may be spliced somewhat independently; they can be at different stages of maturation. (r-strand transcripts.)
16. Splice signals in separate transcription units can faithfully recombine if transcription read-through puts them in the same RNA molecule. (Hybrid RNAs from region 4-region 2; region 1A-region 1B; late r-strand unit-early region 3.)
17. Many splice junction sequences can have alternative 5' or 3' splice partners. (All early regions; major late r-strand transcripts.)
18. When a message switches from an early to a late promoter, splicing within the main body can continue in the same pattern despite different leaders. (Early regions 2 and 3.)
19. Splices that occur efficiently in one host may be inefficient in another host. (Region 3 and fiber: human HeLa vs. monkey CV-1 cells.)
20. The rate or efficiency of splicing can depend on cellular growth conditions. (Spliced late RNAs present in spinner cultures compared with those in monolayer cells.)
21. Cycloheximide, an inhibitor of protein synthesis, depresses the appearance of the more extensively deleted members within a family of transcripts. (Early regions 1A, 1B, and 4.)
22. Adenoviral RNA splicing is catalyzed by host proteins, but some viral proteins play some role in the process at intermediate to late times after infection.

TABLE 2 (*continued*)

GENERAL PATTERNS OF ADENOVIRAL RNA SPLICING

23. The designations "intron" and "exon" are inappropriate for adenoviral transcripts since most segments can serve both roles, depending on the message.

Conclusions are based on the data discussed in the text.
* Akusjärvi and Pettersson, 1979; Zain et al., 1979; Breathnach et al., 1978; Seif et al., 1979.
† The late and early leaders of region 2 transcripts are usually not present in the same RNA (Chow et al., 1979a).

there is a continuous long third leader from coordinate 22.0 to 26.8. In another type this long third leader or the normal mature third leader sequences are not spliced to a downstream coding region but instead extend continuously into the coding regions. Except for RNAs with the i leader, these structures are very rare in the cytoplasm and most likely represent nuclear RNAs that have leaked into the cytoplasm. They are detected by exhaustive electron microscopic survey. In contrast, RNAs with the i leader can be the abundant or predominant late species in cytoplasmic RNA population. This occurs at intermediate times after infection, when adenoviruses infect HeLa cells grown in monolayers rather than in suspensions (Chow, Grodzicker, Sambrook, Lewis, and Broker, unpublished observations), when Ad2 infects monkey cells (as described previously), and when infections are carried out in the presence of cytosine arabinoside, an inhibitor of DNA replication. Analogous leader segments are also seen in cytoplasmic RNAs from Class B adenoviral serotypes Ad-3 and Ad-7 (Kilpatrick et al., 1979). The common denominator of these various conditions seems to be a prolonged infection during which late transcription is proceeding at a slower rate than that at 22 to 30 hours in infected suspension cultures. The joining of leader segments seems to be particularly susceptible to the apparent limitation in the splicing machinery. The preservation of i leader sequences after downstream sequences have been removed, the variable lengths of the third leader segments, and the joining of the incompletely matured leader sequences to the main bodies support the conclusions that the deletion of a single intervening sequence can be multistep, that splicing is not processive and some downstream splices can precede others upstream, and that there are alternative pathways to the same final mRNA (table 2).

Examination of the cytoplasmic RNAs prepared at intermediate times after infection or in the presence of cytosine arabinoside reveals a bias in the distribution of the r-strand products (Chow et al., 1979a, b). The late r-strand mRNA coding for two related proteins of 52K and 55K molecular weights has a main body sequence between coordinates 30.5 and 39.0 (Chow and Broker, 1978; Miller et al., in press; J. E. Smart and J. B. Lewis, personal communication). At intermediate times, the proportion of late r-strand transcripts represented by the 52K-55K RNA is about 19%, which is higher by a factor of about 2 relative to its production at late times. But in the presence of cytosine arabinoside, 84% of the r-strand transcripts in the cytoplasm are the 52K-55K RNA. Since the distribution of the primary transcripts into particular families is dictated at the level of RNA cleavage and polyadenylation during transcription (Nevins and Darnell, 1978b), araC must affect this process and result in either premature termination or preferential cleavage and polyadenylation at coordinate 39.0. When the late RNAs in general, and the 52K-55K RNA in particular, were examined for the presence of a four-part leader (the tripartite plus the i leader segments), again a strong bias was observed. At intermediate times about 50% of the 52K-55K messages have the i leader, whereas 19% of all other RNAs have this leader segment. This distribution is not substantially altered by the inclusion of cycloheximide from 1 hr post infection. But when cytosine arabinoside is present, 88% of the 52K-55K RNA and of all other products of the r-strand transcription unit have the i leader. This is consistent with our previous notion that the i leader is often seen in infections that are retarded because of unfavorable or slower growth conditions.

One other intermediate and late mRNA has been detected and it, too, has a spliced structure. It is transcribed from the l-strand and encodes the protein IVa_2 (Lewis et al., 1977), involved in virion maturation (Persson et al., 1979a). There is a single 5' leader from coordinates 16.1 to 15.7 joined to the main body sequences from 15.1 to 11.2 (Chow and Broker, 1978). Thus, transcription diverges leftward and rightward from the interval at 16.1–16.5. Recently the late promoter sequences in this interval have been recognized by DNA analysis (Ziff and Evans, 1978; Baker and Ziff, 1979). As indicated by our R-loop mapping (Chow et al., 1977b), the IVa_2 transcript and the early region 1B and peptide IX transcripts converge almost precisely at coordinate 11.2. The DNA sequence analysis also shows that the nucleotide sequence AATAAA, which signals RNA transcription termination, is present in both the r- and l-strands near coordinate 11.2 (van der Eb et al., 1979).

Figure 2 shows the complete transcription maps at early, intermediate, and late times after infection. 5'-ends (promoters), 3' polyadenylation sites, and internal splice junctions of different mRNAs are located within narrow intervals, usually between coding regions either on the same strand or on opposite strands. In several cases, notably with the DNA binding protein message, overlapping processing signals and coding sequences are evident. The presumed promoter and 5′ leader segment for the major late DBP RNA from coordinates 72.0–71.9 on the l-strand are complementary to the coding sequences for the 100K protein (approximately 66–76 on the r-strand) (Galibert et al., 1979). Similarly, the minor late promoter and leader for the DBP message from coordinates 86.7–86.2 overlaps fiber coding sequences (86.2–91.3) on the r-strand (Zain and Roberts, 1979). Furthermore, the common internal leader segment of the DBP messages from 68.5–68.3 also overlaps the coding region for the 100K RNA. One case of overlapping coding sequences is evident in the late r-strand transcriptional unit; a 33K protein has been found to correspond to the carboxy-terminus of the 100K protein (Axelrod, 1978; Miller et al. submitted), compatible with the structures of the RNAs observed in that region (Chow et al., 1977a; Chow and Broker, 1978; Chow et al., 1979a). Combining these observations on late RNAs with the alternative splicing patterns in early region transcripts, there are several types of overlapping genes or genetic intervals in the adenovirus chromosome. These coincidences are emphasized in figures 8 and 9, as discussed by Broker and Chow (1979).

CONCLUSIONS

RNA splicing was discovered with the adenovirus system, which has continued to provide examples of many variations on RNA splicing, including the splicing of leader segments and splicing within coding regions. Splicing intermediates and pathways of several types are becoming

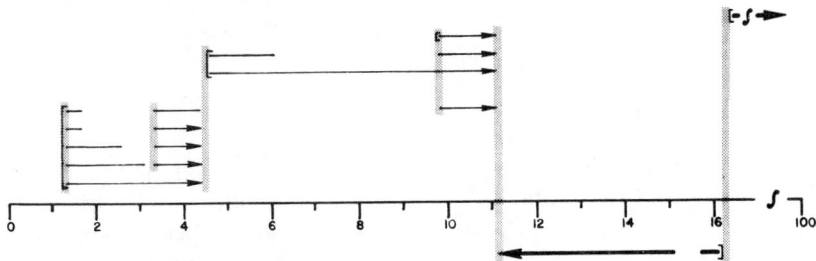

Fig. 8. Coincident RNA processing sites in the Ad2 chromosome region 0.0–17.0. The conventions used in figure 2 to represent spliced early RNAs (thin arrows) and late RNAs (thick arrows) are maintained. Short DNA or RNA regions that serve multiple roles in the synthesis and maturation of different mRNAs are indicated by stippled bars (Broker and Chow, 1979).

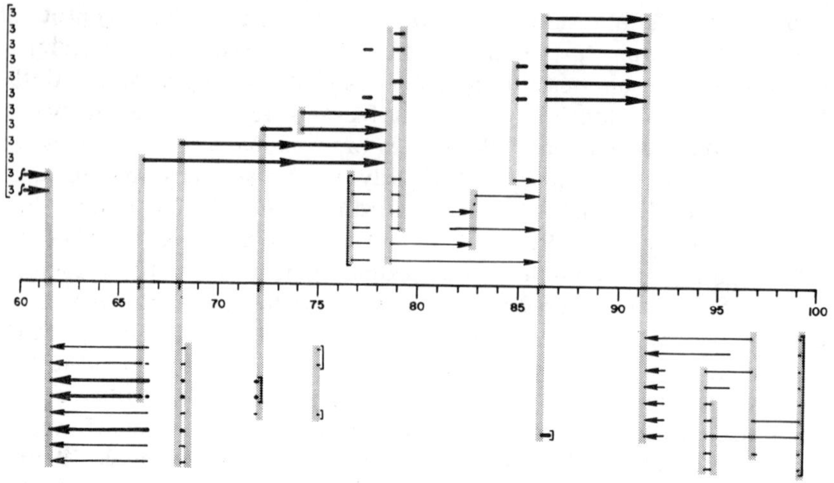

Fig. 9.   Coincident RNA processing events in the Ad2 chromosome region 60.0–100.0. The conventions are described in the legends to figures 2 and 7. All late r-strand RNAs have the tripartite leader (3) derived from coordinates 16.5–16.6/19.5–19.7/26.5–26.8. Each DNA binding protein message (l-strand) can be observed in an alternative form with a small splice from 66.3–66.2 (Chow et al., 1979a). Only two examples are depicted.

evident. Splicing patterns change during the course of infection, either reflecting or causing the stages of the infectious cycle. Because of the great number of transcripts due to variations in splicing, the adenoviruses are exquisitely sensitive probes of the host cells, of cellular growth conditions, and of the effects of various pharmaceutical drugs that affect biochemical processes. The availability of RNA transcription and splicing maps, DNA sequences, RNA sequences, and considerable information on protein sizes and compositions have already allowed tight correlations that reveal a great deal of information about how and why RNA splicing occurs and what it achieves for the genetic expression of the virus.

ACKNOWLEDGMENTS

We wish to thank our colleagues in this research at Cold Spring Harbor Laboratory for providing RNA and DNA samples and stimulating discussions: James Lewis, Richard Gelinas, Richard Roberts, Daniel Klessig, Terri Grodzicker, Joe Sambrook, Jeffrey Engler, and Bill Kilpatrick. The manuscript was skillfully prepared by Marie Moschitta. These studies were supported by Cancer Center Grant CA13106 from the National Cancer Institute.

REFERENCES

Akusjärvi, G., and U. Pettersson. 1979. Sequence analysis of adenovirus DNA: complete nucleotide sequence of the spliced 5′ noncoding region of adenovirus hexon messenger RNA. Cell 16:841–50.

Anderson, C. W., P. R. Baum, and R. F. Gesteland. 1973. Processing of adenovirus 2-induced proteins. J. Virol. 12:241–52.

Anderson, C. W., J. B. Lewis, J. F. Atkins, and R. F. Gesteland. 1974. Cell-free synthesis of adenovirus-2 proteins programmed by fractionated mRNA: a comparison of polypeptide products and mRNA lengths. Proc. Nat. Acad. Sci. USA 71:2756–60.

Axelrod, N. 1978. Phosphoproteins of adenovirus 2. Virology 87:366–83.

Bablanian, R., and W. C. Russell. 1974. Adenovirus polypeptide synthesis in the presence of non-replicating poliovirus. J. Gen. Virol. 24:261–79.

Bachenheimer, S., and J. E. Darnell. 1975. Adenovirus-2 mRNA is transcribed as part of a high-molecular-weight precursor RNA. Proc. Nat. Acad. Sci. USA 72:4445–49.

Bachenheimer, S., and J. E. Darnell. 1976. Hybridization of mRNA from adenovirus-transformed cells to segments of the adenovirus genome. J. Virol. 19:286–89.

Baker, C., J. Hérissé, G. Courtois, F. Galibert, and E. Ziff. 1979. Messenger RNA for the Ad-2 DNA binding protein: DNA sequences encoding the first leader and heterogeneity at the mRNA 5′ end. Cell 18:569–80.

Baker, C. C., and E. B. Ziff. 1979. The biogenesis, structures, and sites of encoding of the 5′ termini of adenovirus 2 messenger RNAs. Cold Spring Harbor Symp. Quant. Biol. 44:415–28.

Berget, S. M., C. Moore, and P. A. Sharp. 1977. Spliced segments at the 5′ terminus of adenovirus-2 late mRNA. Proc. Nat. Acad. Sci. USA 74:3171–75.

Berget, S. M., and P. A. Sharp. 1977. A spliced sequence at the 5′ terminus of adenovirus late mRNA. *In* C. W. Anderson (ed.), Genetic interaction and gene transfer, pp. 332–44. Brookhaven Symposia in Biology, Vol. 29.

Berget, S. M., and P. A. Sharp. 1979. Structure of late adenovirus 2 heterogeneous nuclear RNA. J. Mol. Biol. 129:547–65.

Berk, A. J., and P. A. Sharp. 1977. Ultraviolet mapping of the adenovirus 2 early promoters. Cell 12:45–55.

Berk, A. J., and P. A. Sharp. 1978. Structure of the adenovirus 2 early mRNAs. Cell 14:695–711.

Birnstiel, M. L., B. H. Sells, and I. F. Purdom. 1972. Kinetic complexity of RNA molecules. J. Mol. Biol. 63:21–39.

Blanchard, J.-M., J. Weber, W. Jelinek, and J. E. Darnell. 1978. *In vitro* RNA-RNA splicing in adenovirus-2 mRNA formation. Proc. Nat. Acad. Sci. USA 75:5344–48.

Breathnach, R., C. Benoist, K. O'Hare, F. Gannon, and P. Chambon. 1978. Ovalbumin gene: evidence for a leader sequence in mRNA and DNA sequence at the exon-intron boundaries. Proc. Nat. Acad. Sci. USA 75:4853–57.

Broker, T. R., and L. T. Chow. 1979. Alternative RNA splicing patterns and the clustered transcription and splicing signals of human adenovirus-2. *In* R. Axel, T. Maniatis, and C. F. Fox (eds.), Eukaryotic gene regulation, pp. 611–37. ICN-UCLA Symposium on Molecular and Cellular Biology, Vol. 14. Academic Press, New York.

Broker, T. R., L. T. Chow, A. R. Dunn, R. E. Gelinas, J. A. Hassell, D. F., Klessig, J. B. Lewis, R. J. Roberts, and B. S. Zain. 1977. Adenovirus-2 messengers: an example of baroque molecular architecture. Cold Spring Harbor Symp. Quant. Biol. 42:531–53.

Büttner, W., Z. Veres-Molnár, and M. Green. 1976. Preparative isolation and mapping of adenovirus-2 early messenger RNA species. J. Mol. Biol. 107:93–114.

Casey, J., and N. Davidson. 1977. Rates of formation and thermal stabilities of RNA:DNA and DNA:DNA duplexes at high concentrations of formamide. Nucl. Acids. Res. 4:1539–52.

Chinnadurai, G., H. M. Rho, R. B. Horton, and M. Green. 1976. mRNA from the transforming segment of the adenovirus 2 genome in productively infected and transformed cells. J. Virol. 20:255–63.

Chow, L. T., and T. R. Broker. 1978. The spliced structures of adenovirus-2 fiber message and the other late mRNAs. Cell 15:497–510.

Chow, L. T., R. E. Gelinas, T. R. Broker, and R. J. Roberts. 1977a. An amazing sequence arrangement at the 5′ ends of adenovirus-2 messenger RNA. Cell 12:1–8.

Chow, L. T., J. M. Roberts, J. B. Lewis, and T. R. Broker. 1977b. A map of cytoplasmic RNA transcripts from lytic adenovirus type 2, determined by electron microscopy of RNA:DNA hybrids. Cell 11:819–36.

Chow, L. T., T. R. Broker, and J. B. Lewis. 1979a. Complex splicing patterns of RNAs from the early regions of adenovirus-2. J. Mol. Biol. 134:265–304.

Chow, L. T., J. B. Lewis, and T. R. Broker. 1979b. RNA transcription and splicing at early and intermediate times after adenovirus-2 infection. Cold Spring Harbor Symp. Quant. Biol. 44:401–14.

Craig, E. A., and H. J. Raskas. 1976. Nuclear transcripts larger than the cytoplasmic mRNAs are specified in segments of the adenovirus genome coding for early functions. Cell 8:205–13.

Dixon, E., and H. D. Robertson. 1976. Potential regulatory roles for RNA in cellular development. Cancer Res. 36:3387–93.

Dunn, A. R., and J. A. Hassell. 1977. A novel method to map transcripts: evidence for homology between an adenovirus mRNA and discrete multiple regions of the viral genome. Cell 12:23–36.

Dunn, A. R., M. B. Mathews, L. T. Chow, J. Sambrook, and W. Keller. 1978. A supplementary adenoviral leader sequence and its role in messenger translation. Cell 15:511–26.

Eggerding, F., and H. Raskas. 1978. Effect of protein synthesis inhibitors on viral mRNAs synthesized early in adenovirus type 2 infection. J. Virol. 25:453–58.

Evans, R. M., N. Fraser, E. Ziff, J. Weber, M. Wilson, and J. E. Darnell. 1977. The initiation sites for RNA transcription in Ad2 DNA. Cell 12:733–40.

Everitt, E., L. Lutter, and L. Philipson. 1975. Structural proteins of adenoviruses. XII. Location and neighbor relationship among proteins of adenovirus type 2 as revealed by enzymatic iodination, immunoprecipitation, and chemical cross-linking. Virology 67:197–208.

Farber, M. S., and S. G. Baum. 1978. Transcription of adenovirus RNA in permissive and nonpermissive infections. J. Virol. 27:136–48.

Flint, S. J. 1977. Two "early" mRNA species in adenovirus type 2-transformed rat cells. J. Virol. 23:44–52.

Flint, S. J., P. H. Gallimore, and P. A. Sharp. 1975. Comparison of viral RNA sequences in adenovirus-2 transformed and lytically infected cells. J. Mol. Biol. 96:47–68.

Flint, S. J., and P. A. Sharp. 1976. Adenovirus transcription. V. Quantitation of viral RNA sequences in adenovirus 2-infected and transformed cells. J. Mol. Biol. 106:749–71.

Frost, E., and J. F. Williams. 1978. Mapping of temperature-sensitive and host-range mutants of adenovirus type 5 by marker rescue. Virology 91:39–50.

Fraser, N., and E. Ziff. 1978. RNA structures near poly(A) of adenovirus-2 late messenger RNAs. J. Mol. Biol. 124:27–51.

Galibert, F., J. Hérissé, and G. Courtois. Nucleotide sequence of the *Eco*RI-F fragment of adenovirus 2 genome. Gene 6:1–22.

Galos, R. S., J. Williams, M.-H. Binger, and S. J. Flint. 1979. Location of additional early gene sequences in the adenoviral chromosome. Cell 17:945–56.

Gallimore, P. H., P. A. Sharp, and J. Sambrook. 1974. Viral DNA in transformed cells. II. A study of the sequences of adenovirus 2 DNA in nine lines of transformed rat cells using specific fragments of the viral genome. J. Mol. Biol. 89:49–72.

Garon, D. F., K. W. Berry, J. C. Hierholzer, and J. A. Rose. 1973. Mapping of base sequence heterologies between genomes from different adenovirus serotypes. Virology 54:414–26.

Gelinas, R. E., L. T. Chow, R. J. Roberts, T. R. Broker, and D. F. Klessig. 1977. The structure of late adenovirus type 2 messenger RNAs. *In* C. W. Anderson (ed.), Genetic interaction and gene transfer, pp. 345–47. Brookhaven Symposia in Biology, Vol. 29.

Gelinas, R. E., and R. J. Roberts. 1977. One predominant 5'-undecanucleotide in adenovirus 2 late messenger RNAs. Cell 11:533–44.

Gilead, Z., Y.-H. Jeng, W. S. M. Wold, K. Sugawara, H. M. Rho, M. L. Harter, and M. Green. 1976. Immunological identification of two adenovirus-2 induced early proteins possibly involved in cell transformation. Nature 264:263–66.

Goldberg, S., J. Weber, and J. E. Darnell, Jr. 1977. The definition of a large viral transcription unit late in Ad2 infection of HeLa cells: mapping by effects of ultraviolet irradiation. Cell 10:617–21.

Goldenberg, C. J., and H. J. Raskas. 1979. Splicing patterns of nuclear precursors to the mRNA for adenovirus 2 DNA binding protein. Cell 16:131–38.

Graham, F. L., P. S. Abrahams, C. Mulder, H. L. Heijneker, S. O. Warnaar, F. A. J. deVries, W. Fiers, and A. J. van der Eb. 1974. Studies on *in vitro* transformation by DNA and DNA fragments of human adenoviruses and simian virus 40. Cold Spring Harbor Symp. Quant. Biol. 39:637–50.

Green, M., J. T. Parsons, M. Piña, K. Fujinaga, H. Caffier, and I. Landgraf-Leurs. 1970. Transcription of adenovirus genes in productively infected and in transformed cells. Cold Spring Harbor Symp. Quant. Biol. 35:803–18.

Green, M., W. S. M. Wold, K. H. Brackmann, and M. A. Cartas. 1979. Identification of families of overlapping polypeptides coded by early "transforming" gene region 1 of human adenovirus type 2. Virology 97:275–86.

Grodzicker, T., J. Williams, P. Sharp, and J. Sambrook. 1974. Physical mapping of temperature-sensitive mutations of adenoviruses. Cold Spring Harbor Symp. Quant. Biol. 39:439–46.

Harter, M. L., and J. B. Lewis. 1978. Adenovirus type 2 early proteins synthesized *in vitro* and *in vivo*: identification in infected cells of the 38,000- to 50,000-molecular-weight protein encoded by the left end of the adenovirus type 2 genome. J. Virol. 26:736–49.

Harter, M. L., G. Shanmugam, W. S. M. Wold, and M. Green. 1976. Detection of adenovirus type 2-induced early polypeptides using cycloheximide pretreatment to enhance viral protein synthesis. J. Virol. 19:232–42.

Horwitz, M. S., L. M. Kaplan, M. Abboud, J. Maritato, L. T. Chow, and T. R. Broker. 1978. Adenovirus DNA replication in soluble extracts of infected cell nuclei. Cold Spring Harbor Symp. Quant. Biol. 43:769–80.

Johansson, K., H. Persson, A. M. Lewis, U. Pettersson, C. Tibbetts, and L. Philipson. 1978. Viral DNA sequences and gene products in hamster cells transformed by adenovirus type 2. J. Virol. 27:628–39.

Kitchingman, G. R., S.-P. Lai, and H. Westphal. 1977. Loop structures in hybrids of early RNA and the separated strands of adenovirus DNA. Proc. Nat. Acad. Sci. USA 74:4392–95.

Kilpatrick, B. A., R. E. Gelinas, T. R. Broker, and L. T. Chow. 1979. Comparison of late mRNA splicing among class B and class C adenoviruses. J. Virol. 30:899–912.

Klessig, D. F., and C. W. Anderson. 1975. Block to multiplication of adenovirus serotype 2 in monkey cells. J. Virol. 16:1650–68.

Klessig, D. F. 1977a. Isolation of a variant of human adenovirus serotype 2 that multiplies efficiently on monkey cells. J. Virol. 21:1243–46.

Klessig, D. F. 1977b. Two adenovirus mRNAs have a common 5′ terminal leader sequence encoded at least 10 kb upstream from their main coding sequences. Cell 12:9–21.

Klessig, D. F., and T. Grodzicker. 1979. Mutations that allow human Ad2 and Ad5 to express late genes in monkey cells map in the viral gene encoding the 72K DNA binding protein. Cell 17:957–66.

Klessig, D. F., and L. T. Chow. 1980. Deficient accumulation and incomplete splicing of several late viral RNAs in monkey cells infected by human adenovirus type 2. J. Mol. Biol. 139:221–42.

Knapp, G., J. S. Beckmann, P. F. Johnson, S. A. Fuhrman, and J. Abelson. 1978. Transcription and processing of intervening sequences in yeast tRNA genes. Cell 14:221–36.

Kozak, M. 1978. How do eukaryotic ribosomes select initiation regions in messenger RNA? Cell 15:1109–23.

Lechner, R. L., and T. J. Kelly, Jr. 1977. The structure of replicating adenovirus 2 DNA molecules. Cell 12:1007–20.

Levinson, A. D., and A. J. Levine. 1977. The group C adenovirus tumor antigens: identification in infected and transformed cells and a peptide map analysis. Cell 11:871–79.

Lewis, J. B., C. W. Anderson, and J. F. Atkins. 1977. Further mapping of late adenovirus genes by cell-free translation of RNA selected by hybridization to specific DNA fragments. Cell 12:37–44.

Lewis, J. B., J. F. Atkins, C. W. Anderson, P. R. Baum, and R. F. Gesteland. 1975. Mapping of late adenovirus genes by cell-free translation of RNA selected by hybridization to specific DNA fragments. Proc. Nat. Acad. Sci. USA 72:1344–48.

Lewis, J. B., J. F. Atkins, P. R. Baum, R. Solem, R. F. Gesteland, and C. W. Anderson. 1976. Location and identification of the genes for adenovirus type 2 early polypeptides. Cell 7:141–51.

Lewis, J. B., H. Esche, J. E. Smart, B. Stillman, M. L. Harter, and M. B. Mathews. 1979. The organization and expression of the left third of the adenovirus genome. Cold Spring Harbor Symp. Quant. Biol. 44:493–504.

Maat, J., and H. van Ormondt. 1979. The nucleotide sequence of the transforming *Hind*III-G fragment of Ad5 DNA. The region between map positions 4.5 (*Hpa*I site) and 8.0 (*Hind*III site). Gene 6:75–90.

Mathews, M. B. 1975. Genes for VA-RNA in adenovirus 2. Cell 6:223–30.

McGrogan, M., and H. J. Raskas. 1978. Two regions of the adenovirus-2 genome specify families of late polysomal RNAs containing common sequences. Proc. Nat. Acad. Sci. USA 75:625–29.

Meyer, J., P. D. Neuwald, S.-P. Lai, J. V. Maizel, Jr., and H. Westphal. 1977. Electron microscopy of late adenovirus type 2 mRNA hybridized to double-stranded viral DNA. J. Virol. 21:1010–18.

Miller, J. S., R. P. Ricciardi, B. E. Roberts, B. M. Paterson, and M. B. Mathews. The arrangements of messenger RNAs and protein coding regions in the major late transcription unit of adenovirus-2. J. Mol. Biol. (in press).

Moss, B., and F. Koczot. 1976. Sequence of methylated nucleotides at the 5' terminus of adenovirus-specific RNA. J. Virol. 17:385–92.

Neuwald, P. D., J. Meyer, J. V. Maizel, Jr., and H. Westphal. 1977. Early gene expression of adenovirus type 2: R-loop mapping of mRNA and time course of viral DNA, mRNA, and protein synthesis. J. Virol. 21:1019–30.

Nevins, J. R., and J. E. Darnell. 1978a. Groups of adenovirus type 2 mRNAs derived from a large primary transcript: probable nuclear origin and possible common 3' ends. J. Virol. 25:811–23.

Nevins, J. R., and J. E. Darnell, Jr. 1978b. Steps in the processing of Ad2 mRNA: poly(A)$^+$ nuclear sequences are conserved and poly(A) addition precedes splicing. Cell 15:1477–93.

O'Farrell, P. Z., B. Cordell, P. Valenzuela, W. J. Rutter, and H. M. Goodman, 1978. Structure and processing of yeast precursor tRNAs containing intervening sequences. Nature 274:438–45.

Paterson, B. M., B. E. Roberts, and E. L. Kuff. 1977. Structural gene identification and mapping by DNA-mRNA hybrid-arrested cell-free translation. Proc. Nat. Acad. Sci. USA 74:4370–74.

Peebles, C. L., R. C. Ogden, G. Knapp, and J. Abelson. 1979. Splicing of yeast tRNA precursors: a two-stage reaction. Cell 18:27–35.

Perricaudet, M., G. Akusjärvi, A. Virtanen, and U. Pettersson. 1979. Structure of two spliced mRNAs from the transforming region of human subgroup C adenoviruses. Nature 281:694–96.

Perry, R. P., and D. E. Kelley. 1976. Kinetics of formation of 5' terminal caps in mRNA. Cell 8:433–42.

Persson, H., B. Öberg, and L. Philipson. 1978. Purification and characterization of an early protein (E14K) from adenovirus type 2-infected cells. J. Virol. 28:119–39.

Persson, H., B. Mathiesen, L. Philipson, and U. Pettersson. 1979a. A maturation protein in adenovirus morphogenesis. Virology 93:198–208.

Persson, H., C. Signäs, and L. Philipson. 1979b. Purification and characterization of an early glycoprotein from adenovirus type 2-infected cells. J. Virol. 29:938–48.

Peterlin, B. M., M. Sullivan, H. Westphal, and J. V. Maizel, Jr. 1978. Secondary structure map of psoralen-crosslinked adenovirus DNA studied by electron microscopy. Virology 86:391–97.

Pettersson, U., and M. B. Mathews. 1977. The gene and messenger RNA for adenovirus polypeptide IX. Cell 12:741–50.

Pettersson, U., C. Tibbetts, and L. Philipson. 1976. Hybridization maps of early and late messenger RNA sequences on the adenovirus type 2 genome. J. Mol. Biol. 101:479–501.

Philipson, L., R. Wall, G. Glickman, and J. E. Darnell. 1971. Addition of polyadenylate sequences to virus-specific RNA during adenovirus replication. Proc. Nat. Acad. Sci. USA 68:2806–9.

Puckett, L., and J. E. Darnell. 1977. Essential factors in the kinetic analysis of RNA synthesis in HeLa cells. J. Cell Physiol. 90:521–34.

Rabson, A. S., G. T. O'Conor, L. K. Berezesky, and F. J. Paul. 1964. Enhancement of adenovirus growth in African green monkey kidney cell cultures by SV40. Proc. Soc. Exp. Biol. Med. 116:187–90.

Ross, S., and A. J. Levine. 1979. The genomic map position of the adenovirus type 2 glycoprotein. Virology 99:427–30.

Russell, W. C., and J. J. Skehel. 1972. The polypeptides of adenovirus-infected cells. J. Gen. Virol. 15:45–57.

Saborio, J. L., and B. Öberg. 1976. *In vivo* and *in vitro* synthesis of adenovirus type 2 early proteins. J. Virol. 17:865–75.

Sehgal, P., N. W. Fraser, and J. E. Darnell, Jr. 1979. Early Ad2 transcription units: only promoter-proximal RNA continues to be made in the presence of DRB. Virology 94:185–91.

Seif, I., G. Khoury, and R. Dhar. 1979. BKV splice sequences based on analysis of preferred donor and acceptor sites. Nucl. Acids Res. 6:3387–98.

Sharp, P. A., P. H. Gallimore, and S. J. Flint. 1974. Mapping of adenovirus 2 RNA sequences in lytically infected cells and transformed cell lines. Cold Spring Harbor Symp. Quant. Biol. 39:457–74.

Shatkin, A. 1976. Capping of eukaryotic mRNAs. Cell 9:645–53.

Söderlund, H., U. Pettersson, B. Vennström, L. Philipson, and M. B. Mathews. 1976. A new species of virus-coded low molecular weight RNA from cells infected with adenovirus type 2. Cell 7:585–93.

Sommer, S., M. Salditt-Georgieff, S. Bachenheimer, J. E. Darnell, Y. Furuichi, M. Morgan, and A. J. Shatkin. 1976. The methylation of adenovirus-specific nuclear and cytoplasmic RNA. Nucl. Acids Res. 3:749–65.

Spector, D. J., D. N. Halbert, L. D. Crossland, and H. J. Raskas. 1979. Expression of genes from the transforming region of adenovirus. Cold Spring Harbor Symp. Quant. Biol. 44: 437–45.

Spector, D. J., M. McGrogan, and H. J. Raskas. 1978. Regulation of the appearance of cytoplasmic RNAs from region 1 of the adenovirus 2 genome. J. Mol. Biol. 126:395–414.

Tal, J., E. A. Craig, and H. J. Raskas. 1975. Sequence relationships between adenovirus-2 early RNA and viral RNA size classes synthesized at 18 hours after infection. J. Virol. 15:137–44.

Thomas, M., R. L. White, and R. W. Davis. 1976. Hybridization of RNA to double-stranded DNA: formation of R-loops. Proc. Nat. Acad. Sci. USA 73:2294–98.

Valentine, R. C., and H. G. Pereira. 1965. Antigens and structure of the adenovirus. J. Mol. Biol. 13:13–20.

van der Eb, A. J., H. van Ormondt, P. I. Schrier, J. H. Lupker, H. Jochemsen, P. J. van den Elsen, R. J. DeLeys, J. Maat, C. P. van Beveren, R. Dijkema, and A. deWaard. 1979. Structure and function of the transforming genes of human adenoviruses and SV40. Cold Spring Harbor Symp. Quant. Biol. 44:383–99.

van der Vliet, P. C., and A. J. Levine. 1973. DNA binding proteins specific for cells infected by adenovirus. Nature New Biol. 246:170–74.

van der Vliet, P. C., A. J. Levine, M. J. Ensinger, and H. S. Ginsberg. 1975. Thermolabile DNA binding proteins from cells infected with a temperature-sensitive mutant of adenovirus defective in viral DNA synthesis. J. Virol. 15:348–54.

van Ormondt, H., J. Maat, A. deWaard, and A. J. van der Eb. 1978. The nucleotide sequence of the region of the transforming *Hpa*I/E fragment of Ad5 DNA. Gene 4:309–28.

Wall, R., L. Philipson, and J. E. Darnell. 1972. Processing of adenovirus-specific nuclear RNA during virus replication. Virology 50:27–34.

Walter, G., and J. V. Maizel, Jr. 1974. The polypeptides of adenovirus. IV. Detection of early and late virus-induced polypeptides and their distribution in subcellular fractions. Virology 57:402–8.

Weber, J., W. Jelinek, and J. E. Darnell, Jr. 1977. The definition of a large viral transcription unit late in Ad2 infection of HeLa cells: mapping of nascent RNA molecules labeled in isolated nuclei. Cell 10:611–16.

Westphal, H. and S.-P. Lai. 1977. Quantitative electron microscopy of early adenovirus RNA. J. Mol. Biol. 116:525–48.

White, R. L., and D. S. Hogness. 1977. R loop mapping of the 18S and 28S sequences in the long and short repeating units of *Drosophila melanogaster* rDNA. Cell 10:177–92.

Williams, J. F., C. S. H. Young, and P. E. Austin. 1974. Genetic analysis of human adenovirus type 5 in permissive and nonpermissive cells. Cold Spring Harbor Symp. Quant. Biol. 39: 427–37.

Wilson, M. C., S. G. Sawicki, M. Salditt-Georgieff, and J. E. Darnell. 1978. Adenovirus type 2 mRNA in transformed cells: map positions and difference in transport time. J. Virol. 25:97–103.

Wilson, M. C., N. W. Fraser, and J. E. Darnell. 1979a. Mapping of RNA initiation sites by high doses of UV irradiation: evidence for three independent promoters within the left 11% of the Ad-2 genome. Virology 94:175–84.

Wilson, M. C., J. R. Nevins, J.-M. Blanchard, and J. E. Darnell, Jr. 1979b. The metabolism of mRNA from the transforming region of adenovirus type 2. Cold Spring Harbor Symp. Quant. Biol. 44:447–55.

Wu, M., R. J. Roberts, and N. Davidson. 1977. Structure of the inverted terminal repetition of adenovirus type 2. J. Virol. 21:766–77.

Zain, B. S., and R. J. Roberts. 1979. Sequences from the beginning of the fiber messenger RNA of adenovirus-2. J. Mol. Biol. 131:341–52.

Zain, S., J. Sambrook, R. J. Roberts, W. Keller, M. Fried, and A. R. Dunn. 1979. Nucleotide sequence analysis of the leader segments in a cloned copy of adenovirus 2 fiber mRNA. Cell 16:851–61.

Ziff, E. B., and R. M. Evans. 1978. Coincidence of the promoter and capped 5' terminus of RNA from the adenovirus 2 major late transcription unit. Cell 15:1463–75.

PHILIP S. PERLMAN, NANCY J. ALEXANDER,
DEBORAH K. HANSON, AND HENRY R. MAHLER

Mosaic Genes in Yeast Mitochondria

10

The number of investigators studying the genetics and biogenesis of mitochondria, especially in yeast and other lower eukaryotes, is large and growing. Increasingly the mitochondrial system is seen as an excellent model for probing fundamental aspects of gene structure and regulation. However, many of the early observations were complex and confusing, and they failed to yield simple interpretations with significance and interest for the general scientific community. Within the last year it has become clear that many of the confusing aspects of yeast mitochondrial genetics result from the unusual structure of some mitochondrial genes. A substantial fraction of the mitochondrial genome of *Saccharomyces cerevisiae* consists of eukaryote-like mosaic genes, in which the base sequences (exons) that code for a given protein are interrupted by intervening sequences (introns). Moreover, the number of introns in some genes is not always the same in different laboratory stocks; for example, the gene coding for the large rRNA contains a large intron called ω, which is present in some stocks (ω^+) but not in others (ω^-). The ω intron appears to be a focus for one-way gene conversion of genetic markers, which complicated the analysis of crosses between ω^+ and ω^- strains. The confusing diversity of gene *structure* among yeast strains is in marked contrast to the absolute uniformity of gene *order* on maps of the mitochondrial DNA molecule.

The powerful advantage yeast offers the investigator of gene structure and function is the ready availability of specific mutants in any portion of the mitochondrial genome, including intron mutants that are not available for mosaic genes in other systems. Furthermore, the ease with which

Philip S. Perlman, Department of Genetics, Ohio State University, Columbus, Ohio 43210 mutants can be manipulated genetically permits detailed studies of the correlation of structure and function. In this review we will show how these tools are being applied to the study of the mosaic *cob-box* gene, which codes for the apoprotein of cytochrome *b*.

Saccharomyces cerevisiae is a facultative anaerobe: it will grow on glucose without benefit of mitochondrial function, whereas its growth on nonfermentable carbon sources (e.g., ethanol or glycerol) is strictly dependent on mitochondrial respiration. This property allows the isolation and propagation of *petite* and *mit⁻* mutants that are deficient in mitochondrial function. The *petite* mutation results from extensive deletions of the mitochondrial genome (ρ^-), including genes coding for rRNA or tRNA, or from the complete loss of mitochondrial DNA (ρ^0). The *mit⁻* mutations are point mutations or relatively small deletions in genes essential for mitochondrial function. *Mit⁻* mutants can be differentiated from *petites* by a variety of genetic and biochemical tests including (1) presence of mitochondrial protein synthesis; (2) ability to revert to respiratory competence; and (3) the *petite* restoration test. The latter refers to the production of respiratory-sufficient progeny when a *mit⁻* mutant is crossed to a specific *petite* retaining the segment of the mitochondrial genome harboring the wild-type DNA sequence corresponding to the mutant sequence in the *mit⁻* mutant. Given a *petite* mutant well characterized either genetically or physically by restriction endonuclease analysis, an unlimited number of *mit⁻* mutants can be isolated, all mapping in that portion of the genome retained by the *petite*.

The circular mitochondrial DNA molecule encodes only a small number of gene products that can be classified into two functional groups. The first is composed of the protein products that form parts of the respiratory chain including three subunits of cytochrome oxidase, the apoprotein of cytochrome *b*, and at least two and possibly four subunits of the oligomycin-sensitive ATPase. The other class is composed of gene products involved in translation: the large (21S) and small (14S) rRNAs; at least 25 tRNAs representing acceptor species for all 20 amino acids; and one protein constituent (var1) of the small mitoribosomal subunit. If more products remain to be discovered, they are most likely minor species, which nevertheless could have important regulatory functions.

The genome size of yeast mitochondria is, however, much larger than expected on the basis of the number of genes identified. With a genome of 25 μm, five times as large as that of animal mitochondria, about 50 average-sized proteins could be encoded. The discrepancy between the potential coding capacity and that actually present rests partly on the unusual base composition, which is 82% A+T (Carnevali et al., 1969). Early studies of nearest neighbor frequencies showed that the G+C present was more clustered than predicted by chance and that approximately 50% of the genome was comprised of stretches of nearly pure A+T with a mean length of 300 bp (Bernardi et al., 1972). Subsequently, DNA sequencing of selected portions of the genome revealed stretches of A+T greater than 2,000 bp in length interspersed only infrequently by G+C (Tzagoloff et al., 1979). A small fraction of the genome consists of short stretches (less than 50 bp) of nearly pure G+C, whose function is unknown; all those observed to date are located within the long A+T stretches. This genome organization has led to the hypothesis that the long runs of nearly pure A+T are spacers between genes relatively rich in G+C (Prunell and Bernardi, 1977).

In the last few years, significant and unexpected insights into mitochondrial genome organization have come from comparisons of restriction endonuclease maps of mtDNA from different wild-type yeast strains (Sanders et al., 1977; Morimoto and Rabinowitz, 1979). A map of strain ID41-6/161 used in our laboratory is shown in figure 1 (see also Vincent et al., 1980). The strains may differ in the location of given restriction sites; these differences largely result from deletions and insertions of fewer than 1,000 bp scattered throughout the genome. The linear order of genes, however, does not appear to vary. In addition to these small deletions and insertions, there are significant differences in the total molecular weights of the mtDNAs of some strains. These are largely explained by three DNA segments, 1,000–3,000 bp long, that are present or absent in specific regions of the genome. One of these is the ω^+ intron in the 21S rRNA gene mentioned above; another is an insertion in the *cob-box* gene; and the third is in or near the *oxi3* gene that codes for coxI, a subunit of cytochrome oxidase. In contrast to every other mitochondrial genome analyzed, the structural genes for the two rRNAs of yeast are not contiguous; they are separated by nearly half the genome, and a number of structural genes have been identified in the region between them. As we will describe below, a combination of physical and genetic analysis has shown at least two genes in the yeast mitochondrion are mosaic, containing introns.

THE 21S rRNA GENE

The first clear-cut example of a gene on yeast mtDNA with a mosaic structure was provided by the gene coding for the large (21S) rRNA species (Bos et al., 1978). This gene is located between the *var1* locus and a cluster of tRNA genes and includes the unique *Sal*I site that is used to orient the physical and genetic map of yeast mtDNA (figures 1 and 2). As is the case

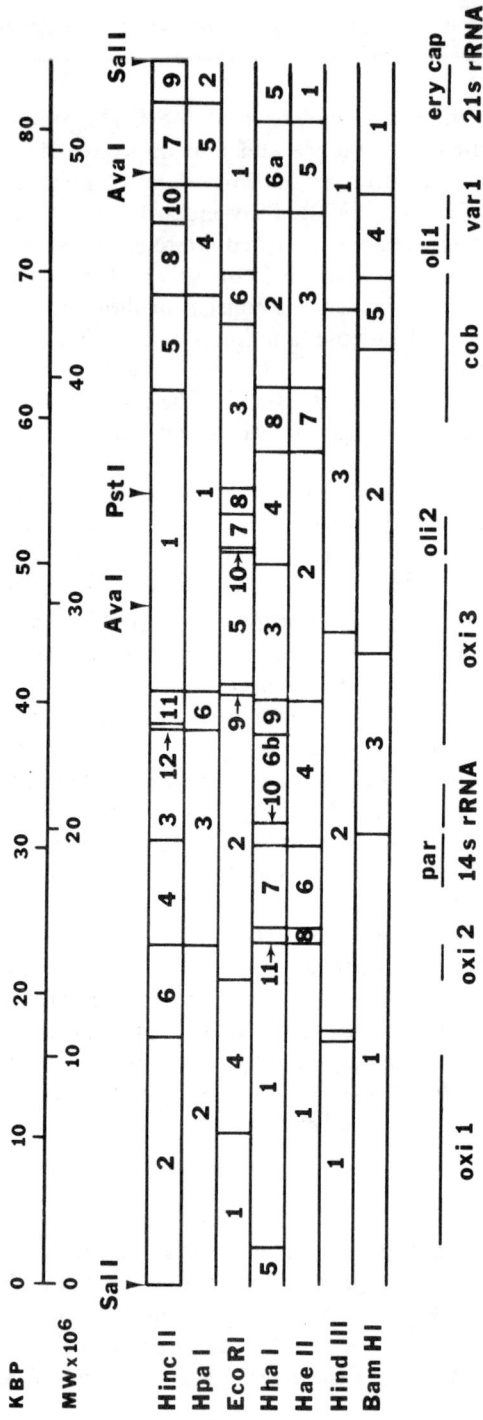

Fig. 1. Restriction fragment map of mtDNA from mit⁺ ID41-6/161. mtDNA was digested with restriction endonucleases (*Hinc*II, *Hha*I, *Eco*RI, *Bam*HI, *Ava*I, *Sal*I, *Pst*I, *Hind*III, *Hpa*I, and *Hae*II) singly and in various combinations. The location of genes is based upon information from the restriction analysis of *petite* DNA. (Information for this map supplied by R. D. Vincent)

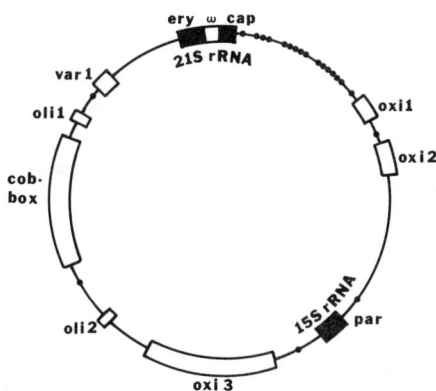

Fig. 2.   Map of the mitochondrial genome. Loci for rRNA subunits (shaded boxes), mitochondrial membrane component subunits, var1 protein, antibiotic resistance, and tRNAs (•) are arbitrarily spaced (approximate location) on the circular genome based on results from recombination and petite deletion mapping studies.

for mitochondrial genomes from all other species studied to date, there is a single copy of the 21S rRNA gene per genome and all genomes in each cell are the same.

Several lines of evidence show that the 21S rRNA gene in mtDNA from some yeast strains contains an intervening sequence of roughly 1,100 base pairs. Initially, the 21S rRNA gene was physically mapped largely within a *Hae*III fragment of roughly 4,500 BP (figure 3) (Jacq et al., 1977). Once a detailed restriction site map of that fragment became available (Heyting and Menke, 1979; Faye et al., 1979), labeled 21S rRNA was found to hybridize to restriction fragments on both sides of the *Hind*III fragment contained within the larger *Hae*III fragment. No labeled rRNA hybridized to the 600 bp *Hind*III fragment, suggesting that this fragment was part of an intervening sequence within the 21S rRNA gene. Further support for this hypothesis was provided by Bos and colleagues (1978), who observed by electron microscopy loops in heteroduplexes between the 21S rRNA and DNA fragments containing the gene for the 21S rRNA.

Similar studies of rRNA genes in mtDNA from *Neurospora crassa* (see Manella et al., 1979) and cpDNA from *Chlamydomonas reinhardtii* (Rochaix and Malnoe, 1978) have also identified a single long insertion within the large rRNA gene. As in yeast, each insertion is located near the 3' end of the rRNA gene. Intervening sequences in chromosomal rRNA genes have also been reported in several organisms (e.g., Din et al., 1979; Glover and Hogness, 1977).

Although it is known that the physical maps and total lengths of mitochondrial genomes from a number of yeast strains vary considerably

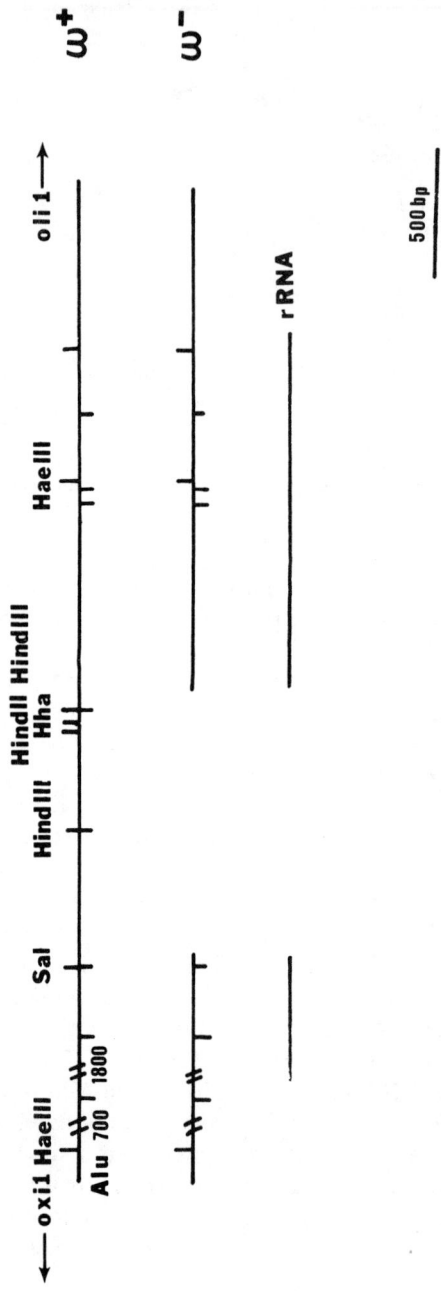

Fig. 3. Fine-structure of the 21s ribosomal region of the mtDNA, 21s RNA hybridizes on either side of the *HindIII* fragment contained within the larger *HaeIII* fragment (Bos et al., 1978; Faye et al., 1979). No rRNA hybridizes to the *HindIII* fragment in ω⁺ strains, suggesting that this fragment is an intervening sequence.

(Sanders et al., 1977; Morimoto and Rabinowitz, 1979), the order of their genes is highly conserved. The main differences in physical maps are due to the presence or absence of continuous base sequences (50-3,000 bp in length) located in several discrete positions on the genome. As described above for the 21S rRNA gene, in some yeast strains an intervening sequence of 1,100 bp has been identified between the genetic markers *cap^r* and *ery^r*. Other yeast strains are missing all (or virtually all) of this 21S rRNA intron (fig. 3). Therefore, this intron is not likely to provide an essential mitochondrial function or participate in any essential regulatory circuits since its absence does not result in a defect in cellular functions. However, it is very likely that some base sequences within the intron participate in the RNA splicing event(s) required to form the intact rRNA. It is conceivable that the intron codes for a protein that participates in the processing events; however, if that is the case, then the protein cannot be essential for splicing any other gene transcript.

As will be noted later, at least one other gene (the *cob-box* gene) on yeast mtDNA exists in alternate forms where at least one intron present in the gene in one strain is absent from the homologous gene in another strain. To some extent this unusual feature of the yeast system should simplify the task of locating splice points. It is clear that the introns are very stable during vegetative growth of cells, but we do not know the mode of origin of the alternate forms of these several genes. It is likely that laboratory stocks of *S. cerevisiae* have mitochondrial and nuclear genes derived from a variety of interfertile species of *Saccharomyces* and that the alternate forms originated during speciation and were first reassorted in the early days of yeast genetics. Unfortunately little is known about the origin of those species so that we do not know which form of a given gene is more ancient.

Some aspects of the 21S rRNA intron in yeast mtDNA should be mentioned here. Some years ago, an unusually efficient case of polar recombination (gene conversion) was documented involving genetic markers affecting the sensitivity to the antibiotics chloramphenicol, erythromycin, and spiramycin, all now known to be located in the 21S rRNA gene. It was postulated that an interaction between alleles of a nearby genetic locus termed ω is responsible for the distortion in transmission of markers (reviewed in Dujon et al., 1974; Perlman and Birky, 1974; Gillham, 1978). The two original allelic forms were called ω^+ and ω^-, and in crosses between ω^+ and ω^- strains, ω^- alleles were shown to be converted to ω^+ alleles. It was also noted that nearby drug markers in the ω^- strain are converted to the allele of the ω^+ parent with a gradient of efficiency related to the distance of the marker from ω. It is now clear that ω^+ strains are those having the ribosomal intron and ω^- strains are those lacking it. Although it is not known whether the base sequences necessary for splicing are the same as those involved in gene conversion, it appears that the ω^- genomes converted to ω^+ acquire the entire insert (Heyting and Menke, 1979), and no shorter or longer forms of ω have been observed in wild type strains.

NOT ALL GENES ON YEAST mtDNA ARE MOSAIC

Although it is clear that the mitochondrial gene coding for the 21S rRNA in yeast contains an intron in ω^+ strains, some other genes on yeast mtDNA are not interrupted. Based on direct nucleotide and protein sequence determinations, the gene coding for the proteolipid (DCCD-binding protein) of the oligomycin-sensitive ATPase (the *oli1/pho2* gene) has been found to be physically colinear with its protein product (Tzagoloff et al., 1979; Hensgens et al., 1979) and so contains no intervening sequences. Although subunit II of cytochrome *c* oxidase from yeast has not been sequenced, preliminary evidence suggests that the structural gene coding for this protein (*oxi2* gene) also is not split (T. Fox, personal communication). Several groups have reported the nucleotide sequence for a number of mitochondrial tRNA genes (Tzagoloff et al., 1979; Grivell et al., 1979); and although most of these genes have some unusual properties, none was found to be split. This result is especially intriguing since some tRNA genes in the yeast nucleus do contain short intervening sequences.

Analysis of mitochondrial genes by DNA sequencing, RNA/DNA hybridization, and protein sequencing has provided definitive information concerning the structure of several genes. Despite the lack of such detailed information for most mitochondrial genes, three others appear to be mosaic, based on genetic and physical mapping data. That conclusion is still tentative for the *oxi3* (Carignani et al., 1979) and *var1* (Vincent et al., 1980) genes but is almost certain for the *cob-box* gene. The remainder of this paper will discuss the evidence that supports the hypothesis that the *cob-box* gene of yeast mtDNA has a mosaic structure.

TYPES OF MITOCHONDRIAL MUTANTS

The basic approach to the identification and structural analysis of genes on mitochondrial DNA has been a rather traditional one in which many independently isolated mutants were analyzed genetically and biochemically. Four main classes of mutants have been identified and used extensively, especially to study the *cob-box* gene.

Antibiotic-Resistant Mutants

Until recentlyc mutants resistant to a number of antibiotics that inhibit the growth of wild-type (antibiotic-sensitive) yeast cells on nonfermentable carbon sources have been a main focus of mitochondrial genetics (reviewed in Gillham, 1978). There are mutants resistant to the mitochondrial protein synthesis inhibitors chloramphĕnicol, erythromycin, spiramycin, and paromomycin; others are resistant to oligomycin and venturicidin, which inhibit ATP synthetase in sensitive strains.

Recently a number of mutants have been found that are resistant to the inhibitors of complex III of the mitochondrial respiratory chain: antimycin A, diuron, mucidin, and funiculosin (Michaelis and Pratje, 1977; Subik et al., 1977: Burger et al., 1976; Colson et al., 1977; Colson and Slonimski, 1979). In general, a mutant selected for resistance to one of these drugs is not cross-resistant to the others. These mutants represent a number of distinct alleles all mapping within a single, but fairly large, portion of the mitochondrial genome. In the earliest reports (e.g., Burger et al., 1976; Michaelis and Pratje, 1977), mutants resistant to antimycin A (*ana*[r] mutants) were genetically mapped into two physically distinct clusters, and thus the literature reflects the possibility that there were two (or more) genes affecting cytochrome *b* function. As additional drugs were used as probes of cytochrome *b* function, clusters of mutant sites were found to be distributed quite widely within the *cob-box* region, necessitating further subdivision into more than two loci (Colson and Slonimski, 1979). As is the case with many drug-resistant mutants in a variety of systems, these are most likely missense mutations, located in a region of the DNA coding for a protein, that result in altered drug-binding properties of the target enzyme (Johnson, this volume, describes a significant exception to this mechanism of drug resistance). The large physical separation in the map position of such drug-resistant mutants all affecting cytochrome *b* functions suggests that the *cob-box* region contains several genes or one mosaic gene.

Mit⁻ Mutants

In 1975 Tzagoloff and colleagues first characterized a class of mitochondrially inherited respiration-deficient mutants termed *mit*⁻ mutants (Slonimski and Tzagoloff, 1976). These differ from the previously available cytoplasmic *petite* mutants (see below) in several important aspects: (1) most *mit*⁻ mutants eliminate only a portion of the respiratory chain; (2) all retain mitochondrial protein synthesis; and (3) most can revert to respiratory sufficiency. *Mit⁻* mutants are most usefully thought of as typical loss-of-function mutants although, as we will note later, some of them eliminate respiratory function by affecting regulatory rather than purely functional sites.

One reason for some of the confusion in the scientific community concerning the nature of mitochondrial genes is that several different nomenclatures have been used to categorize the mutants used to define genes on mtDNA. Originally, mutants affecting cytochrome *b*-linked activities were called *cob* mutants to distinguish them from *oxi* mutants, which lack cytochrome *c* oxidase (Slonimski and Tzagoloff, 1976) or from *pho* mutants, which have a reduced phosphorylation activity (oligomycin-sensitive ATPase) (Foury and Tzagoloff, 1976a). However, Linnane's group has used the terms *cyb* and *cya* in place of *cob* and *oxi*, respectively (Cobon et al., 1976).

When the total number of available *cob* (*cyb*) mutants was small, the region containing them was divided into two segments, *cob1* and *cob2*, because all mutants appeared to map in one of these two easily separable clusters (Tzagoloff et al., 1976). As more cytochrome *b*-deficient mutants were analyzed biochemically, it became clear that some mutants clearly mapping in the *cob* region were simultaneously deficient in cytochromes *b* and *aa₃* (Pajot et al., 1976; Cobon et al., 1976; Claisse et al., 1978; Alexander et al., 1979). This more complex situation has been termed the "*box* phenotype" (cytochrome *b*- and *oxi*dase-deficient) and such mutants were called *box* mutants. Since *cob* and *box* (and drug-resistant) mutants are interspersed on the genetic map (Slonimski et al., 1978b), a number of laboratories have now agreed to call the region *cob-box* (Slonimski et al., in preparation). As will be described later, the region has now been divided into nine segments and the *cob1/cob2* nomenclature is no longer used by us.

Conditional Mutants

Mutants have also been isolated that are capable of growth on glycerol (respiratory-sufficient) at one temperature but not at another (Lancashire, 1976; Bolotin-Fukuhara, 1977). Both heat- and cold-sensitive mutants have been isolated, and a number have been mapped in the *cob-box* region (Haid et al., 1979). Most of the conditional *cob-box* mutants grow at the permissive temperature but fail to make a functional product when shifted to the restrictive temperature even though they are capable of undergoing a limited number of cell generations.

Deletion Mutants

Two types of deletion mutants have been useful in the characterization of mitochondrial genes: cytoplasmic *petite* (ρ^-) and deletion *mit⁻* mutants. Cytoplasmic *petite* mutants were first reported in 1949 (Ephrussi et al., 1949), and have been extensively used since then to study mitochondrial inheritance and biogenesis. In the last five years, their great value as tools to study the mitochondrial genome has been more fully recognized. *Petites* are a frequently obtained class of respiratory-deficient mutants that are characterized by being deficient in oligomycin-sensitive ATP synthetase, cytochromes aa_3 and b and the related enzyme activities, and are also deficient in mitochondrial protein synthesis (reviewed in Gillham, 1978). They are the result of large deletions in the mitochondrial genome and thus do not revert to respiratory sufficiency. Some are deleted for the entire mitochondrial genome (ρ^0 mutants), and others have mtDNA (ρ^- mutants) retaining as little as 70 bp, or as much as roughly 50% of the wild-type genome (see Locker et al., 1979). Mutants retaining roughly 10% of the wild-type genome are quite common, and it is possible to obtain a library of such mutants that in its aggregate represents the entire genetic map. Such mutants have been very useful for ordering point mutations on this map. Although the mtDNA of some *petite* mutants may contain inversions, duplications, or internal deletions (Locker et al., 1974; Heyting et al., 1979; Lewin et al., 1978), many other mutants have been shown to retain a single continuous segment of the wild-type map (reiterated in a head-to-tail or head-to-head fashion). These "simple *petites*" have been very useful both for constructing the complete physical map of the wild-type mtDNA and in obtaining a detailed physical map of selected regions of the genome (Sanders et al., 1977; Slonimski et al., 1978b; Morimoto and Rabinowitz, 1979; Alexander et al., 1979; Vincent et al., 1980). Examples of how *petite* mutants can be used to localize specific mutants sites on the physical map will be presented in later sections.

Several groups have noted that some *mit⁻* mutants never revert to *mit⁺*, and at least some of those mutants have been shown to be deleted for a portion of the mitochondrial genome. One such mutant is deleted for a single continuous segment 3,300 bp long (Alexander et al., 1979). Another type of *mit⁻* mutant, a missense mutant, has helped to identify the location of the *oxi1* gene (Fox, 1979).

STRUCTURE OF THE COB-BOX GENE

Three separate groups have studied the structure of the *cob-box* region using separate collections of mutants, and all have independently reached the conclusion that this region contains a single gene coding for the primary sequence of the apoprotein of cytochrome *b*, a protein of roughly 30,000 daltons (p30) (Slonimski et al., 1978a; Haid et al., 1979; Alexander et al., 1979). Furthermore, each group has concluded that the *cob-box* region consists of approximately 8,000 bp and has a mosaic structure consisting of coding and noncoding sequences.

Figure 4 is a schematic diagram of the current model of the *cob-box* gene. This model is a result of the collaboration among the three investigative groups. All of the mutants from the three laboratories were compared genetically and biochemically, and though the results of the collaboration will be presented elsewhere (Slonimski et al., in preparation), this paper will describe the model and summarize how the model was deduced, emphasizing the data from the Columbus-Bloomington group.

The *cob-box* region has been located on the mitochondrial genome between the *oli1* (*pho2*) and *oli2* (*pho1*) genes (figs. 1 and 2). It consists of

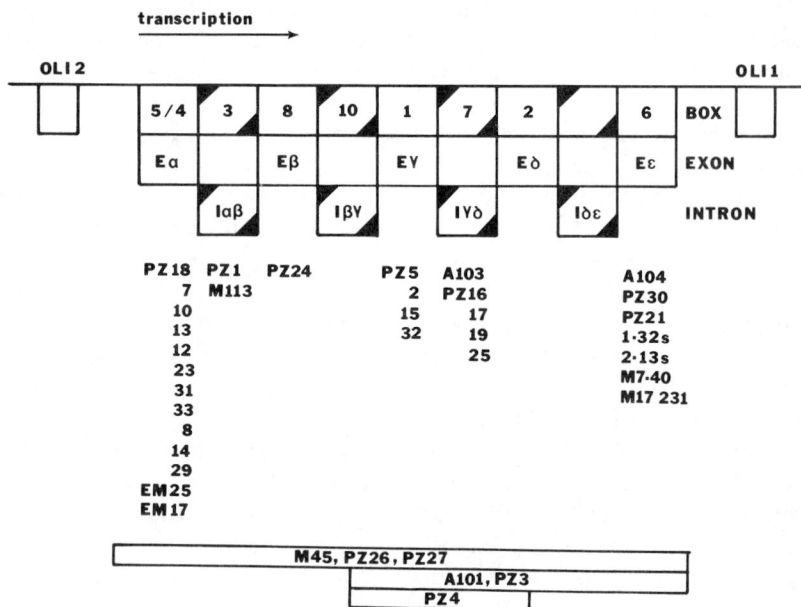

transcription →

OLI 2										OLI 1
	5/4	3	8	10	1	7	2		6	BOX
	Eα		Eβ		Eγ		Eδ		Eε	EXON
		Iαβ		Iβγ		Iγδ		Iδε		INTRON

PZ18	PZ1	PZ24		PZ5	A103		A104
7	M113			2	PZ16		PZ30
10				15	17		PZ21
13				32	19		1·32s
12					25		2·13s
23							M7·40
31							M17 231
33							
8							
14							
29							
EM25							
EM17							

M45, PZ26, PZ27
A101, PZ3
PZ4

Fig. 4.   Schematic diagram of the cob-box region. The *cob-box* region is located between the markers *oli1* and *oli2* and has been subdivided into 4 introns (spacers) and 5 exons (coding for cytochrome *b*) (see text for further discussion). *Mit⁻* mutants used in our laboratory are listed below the appropriate exon or intron to which they have been genetically mapped. Deletion *mit⁻* mutants M45, PZ26, PZ27, A101, PZ3, and PZ4 are drawn to indicate the region of deletion.

five exons and four introns, though it is likely that some portions of the gene (e.g., promoter, mRNA leader, etc.) remain to be identified at the *oli2* proximal end of the gene. It is transcribed from the *oli2* proximal end toward the *oli1* proximal end.

The nomenclature used to identify the exons and introns was first introduced by Slonimski and coworkers (Kotylak and Slonimski, 1976), and the mutants were assigned "*box*" loci (fig. 4, top). The *box* loci were assigned numbers in the order of their identification, and therefore the numbers do not correspond to their actual map order. In fact, several *box* loci were defined only recently, with the result that several versions of the map of *box* loci have appeared in the literature. For greater clarity a new, and we hope, final, nomenclature has been adopted (Slonimski et al., in preparation) and is also shown in figure 4. Each gene segment shown to be an exon was assigned a Greek letter ($\alpha \rightarrow \epsilon$) where α is the first exon, β the second, and so on. The four intron segments are named according to the exons they separate; for example, $I\alpha\beta$ separates exon α from exon β. We note here that $I\delta\epsilon$ does not appear to coincide with any previously described *box* locus, and its inclusion here is based chiefly on the results of Grivell and colleagues (1979) (see below: "Transcripts of the Cob-Box Region"), and the finding that exon δ and exon ϵ are not physically or genetically contiguous.

PHENOTYPIC ANALYSIS OF COB-BOX MUTANTS

All the *cob-box* mutants are deficient in cytochrome *b*-linked enzyme activities, and nearly all of them lack the cytochrome *b* absorption band. Some mutants are also deficient in cytochrome *c* oxidase activity and cytochrome aa_3; this is an unexpected phenotypic property shared by nearly all the intron mutants (Claisse et al., 1978; Haid et al., 1979). This novel intron phenotype, however, plays little role in the characterization of the structure of the *cob-box* gene, and so this interesting result will be discussed in detail only at the end of this paper.

Since all *cob-box* mutants retain mitochondrial protein synthesis, useful information has been obtained from the analysis of proteins made on their mitochondrial ribosomes (Mahler et al., 1978; Claisse et al., 1978; Alexander et al., 1979; Haid et al., 1979). Most groups use a version of the technique of Douglas and Butow (1975) (see also Douglas et al., 1979) to visualize those products of mitochondrial protein synthesis. Briefly, cells are labeled with $^{35}SO_4$ in the presence of cycloheximide so that virtually all of the label is found in proteins synthesized in the mitochondria. The mitochondria are then isolated, solubilized with SDS, and the proteins

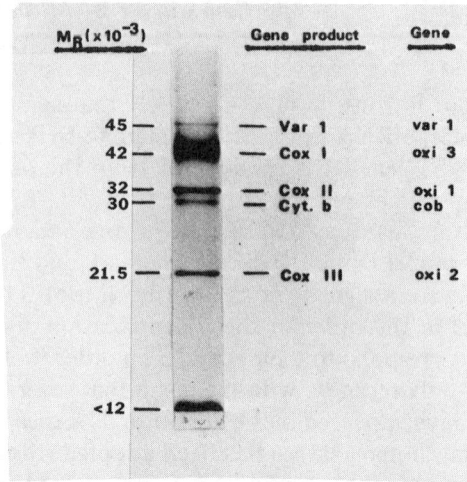

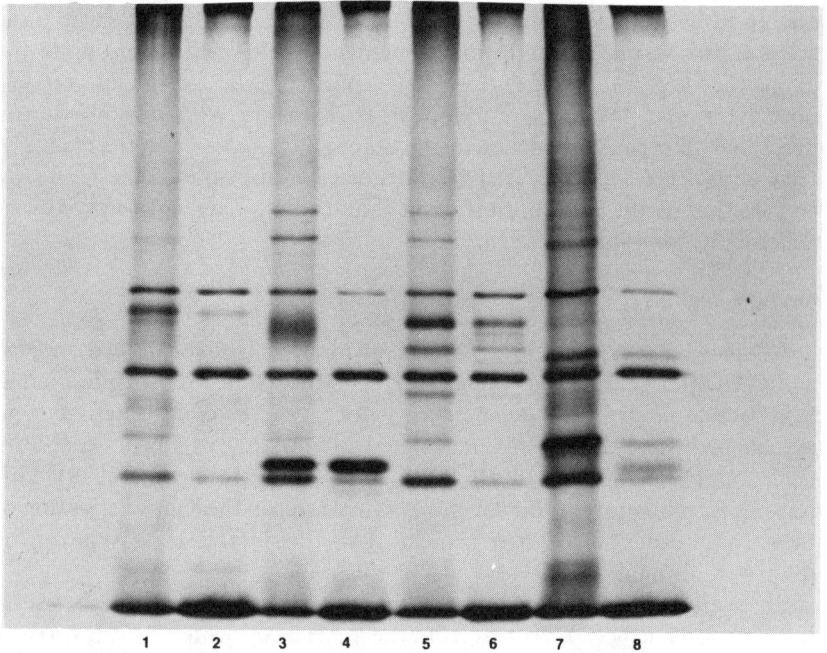

Fig. 5.   Mitochondrial protein synthetic products. Cells (*in vivo*) or isolated mitochondria (*in vitro*) were labeled with $^{35}SO_4$ in the presence of cycloheximide and the proteins separated on an SDS polyacrylamide slab gel. (*above*) Mitochondrial genes and gene products from wild-type strain ID41-6/161 (*in vivo*). (*below*) Mitochondrial protein products from *mit*⁻ mutants labeled *in vivo* or *in vitro, lane 1*: PZ1 *in vivo; lane 2*: PZ1 *in vitro; lane 3*: PZ 30 *in vivo; lane 4*: PZ30 *in vitro; lane 5*: M44 *in vivo* (*oxi3 mit*⁻); *lane 6*: M44 *in vitro; lane 7*: A 103 *in vivo; lane 8*: A 103 *in vitro.*

separated on a polyacrylamide slab gel. Figure 5A is a typical autoradiograph of proteins made by the mitochondria of a wild-type strain. When *cob-box* mutants are analyzed in this fashion, it is found that many of them lack apocytochrome *b* (p30) and have one or more extra proteins larger and/or smaller than p30. Fortunately, it has been possible to substantiate these findings with experiments using isolated mitochondria *in vitro*. In studies performed by J. Steinkeler (Steinkeler and Mahler, 1980), some of which are shown in figure 5B, the same products are formed by the wild type and all mutants investigated so far when cells are labeled *in vivo* or mitochondria *in vitro*.

PHYSICAL MAP OF THE COB-BOX REGION

Since we are interested in presenting an over-all view of the *cob-box* region, it is necessary to compare the mtDNA from wild-type stocks used by the different groups of investigators. Using standard restriction site mapping procedures, a detailed physical map of the *cob-box* region (fig. 6) was constructed for the wild-type strain used by us, strain ID41-6/161 (Alexander et al., 1979; Vincent et al., 1980; Perlman, unpublished). A similar map of the *cob-box* region in strain 777-4A was prepared by Slonimski and colleagues (1978b; unpublished), with the two maps being identical. D273-10B, used by Tzagoloff's group (Morimoto and Rabinowitz, 1979), is very similar in the *cob-box* region except that the *Bam*HI and *Bgl*II sites located in *Hinc*II fragment 5 are missing. The *Eco*RI and *Hha*I sites located between the two *Hinc*II sites are retained but are closer together by roughly 3,000bp. However, the three strains appear to be identical between the two *Eco*RI sites.

The genetic map of the *cob-box* region was constructed chiefly by deletion mapping using a collection of *rho⁻* mutants (Slonimski et al., in preparation). Many of the *rho⁻* genomes used for mapping have also been

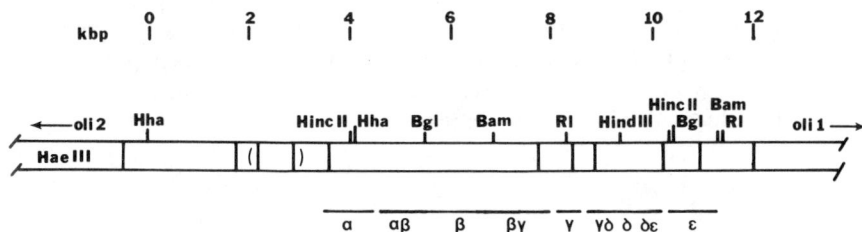

Fig. 6. Physical map of the *cob-box* region in strain ID41-6/161. The approximate location of the exons and introns is shown. The location of the *Hae*III sites (identified within the parentheses) is not precisely known. *Hha* = *Hha*I, *Bgl* = *Bgl*I, *Bam* = *Bam*HI, RI = *Eco*RI. (Information for this map came from P. P. Slonimski, R. D. Vincent, and P. S. Perlman).

analyzed physically (Alexander et al., 1979; Slonimski et al., 1978b), and so it has been possible to map specific clusters of mutations in this way. The localization of all nine *cob-box* segments is not yet complete, but the partial data presented here (fig. 6) will suffice for this discussion. It should be clear at this point that *cob-box* mutants are scattered throughout a long segment of the genome and that the most distant mutations are separated by at least 5,000 bp.

The physical map of several deletion *mit⁻* mutants has also been determined, and those data provide a separate estimate of the size of the region (Alexander et al., 1979). Mutant PZ27 fails to recombine to yield wild-type progeny when crossed with all *cob-box* mutants, and is restored to respiration sufficiency only by those *petite* mutants capable of restoring all mutant sites. It is, therefore, missing the entire *cob-box* region, and was found to be deleted for 9.0×10^3 bp. Another deletion mutant, A101, lacks three exons and two (or three) introns at the *oli1*-proximal end of the region but retains two exons and one intron at the other end. Physical analysis of this mutant indicates it is deleted for 3,300 bp (Alexander et al., 1979).

We therefore see that the *cob-box* region is physically very large (at least 5,000 bp) and could easily contain several genes. As will be shown later, the product of the *cob-box* gene is a protein of 30,000 daltons (p30) that requires a mRNA species of at least 900 nucleotides for its complete specification. Provided that we can show that physically distant mutations affect the primary sequence of a single protein, that is, that the *cob-box* region contains a single gene, then the physical mapping data shown here comprise an important aspect of the demonstration that the gene has an unusual, interrupted structure.

IDENTIFICATION OF EXONS

A number of properties of the mutants mapping in segments $\alpha \rightarrow \epsilon$ support the conclusion that these regions contain the base sequences specifying the primary sequence of apocytochrome *b*. If this hypothesis is correct, we would expect to find both missense and chain termination mutations located within these exons.

Missense Mutants

Several types of mutants have been reported that are probably missense mutants. Haid and colleagues (1979) reported a mutant in exon γ that has an altered cytochrome *b* spectrum; Claisse and colleagues (1978) have reported mutants in exons α amd ε that have a normal spectrum but no associated enzyme activity. Colson and Slonimski (1979) have reported many drug-resistant mutants scattered throughout the *cob-box* region all of which clearly map in the regions designated as exons. All the above mutants synthesize p30, but no alteration in the primary sequence of p30 has been demonstrated for any of them. Thus, we can only hypothesize they are missense mutants.

Also mapping in exons are some *mit⁻* mutants that synthesize a protein that co-migrates with p30 but have no cytochrome *b*-associated enzyme activity. We focused on one of these, EM17 mapping in exon α, and analyzed its p30 protein by fingerprint analysis using the method of Cleveland and colleagues (1977). We observed that the pattern of partial protease fragments of isolated p30 from EM17 resembles that of wild type p30 but was somewhat different (fig. 7), suggestive of an amino acid substitution (Hanson et al., 1979). It should be stressed that the *Staphlococcus aureus* V8 protease used here was selected because it yields a different pattern for p30 species from wild type and the mutant; several other enzymes yield much more similar patterns. When the EM17 mutation was removed by recombination (by mating with an appropriate petite mutant and isolating progeny that can grow on glycerol), the fingerprint pattern of p30 was then identical to that of the original wild-type parent. These results clearly demonstrate that the p30 species from EM17 is closely related to that of wild type and is consistent with the hypothesis that the mutation is a base substitution or small deletion.

We next extended this analysis by isolating a derivative of EM17, strain 1-32, that has a second *cob-box* mutation physically and genetically distant from exon α, that is, in exon ε (Hanson et al., 1979). Strain 1-32 still synthesizes a p30 species, and as shown in figure 7, the *S. aureus* V8 protease pattern of p30 from the double mutant is further modified from that seen with EM17. The second mutation in exon ε, termed 1-32s, was separated from the double mutant by recombination with an appropriate *petite* strain and was found to synthesize a p30 species yielding a third, still different, protease digestion pattern.

None of the peptide patterns in figure 7 appears to contain a mixture of two of the p30 species. A distinct peptide in digests of wild-type p30 is missing from digests of p30 from EM17 (*), and similarly specific peptides are present and then absent from the pattern as the arrangement of mutant sites is manipulated. We conclude that these mutations (EM17 and 1-32s), located at least 5,000 bp apart (see figure 6), affect the primary sequence of p30 and so are most likely missense or small deletion/insertion mutants. They could be frameshift mutations, but it is unlikely that two different frameshift mutations would result in a protein of apparently normal length in all three combinations studied.

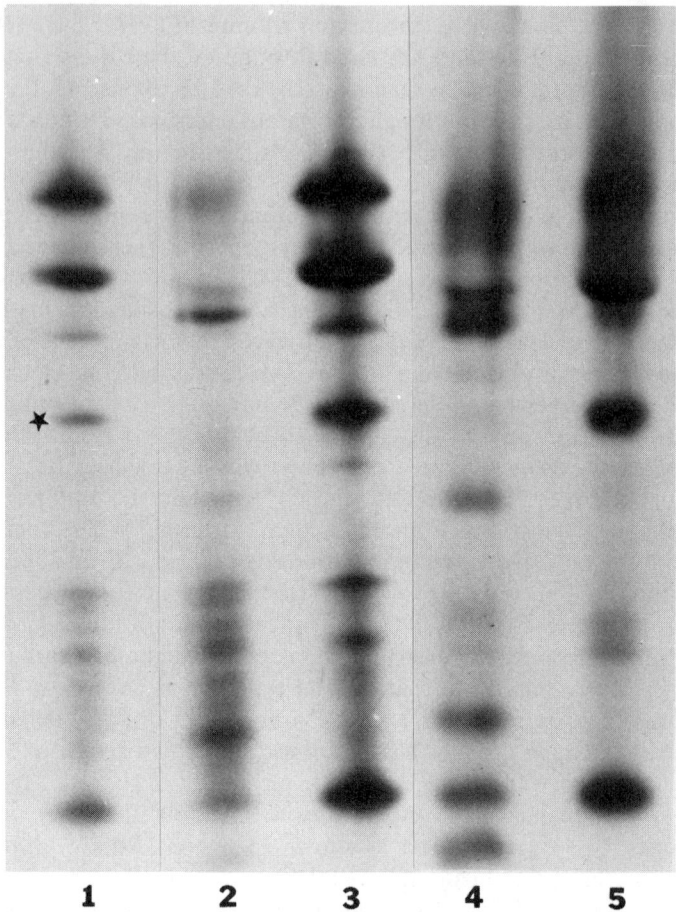

Fig. 7.   Radioautograms of proteolytic digests of isolated p30. p30 was isolated from each strain and digested with 50 μg/well of *S. aureus* V8 protease. The fingerprint pattern is different for EM17 and the double mutant 1-32, suggesting that EM17 and 1-32s are missense mutants (see text for details). *lane 1*: wild type ρ^+ (star) indicates the fragment not present in EM17 or the double mutant 1-32; *lane 2*: EM17 *mit⁻* (Eα); *lane 3*: respiratory-sufficient diploid formed from the cross EM 17 × ρ⁻ (retaining exon α); *lane 4*: double mutant 1-32; *lane 5*: 1-32s (Eε).

Chain Termination Mutants

Most *mit⁻* mutants mapping in segments α→ε make no p30 species but do synthesize a protein smaller than p30 but not present in the wild type. Several groups have reported fingerprint patterns of these extra proteins where clear similarities with digests of wild type p30 were seen, especially for mutants having extra proteins larger than 17,000 daltons (Hanson et al., 1979; Haid et al., 1979). A good example of this point is shown in figure 9 using mutant A104 (exon ε), which makes an extra protein of 22,000 daltons (see figure 7). Fingerprints of shorter novel proteins are more difficult to interpret; however, several groups have recently shown that all these shortened polypeptides cross-react well with antibody to cytochrome *b* (see Grivell et al., 1979; M. Solioz, personal communication). Thus, it is quite clear that many mutants mapping in exons produce fragments of p30. This strongly suggests they are chain termination mutants.

If this hypothesis is correct, one would expect a clear correlation between the map position of the mutation and the size of the fragment. Indeed, that correlation was first reported by Claisse and colleagues (1978) and was more fully documented by Haid and his coworkers (1979). An example of this result is presented in figure 8 for mutants mapping in exons α, γ, and ε.

In figure 6 we showed that the exons α and ε are located at least 3,500 bp apart on the physical map and γ and ε at least 1,600 bp apart. If there were no intervening sequences, then based on our estimates of the difference in fragment sizes of the proteins made in exon mutants, the α→γ distance should be roughly 180 bp and the δ→ε distance roughly 120 bp. Even

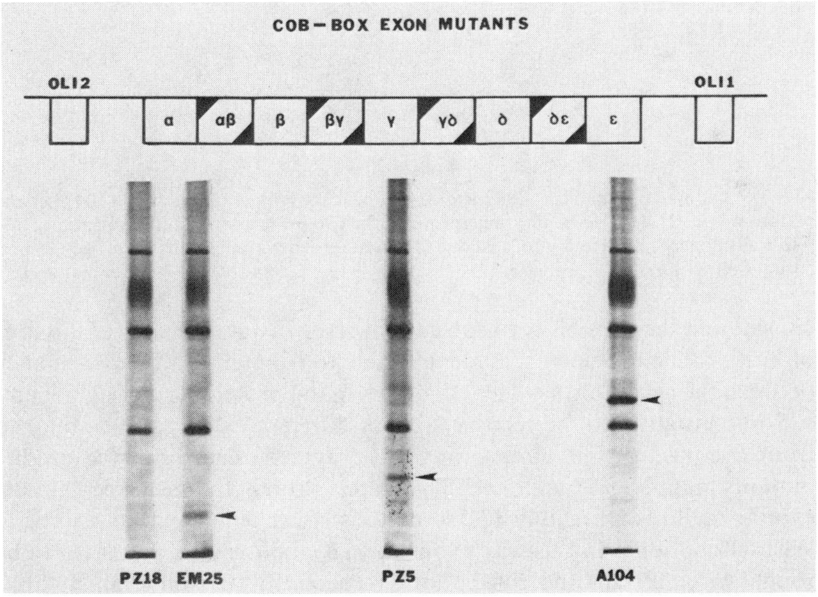

Fig. 8. Radioautograms of mitochondrial translation products of selected *mit⁻* exon mutants. PZ18 and EM25 map in Eα, PZ5 in Eγ, and A104 in Eε. Arrows indicate fragments of p30 formed by the mutants not found in the *mit⁺* parent. All mutants lack p30.

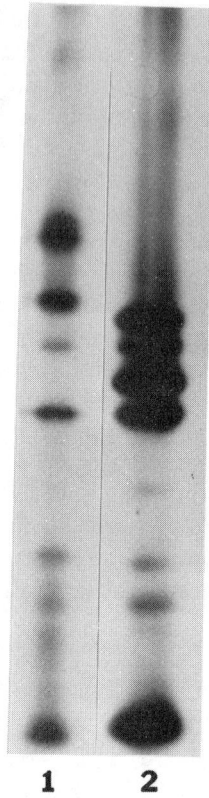

1   2

Fig. 9.   Radioautogram of proteolytic digest of p22 from *mit⁻* exon mutant A104 (Eε) and p30 from *mit⁺* (ID41-6/161). The fingerprints of the two proteins are similar, suggesting that p22 is a fragment of p30 and is formed due to a nonsense mutation. Bands were digested with 50 μg/well *S. aureus* V8 protease.

considering the possibility of substantial error in our estimates of the sizes of the protein fragments made in the exon mutants, this discrepancy between the actual physical length of DNA and protein is most significant.

Some insight into the relative sizes of different exons can be obtained from a consideration of the range of fragment sizes observed among mutants mapping in each exon. The largest protein fragment seen among exon α mutants is roughly 13,500 daltons so we would expect α to be at least 400 bp long. In this way exons β, γ, δ, and ϵ can be estimated to be roughly 75, 90, 120, and 240 bp long respectively (Grivell et al., 1979).

Another important observation is that within three of the exons (α, γ, and ϵ) several independent mutants synthesize p30 fragments of different apparent length. It has been possible to establish the map order of some of those mutants within each exon by deletion mapping. As shown in figure 10, for several mutants mapping in exon ε the fragment size is still correlated with the position of the relevant mutational sites. Given the excellent correlation between protein fragment size and map position of

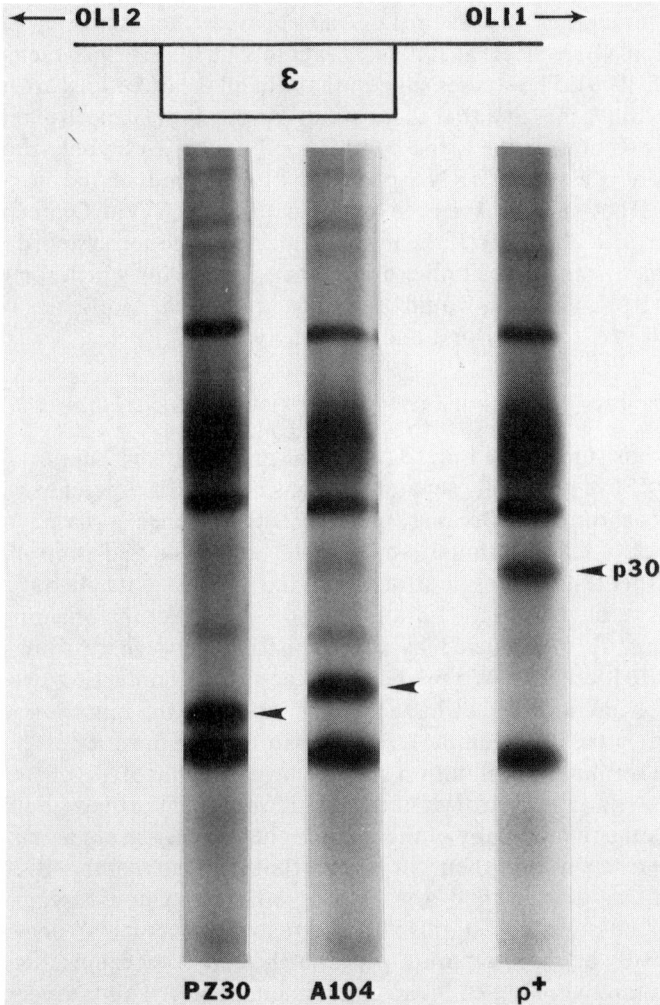

Fig. 10.   Map position and mitochondrial protein products of *mit⁻* mutants from exon ε and wild type (ρ⁺). The fragment size (arrows) of novel proteins produced in these mutants correlates with the position of the relevant mutational site in Eε.

the mutants, it appears that the *cob-box* gene is transcribed from α through ϵ as indicated in figure 4. The actual beginning of the transcriptional unit has not, of course, been identified, and could well lie on the *oli2* side of exon α.

The high frequency of chain termination mutants in the *cob-box* region has been rather puzzling for some time. Recent DNA sequencing studies of two structural genes on yeast mtDNA has provided an interesting possible explanation (Grivell et al., 1979; Tzagoloff et al., 1979; Macino and Tzagoloff, 1979). The studies suggest that not all the 61 codons are used in mitochondrial genes and that there is a striking lack of G and C in the third (wobble) position. In fact, the codons used appear to be only those for which there is a specific tRNA present within the mitochondria. So far, about 25 tRNA species have been localized on mtDNA (Van Ommen et al., 1977; Martin et al., 1977). If the mitochondrial translation system does not permit efficient third position wobble, then codons for which there is no specific tRNA species would serve as additional nonsense codons (Tzagoloff et al., 1979; Borst and Grivell, 1978).

A FURTHER TEST OF THE POLARITY OF THE COB-BOX GENE

The results summarized up to this point provide strong support for the existence of five physically separate regions coding for different segments of apocytochrome *b*. Because much of the evidence is based on the phenotypes of chain termination mutants, we considered it essential to devise further tests of mosaic structure of the *cob-box* gene. As outlined in figure 11, we have used the data in figure 8 to predict the phenotypes of double mutants constructed by recombination between two previously characterized mutants. We predict that each strain containing two chain termination mutations will have the phenotype of the mutation earliest (upstream) in the gene. Similarly, double mutants having one missense and one chain termination mutation should have the phenotype of the chain termination mutant regardless of the relative order of the two mutant sites. If, on the other hand, many of the mutant phenotypes are not actually due to chain termination, then, provided that enough combinations are examined, we should find some cases where the downstream (*oli1*-proximal) mutation is epistatic to the upstream (*oli2*-proximal) one. Also if there are two or more separate genes in the *cob-box* region, then some double mutants should exhibit an additive phenotype. This is especially so for the combination of a missense mutant in exon α with a chain termination mutant in exon ϵ, the most distant parts of the *cob-box* region.

We have constructed and analyzed the phenotypes of double mutants

COB - BOX PROCESSING

mit$^+$

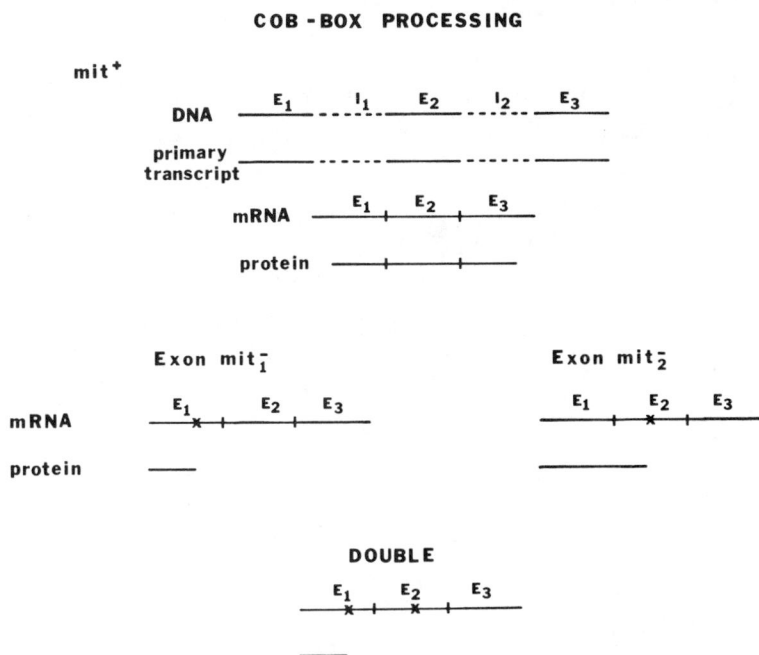

Exon mit$_1^-$ Exon mit$_2^-$

DOUBLE

Fig. 11. Schematic diagram of *cob-box* processing. In *mit*$^+$ cells, intron sequences may be spliced from the primary transcript so that the mRNA contains only exon (coding) sequences. A chain termination mutation in E_1 produces a short protein. A double mutant containing nonsense mutations in E_1 and in E_2 would be expected to have the protein phenotype of the single E_1 mutant.

having chain termination mutations in exons α, β,γ , and ϵ and a missense mutation in exon α (Mahler and Perlman, 1979; Alexander et al., 1980). In all cases the prediction based on the mosaic gene model was confirmed, and a typical result (exon α + exon γ) is shown in figure 12. Clearly the upstream mutation extinguished the synthesis of the fragment of p30 associated with the downstream mutational site. Double mutants between the missense mutant EM17 (exon α) and chain termination mutants in exons β,γ , and ϵ or upstream within exon α all exhibited the predicted phenotype.

These data, taken together with the previous evidence, clearly indicate that the base sequences coding for p30 are scattered over a long (8,000 bp) portion of the yeast mitochondrial genome. The possibility that the region contains two or more genes coding for different p30 species regulated independently or together is rendered most unlikely by the results with double mutants; no additive phenotypes were observed, and the altered p30

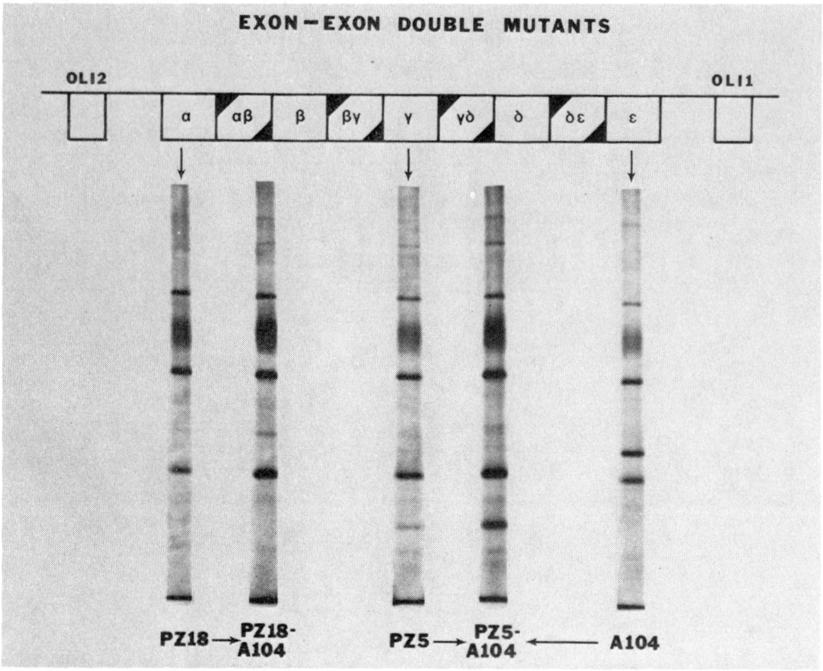

Fig. 12. Exon-exon double mutants. Autoradiograms of mitochondrial protein products from the single exon nonsense mutants, PZ18 (Eα), PZ5 (Eγ), and A104 (Eε) as well as those from the double mutants, PZ18-A104 (Eα-Eε) and PZ5-A104 (Eγ-Eε) are shown. In a double mutant the upstream mutation is epistatic to the downstream mutation.

due to an upstream missense mutation is extinguished by downstream chain termination mutations. The fact that downstream chain termination mutants also extinguish the wild type p30 also supports this conclusion.

ALL EXONS COMPRISE A SINGLE COMPLEMENTATION GROUP

Further support for the mosaic structure of the *cob-box* region comes from the results of complementation tests using the procedure devised recently by Slonimski and colleagues (1978b) and by Foury and Tzagoloff (1978). Unlike the situation for nuclear genes, a stable *trans* configuration cannot be obtained for mitochondrial genes; therefore, complementation must be assessed soon after mating, when most diploid cells are still heteroplasmic, i.e., retain both parental genomes in a common cytoplasm. Since neither mutant parent strain respires, the mating mixture is assayed for the early appearance of respiratory enzymes (as measured by oxygen uptake by whole cells). As shown diagrammatically in figure 13, when the mating is between two mutants mapping in the same gene, no respiration is detectable (no complementation) for up to 12 hours. In those cases where the two mutants used can recombine to yield a few percent wild-type progeny, a low rate of respiration (per diploid cell) is detectable after roughly 20 hours. However, when the mating utilizes two mutants in different genes, a different result is obtained (fig. 13); within 5–10 hours a high rate of respiration becomes manifest and is retained for several hours. Thereafter the rate (per cell) decreases to a final low value. The respiratory rate decreases because the proportion of diploid cells capable of exhibiting complementation (that is, having both mutant genomes) decreases due to the rapid rate of mitotic segregation typical of mitochondrial genes in this system. The low final rate of respiration in these experiments is that characteristic of recombinant cells.

This complementation test has been most thoroughly utilized by Slonimski's group, and their main results have been confirmed recently by Haid and colleagues (1979) and by us. Basically, all mutants mapping in exons comprise a single complementation group: they fail to complement each other but are readily complemented by mutants in other genes (e.g., *oxi3, oxi1, oxi2, var1*). However, mutations mapping in introns between adjacent exons define separate complementation groups; thus, the *cob-box* region contains several complementation groups one of which is broken into at least five pieces. (See below, for further description of intron mutants.)

IDENTIFICATION OF INTRONS

As shown in figure 4, some of our mutants map between pairs of exons. These intervening sequences and a third one are defined genetically more fully by the mutant collections of other workers (Slonimski et al., in

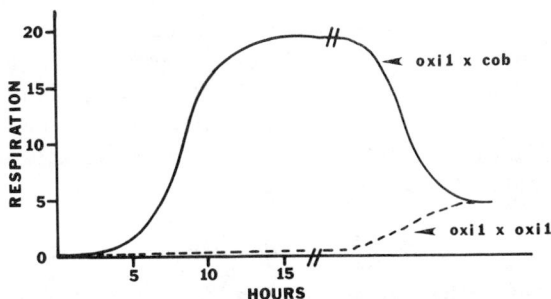

Fig. 13. Kinetics of complementation in zygotes formed between respiratory-deficient mutants. A schematic representation of the results seen for O₂ uptake by zygotes formed by mating *mit⁻* mutants in different genes (*oxi*1 × *cob⁻*) (complementation) and *mit⁻* in the same gene (*oxi*1 × *oxi*1) (no complementation). The final low rate of respiration (similar in both types of matings) is that resulting from recombinant cells.

preparation). In this section we will discuss the properties of these mutants that prompt us to conclude that they are not mutations in coding portions of the *cob-box* gene and thus define the location and properties of several introns.

Phenotypes of Intron Mutants

Intron mutants exhibit phenotypes more complex than those of exon mutants. Typically, they are totally deficient in cytochromes *b* and *aa*₃. When the products of mitochondrial protein synthesis are analyzed on polyacrylamide gels (fig. 14), p30 and subunit I of cytochrome *c* oxidase (see also figure 5) are totally absent. Mutants in I$\alpha\beta$ typically synthesize one protein larger than p30 that is not present in extracts of wild-type cells. An example is PZ1, which synthesizes a protein of 45,000 daltons (p45). Other mutants mapping within this physically large region accumulate novel extra proteins of 40,000 or 35,000 daltons instead of p45 (Claisse et al., 1978; Haid et al., 1979). Mutants mapping I$\gamma\delta$ all show a very complex protein phenotype with several novel proteins both larger and smaller than p30. In figure 14 we show the phenotypes of two such mutants that recombine with each other and are displayed in their map order. They both accumulate an additional protein of roughly 35,000 daltons plus several smaller polypeptides. As shown in figure 4, mutants with phenotypes typical of exons map adjacent to these mutants, but all clusters of intron mutations are clearly separable from exon clusters; eventually, some intron mutants mapping within just a few base pairs of a bona fide exon mutant should be obtained (provided that base sequences near the splice points of mosaic genes are important), but these are not prominent among the available mutants.

The unexpected absence of cytochrome *c* oxidase activity (and the absence of one mitochondrially synthesized subunit of that enzyme, coxI) is a unique property of most intron mutants, one not shared by any of the exon mutants studied to date. Although that phenotypic lesion may turn out to be one of the most interesting properties of the *cob-box* region, it is not germane to the analysis of the structure of the *cob-box* gene but will be discussed separately later in this paper.

Origin of the Extra Proteins Made by Intron Mutants

Novel proteins can be synthesized by mutant strains for any of a number of reasons (see Mahler et al., 1978). Here we are chiefly concerned with the following possibilities. (1) A protein larger than the normal one could arise by a frameshift mutation so that in the altered frame translation extends beyond the normal termination point; by the failure to process a form of a natural, larger protein precursor; or by translation of one or more interverning sequences not excised from the primary transcript because of a mutation within the intron. (2) A protein (or multiple proteins) smaller than the normal one could result from partial proteolysis of a larger protein, or by initiation of translation within a mRNA species at secondary sites not normally used. (3) Both large and small novel proteins could result from enhanced expression of another gene(s), whose protein products are not detected in wild types, as a consequence of a specific defect in the *cob-box* gene. (4) Multiple extra proteins could result quite readily from a combination of several of these factors.

Clearly, the simplest type of intron defect to characterize would be one in which an alteration in base sequence within an intron abolishes the splicing of the adjacent exons. In that case we would expect there to be no p30 and in its place a protein at least as large as those specified by the chain termination mutants in the adjacent upstream exon. The novel proteins synthesized by the intron mutant shown in figure 14 (and other mutants analyzed by other workers) have been analyzed in several ways, and all the evidence supports the conclusion that they are translation products of the *cob-box* region that result from translation across exon-intron boundaries.

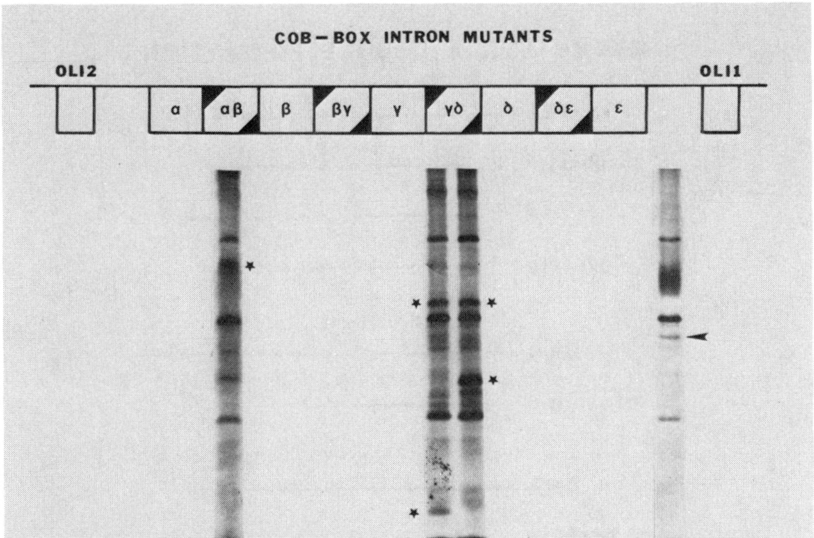

Fig. 14. Radioautograms of mitochondrial translation products of *mit*⁺ (1D41-6/161) and selected *mit⁻* intron mutants. All intron mutants have no coxI and no p30 (arrow) and produce one or more novel proteins (stars). PZ1 (Iαβ) produces p45; PZ16 (Iγδ) produces p35 and p14; A103 (Iγδ) produces p35 and p25.

Several types of evidence show that the intron-specific proteins are related to the absent p30. By fingerprinting we have found that several of the proteins typical of Iγδ mutants (e.g., p25 and p35; see figure 14) closely resemble p30 and the chain termination fragments made by mutants of the preceeding exon, exon γ (Claisse et al., 1980). That test is unconvincing for p45 synthesized by the mutant mapping in Iαβ; that is not too surprising since we would expect only about 14,000 daltons of that protein to be related to p30. More convincing has been the work by others showing that most intron-specific proteins are antigenically related to apocytochrome *b* (see Grivell et al., 1979; M. Solioz, personal communication). Together these two types of analysis indicate that the novel proteins synthesized by intron mutants have properties both related and unrelated to p30 and are consistent with the processing model.

We have constructed and assessed the phenotypes of double mutants containing an intron and a chain termination exon mutation (Alexander et al., 1980). As indicated in figure 15, we would predict that a chain termination mutation upstream from an intron mutation should extinguish the synthesis of the novel protein associated with the intron mutant. That prediction was confirmed for mutants in Iαβ and Iγδ (fig. 16); in the latter case it is significant that *all* the novel proteins were

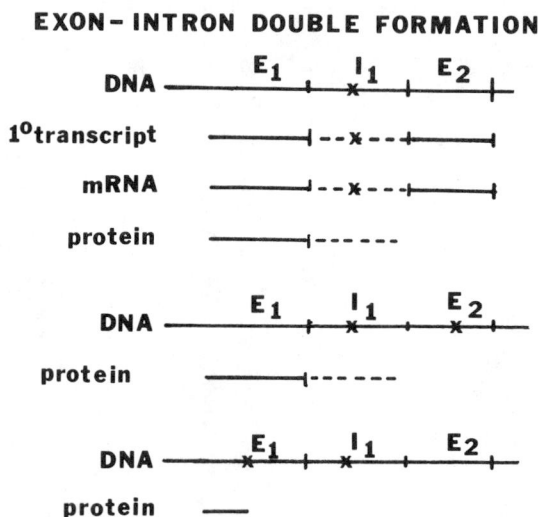

EXON – INTRON DOUBLE FORMATION

Fig. 15. Schematic diagram of predicted phenotype of exon-intron double mutants. An intron mutation may result in the reading of intron sequences normally excised during RNA processing, thus resulting in the formation of a novel protein. A double mutant formed between an exon and an intron mutant would be expected to have the phenotype characteristic of the upstream mutant.

extinguished regardless of whether the chain termination site was nearby (exon γ) or distant (exon α). In this fashion we have shown that the synthesis of intron-specific proteins requires translation through *all* preceding exons.

We next tested whether these novel proteins characteristic of a specific intron mutation arise from translation terminated within that intron or from terminations beyond it in a later intron or exon. We did this by determining the phenotypic consequences of second mutations downstream. In each case such a downstream exon or intron mutation had no effect on the phenotype of the upstream intron mutation (fig. 16), and the phenotype of the downstream mutation was extinguished by the upstream intron. Thus, it is most likely that each intron-specific protein contains the portion of p30 encoded by exons preceding it plus a number of amino acids specified by some intron sequences that are normally removed from the primary transcript.

We cannot currently provide a detailed explanation of the origin of the

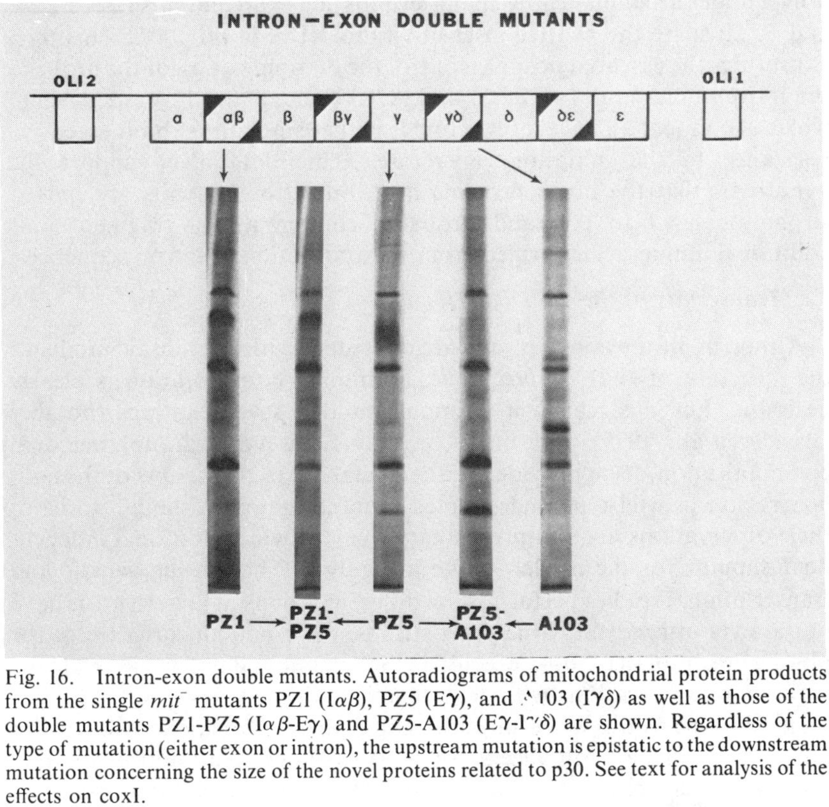

Fig. 16. Intron-exon double mutants. Autoradiograms of mitochondrial protein products from the single *mit⁻* mutants PZ1 (Iαβ), PZ5 (Eγ), and A103 (Iγδ) as well as those of the double mutants PZ1-PZ5 (Iαβ-Eγ) and PZ5-A103 (Eγ-Iⁿδ) are shown. Regardless of the type of mutation (either exon or intron), the upstream mutation is epistatic to the downstream mutation concerning the size of the novel proteins related to p30. See text for analysis of the effects on coxI.

multiple extra proteins typical of I$\gamma\delta$ mutants. We feel that sufficient plausible mechanisms are evident including proteolysis, internal initiation of translation, inefficient reading of rarely-used codons and/or abortive RNA processing.

USE OF NALIDIXIC ACID AS A PROBE

This part of our studies was prompted by some recent reports that nalidixic acid (NDA), a known inhibitor of prokaryotic DNA gyrases (topoisomerases—see e.g., Mizuuchi et al., 1978), can also affect gene expression in a selective manner (Smith et al., 1978; Yang et al., 1979; Sanzey, 1979). We (Mahler and Johnson, 1979) found that NDA at low concentrations (25-50 μg/ml) inhibits mitochondrial gene expession in mit^+ cells in a differential fashion in the order: coxI $>$ var 1 $>$ cox II $=$ coxIII $>$ apocytochrome b. This has permitted the use of NDA as a selective probe for polypeptide products related to apocytochrome b compared with those of other proteins, e.g., coxI. As shown in figure 17, the polypeptides accumulated by chain-terminating exon mutants PZ5, PZ24, and PZ29 share the relative insensitivity to NDA of mit^+ p30. This then constitutes the essential positive control for the application of the probe to the intron mutants in I$\gamma\delta$ PZ25 and A103 (lanes 9 and 13, respectively), which shows that all novel polypeptides in these mutants exhibit extensive resistance to the inhibitor. Therefore, these data also support the hypothesis that the novel proteins made in intron mutants are indeed strongly related to p30 and probably contain a p30 fragment plus additional amino acids formed from the translation of intron sequences.

Transcripts of the Cob-Box Region

Although all the above results are consistent with the mosaic model of the structure of the *cob-box* gene, additional corroboration is clearly desirable. For this reason it is important that several groups (notably Grivell et al., 1979) and more recently Schweyen's group (personal communication) have provided a direct analysis of transcripts of the *cob-box* region in wild-type and selected mutant strains. Although some of their observations are still preliminary, they provide additional independent support for the model shown in figure 4. Though the genetic and transcriptional studies performed by different groups of investigators have not always utilized the same yeast strains, the exact concordance of the physical maps of the *cob-box* region in those strains allows one to relate the two sets of observations with confidence.

In wild-type strains with a physical map of the *cob-box* region like that of our strain, Grivell and colleagues (1979) have detected multiple discrete transcripts that clearly overlap with each other; the smallest and most

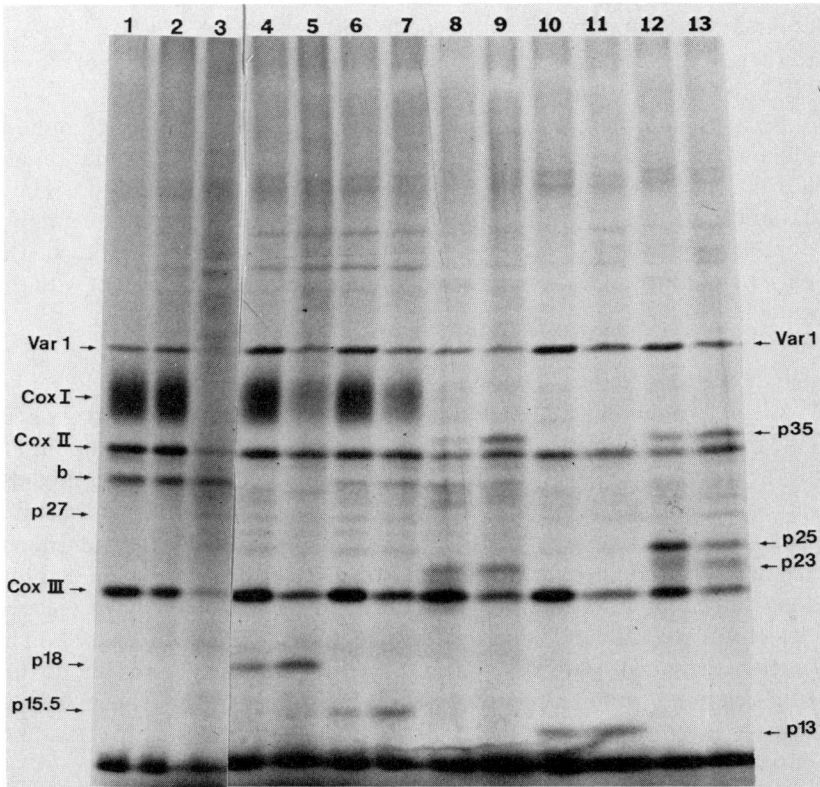

Fig. 17. Nalidixic acid inhibits mitochondrial gene expression. Treatment of *mit⁺* and *mit⁻* cells with nalidixic acid (NDA) (lanes 3, 5, 7, 9, 11, 13) shows a differential effect on gene expression in the order: cox I > varl > coxII = coxIII > apocytochrome *b*. *Lane 1*: ID41-6/161; *lane 2*: ID41-6/161 + 50μl NaOH (solvent for NDA); *lane 3*: ID41-6/161 + NDA (50 μg/ml); *lane 4*: PZ5; *lane 5*: PZ5 + NDA; *lane 6*: PZ24; *lane 7*: PZ24 + NDA; *lane 8*: PZ25; *lane 9*: PZ25 + NDA; *lane 10*: PZ29; *lane 11*: PZ29 + NDA; *lane 12*: A103; *lane 13*: A103 + NDA. All lanes were loaded with equal cpm to permit ready visualization of differential effects.

prominent is an 18S species that is a bit longer than the 900 nucleotides minimally required to specify a protein molecule containing 250 residues (p30). Analysis of DNA-RNA hybrids with the electron microscope clearly reveals that this 18S species is not continuous. In fact, the hybrid spans the whole gene, but exhibits five double-stranded regions of homology interrupted by four single-strand loops. DNA probes containing sequences only from Iαβ fail to hybridize to the 18S species but do hybridize to some of the larger transcripts, which presumably are processing intermediates. More detailed relationships between the different *cob-box* transcripts are only now beginning to emerge, but sufficient evidence has already ac-

cumulated to support the split gene hypothesis for the *cob-box* gene and
to suggest that the main features of its expression resemble those of mosaic
viral and nuclear genes in other eucaryotic systems.

Finally, in order to interpret the results of our studies of mutant
phenotypes, we have made the implicit assumption that exon mutations do
not interfere with processing events, whereas intron mutations do. These
assumptions must be tested directly in order to lay to rest any remaining
doubts about the *cob-box* system; however, preliminary studies of the
pattern of transcripts from intron mutants also appear consistent with the
model.

THE MODEL REVISITED

We have now reviewed the main lines of evidence that support the specific model for the structure of the *cob-box* gene shown ealier in figure 11.
In the course of this discussion, we have considered several other models,
but conclude that the mosaic model is the most probable one. Nine distinct
gene segments have been identified, but it is quite possible that additional
segments extend beyond exon α (toward *oli2*); also, a short exon within a
long intron or a short intron within a long exon could have escaped
detection. We anticipate that the fine structure of the *cob-box* gene will be
elucidated shortly by DNA and protein sequencing and by further
transcriptional mapping studies. In the remainder of this paper, we will
discuss three aspects of the *cob-box* structure and function that are
currently being investigated by us.

THE SPLICING PATHWAY

Transcripts of mosaic mitochondrial genes appear to be more readily
accessible to experimentation than are those of mosaic nuclear genes or
even those of animal viruses. A key feature, not now available with the
other systems, is the analysis of alterations in the processing events caused
by intron mutations. We expect that direct studies of transcripts will define
the processing pathway of the *cob-box* gene; however, we have already
obtained some insight into this important problem from our analysis of
intron-exon double mutants.

As indicated previously, exon phenotypes are consistent with the
conclusion that at least those introns preceeding the chain termination
mutation are processed normally. For example, in the case of mutant PZ5
(exon γ) we have concluded that I$\alpha\beta$ and I$\beta\gamma$ must have been processed
normally to permit the translation of fragment polypeptide p18. If, for
example, processing at I$\alpha\beta$ were defective because of the PZ5 mutation, we
would expect a phenotype resembling that of a mutant in I$\alpha\beta$ (e.g., PZ1,

fig. 14). Similarly, if mutations in a downstream intron, e.g., A103 (I$\gamma\delta$) interfere with processing at upstream introns, then again we would expect the phenotype of A103 to resemble that of PZ1. These considerations apply with equal force to the double mutants; as shown in figure 16, the double mutant containing PZ5 and A103 accumulative p18, typical of PZ5, indicating that failure to process at I$\gamma\delta$ does not interfere with processing events at upstream introns. This approach does not, however, permit us to assess the effects of intron mutations on processing at downstream introns. This information must therefore be obtained directly from studies of the transcripts. We can already conclude that if there is any obligatory sequence of splicing events, I$\gamma\delta$ is processed no sooner than third in the sequence of four steps.

INTERACTIONS BETWEEN MITOCHONDRIAL
GENES: THE "BOX PHENOTYPE"

It was quite unexpected to find some years ago that some cytochrome *b*-deficient mutants also lack cytochrome *c* oxidase. Even before enough was known about the structure of the *cob-box* region to realize that the *box* phenotype was a feature of intron mutations, it was found that the synthesis of only one subunit of cytochrome oxidase was affected, namely, subunit I (coxI). Meanwhile, evidence has accumulated that coxI is a product of the *oxi3* region of the mitochondrial genome and that the *oxi3* gene also appears to have a mosaic structure.

Although the details of this gene interaction are still being investigated, we have made enough progress to warrant some discussion here. We think that our data show that the *cob-box* region exerts positive rather than negative control over the expression of the *oxi3* region and that in wild-type cells the expression of the two genes is coupled through that control circuit. Concerning specific possible negative control mechanisms, we can rule out interference with *oxi3* processing or transcription by unprocessed *cob-box* transcripts by noting that deletion mutants lacking the entire *cob-box* region also make no coxI (Alexander et al., 1979). The intron-specific proteins are also not involved in negative control of *oxi3* because in specific double mutants (e.g., fig. 16) these intron-specific proteins are extinguished by an upstream chain termination mutation while the coxI deficiency persists.

Since coxI deficiencies are exhibited by both intron mutants and deletion mutants, we conclude that the regulation is positive. That is, *oxi3* expression requires something from the *cob-box* gene. We can imagine several possible mechanisms for this regulation; for example, the *cob-box* introns could code for an enzyme required to transcribe or process *oxi3*.

This is a very real possibility since our results with intron-exon double mutants indicate that at least two of the *cob-box* introns can be translated for some distance past exon-intron boundaries in at least one reading frame. Alternatively, specific sequences contained in the *cob-box* primary transcripts, either released or created by splicing, could facilitate processing of *oxi3* transcripts. Slonimski (1979) and Davidson and Britten (1979) have already discussed in general terms how one RNA species may serve as a guide or scaffold for the processing of another RNA species. The guide may be part of a unimolecular or bimolecular interaction, as diagrammed in figure 18. Indeed, evidence that bimolecular interactions at the RNA level may be possible has been presented by Slonimski (1979). Complementation between intron and exon mutants of the *cob-box* region appears to be a consequence of an intragenic *trans* interaction. According to the guide RNA hypothesis, transcript(s) from the exon mutant act on the incompletely processed transcript of the intron mutant to overcome the block to processing imposed by the intron mutation. By analogy, in the wild type either bi-or unimolecular interactions could effect the same processing event. Whether such RNA-RNA interactions do in fact occur, either for intra- or intergenic splicing events, is under active investigation.

ALTERNATE FORMS OF THE COB-BOX GENE

Several years ago Foury and Tzagoloff(1976b) reported that none of the *cob* mutants derived from their wild-type strain, D273-10B, exhibited the "*box* phenotype" that had already been reported by several groups of workers. Several mutants of D273-10B that lacked cytochrome *b* and *aa*$_3$ were shown to be double mutants with separable mutations in *cob* and in one of the *oxi* loci. On the other hand, it is now established that none of the well-characterized *box* mutants obtained from our strain, or the strain used by Slonimski and colleagues (1978b) or Haid and his coworkers (1979), have an expressed mutation outside the *cob-box* region. So, we must

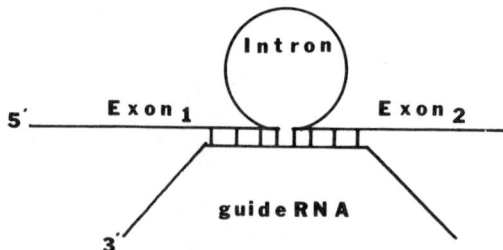

Fig. 18.   Model for RNA used as a guide for splicing of intron sequences.

seriously ask whether the interaction between *cob-box* and *oxi3* is a general feature of mitochondrial genomes or restricted only to particular strains.

It has been known for some time that mtDNA of several yeast strains harbor deletions roughly 3,000 bp in length located near the *cob-box* region (Sanders et al., 1977; Morimoto and Rabinowitz, 1979). Now that the physical map of the *cob-box* region is known (see figure 6), it is quite clear that D273-10B has such a deletion wholly within the *cob-box* region (more likely two deletions separated by exon β, which is retained; Grivell et al., 1979). Morimoto and Rabinowitz (1979) have already shown that mtDNA from this strain lacks the *Bam*HI, *Bgl*II and *Xba*I sites present in fragment *Hae*III-1 and that the flanking retained sites (*Eco*RI and *Hha*I) are closer together by some 3,000 bp (see figure 6). The segment between the two *Eco*RI sites is identical in the two strains, and that fragment clearly contains two introns in our strain. Consistent also with the data of Grivell and colleagues (1979), we conclude that the *cob-box* gene of D273-10B consists of three exons and two introns; it lacks introns Iαβ and Iβγ and exons Eα, Eβ, and Eγ are fused to form one single, long exon. The inference that the coding regions are the same, although their spatial arrangement is different in the first half of the gene, is based in part on our finding that the p30 species synthesized by strains having either structure of the *cob-box* gene are identical in electrophoretic mobility and fingerprint pattern. Also, some of the mutants from Tzagoloff's collection have already been mapped among our mutants; for example, mutant M7-40 appears to represent a typical chain termination mutant in exon ε.

We wished to learn whether the regulation of coxI synthesis seen in our strain is qualitatively different in D273-10B. This could be accomplished in two ways: (1) isolate mutants in D273-10B that map in intron Iγδ and analyze their phenotype; or (2) from our collection select a mutant harboring a lesion mapping in intron Iγδ and cross that mutant site into D273-10B. The latter method offers the advantage that the constructed mutant of D273-10B, a strain with the deletion in the *cob-box* gene, will contain the same mutant allele of Iγδ as one of our mutants, a strain without that deletion. Thus, by comparing the protein phenotype of these two strains, we could determine if a mutant in Iγδ results in the same *box* phenotype, regardless of the over-all structure of the *cob-box* gene. The construction of such a mutant is described in the following paragraphs and outlined in figure 19 (S. Dhawale, unpublished observations).

Mutant A103 (intron Iγδ) was mutagenized with ethidium bromide, and *petite* derivatives were isolated. Among several hundred *petites* screened, we isolated several that retain the latter 2/3 of the *cob-box* region (from

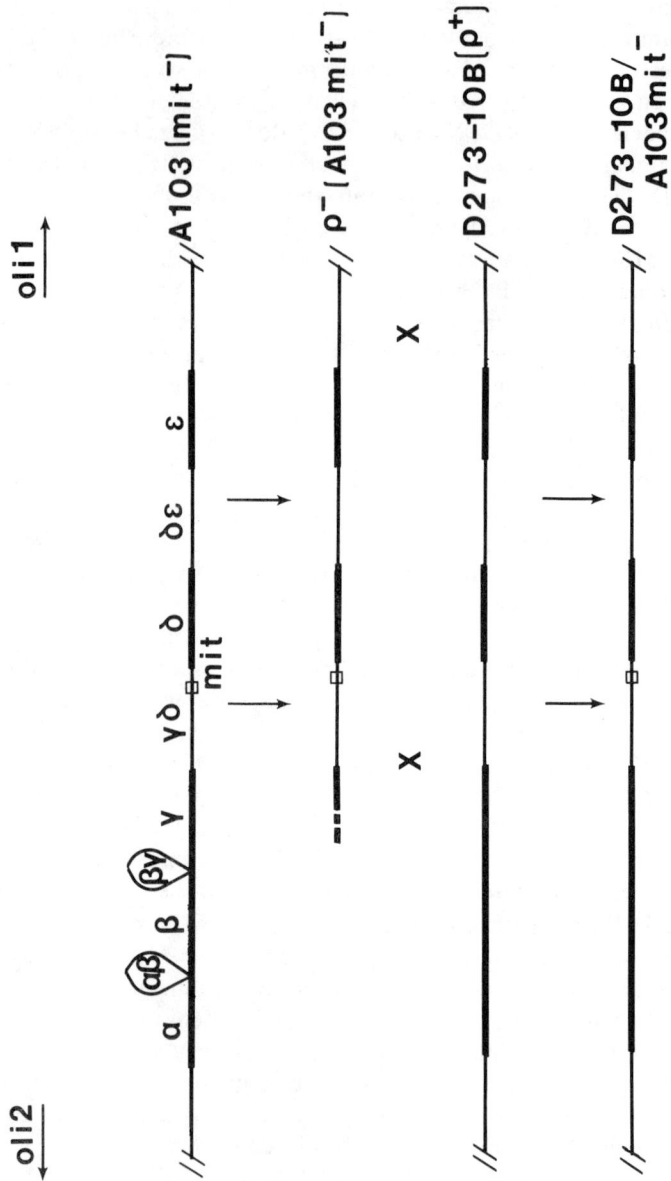

Fig. 19. Schematic diagram for procedure of forming new *mit⁻* mutant. *Mit⁻* mutant A 103 (1γδ) was mutagenized with ethidium bromide, and petites retaining the mutant A 103 site were selected. They were then mated to a wild type *kar I* strain containing D273-10B mito-chondria (D273-10B mito-chondrial genome lacks Iαβ and Iβγ) and *mit⁻* mutants were selected. See text for further analysis. (Information for this scheme supplied by S. Dha-wale.)

exon β through exon ϵ), but are deleted for the first exon and intron. Those petites restore mutants in all retained regions except for A103 since the *petites* themselves harbor the mutant allele at that site. One of those *petites* was then mated to a wild-type *kar1* strain containing mtDNA from D273-10B, and several *mit⁻* recombinants were isolated (in the ID41-6/161 nuclear background). These were shown to be mutant exclusively at the A103 site by the results of appropriate test crosses.

Each constructed *mit⁻* strain was then analyzed phenotypically along with the original mutant (A103), and each had the same complex *box* phenotype as does A103 (illustrated in figure 14). As a further demonstration that our constructed *mit⁻* mutant did indeed have the organization of the *cob-box* gene characteristic of D273-10B, we isolated mtDNA from D273-10B and the mutant (Dhawale, 1980) and compared appropriate restriction enzyme digests with those of mtDNA of a *petite* from our strain that retains the entire *cob-box* region. The constructed *mit⁻* mutant and strain D273-10B were found to have identical restriction fragment patterns. Thus, the *box* phenotype can be obtained in strains lacking two introns. This is an important finding since the *oxi3* regions of the two strains are also different (Sanders et al., 1977; Morimoto and Rabinowitz, 1979). It was therefore possible that the failure to find *box* mutants in D273-10B was a consequence of differences in *cob-box*, in *oxi3*, or in both regions. Although Tzagoloff's group has analyzed quite a few *cob* mutants (Foury and Tzagoloff, 1976b), their failure to obtain clearcut *box* mutants is probably a consequence of the sample size examined. We conclude that the positive regulatory interactions between *cob-box* and *oxi3* are present in their strain and suggest that the properties of those introns present in all strains examined to date (namely I$\gamma\delta$ and I$\delta\epsilon$) should be scrutinized very closely to obtain further insights into the "*box*" phenomenon. Clearly, introns I$\alpha\beta$ and I$\beta\gamma$ do not code for a processing enzyme needed by *oxi3* since a *box* phenotype can be obtained without the presence of those sequences.

We are considering two extreme cases in designing further experiments: (1) each intron in our strain has a specific (or the same) effect on *oxi3* expression; and (2) only one intron influences *oxi3* directly. In the first case, the absence of two introns in D273-10B might necessitate compensating changes in *oxi3*, and in the latter case we might anticipate that the upstream introns must be processed before the downstream ones. We wish to emphasize that the alternate forms of the *cob-box* and *oxi3* genes are valuable experimental probes of this interesting and unusual regulatory interaction between two (probably) mosaic genes that deserve full exploitation.

ACKNOWLEDGMENTS

The authors would like to express their deep appreciation to Drs. P. Slonimski, R. Schweyen, M. Solioz, and Z. Kotylak for sharing unpublished data and for their collaboration and discussions on various aspects of this work. We also would like to thank S. Dhawale for sharing unpublished work, and D. Miller, J. Johnson, R. D. Vincent, and H. P. Zassenhaus for their assistance. This work was supported by Research Grant GM 19607 (to C. W. Birky, Jr., and P. S. P.) and GM 12228 (to H. R. M.) from the National Institute of General Medical Sciences.

REFERENCES

Alexander, N. J., D. K. Hanson, P. S. Perlman, and H. R. Mahler. 1980. Mosaic organization of a mitochondrial gene: evidence from double mutants in the *cob-box* region of *Saccharomyces cerevisiae*. Cell (submitted).

Alexander, N. J., R. D. Vincent, P. S. Perlman, D. H. Miller, D. K. Hanson, and H. R. Mahler. 1979. Regulatory interactions between mitochondrial genes. I. Genetic and biochemical characterization of some mutant types affecting apocytochrome *b* and cytochrome oxidase. J. Biol. Chem. 254:2471-9.

Bernardi, G., G. Piperno, and G. Fonty. 1972. The mitochondrial genome of wild-type yeast cells. I. Preparation and heterogeneity of mitochondrial DNA. J. Mol. Biol. 65:173-89.

Bolotin-Fukuhara, M., G. Faye, and H. Fukuhara. 1977. Temperature-sensitive respiratory-deficient mitochondrial mutations: isolation and genetic mapping. Molec. Gen. Genet. 152:295-305.

Borst, P., and L. A. Grivell. 1978. The mitochondrial genome of yeast. Cell 15:705-23.

Bos, J. L., C. Heyting, P. Borst, A. C. Arnberg, and E. F. T. Van Bruggen, 1978. An insert in the single gene for the large ribosomal RNA in yeast mitochondrial DNA. Nature 275:336-38.

Burger, G., B. Lang, W. Bandlow, R. J. Schweyen, B. Backhaus, and F. Kaudewitz. 1976. Antimycin resistance in *Saccharamyces cerevisiae*: a new mutation of the mitDNA conferring antimycin resistance on the mitochondrial respiratory chain. Biochem. Biophys. Res. Commun. 72:1201-8.

Carignani, G., G. Dujardin, and P. P. Slonimski. 1979. Petite deletion map of the mitochondrial *oxi3* region in *Saccharomyces cerevisiae*. Molec. Gen. Genet. 167:301-8.

Carnevali, F., G. Morpurgo, and G. Tecce. 1969. Cytoplasmic DNA from petite colonies of *Saccharomyces cerevisiae*: A hypothesis on the nature of the mutation. Science 163:1331-33.

Claisse, M., P. P. Slonimski, J. Johnson, and H. R. Mahler. 1980. Mutations within an intron and its flanking sites: patterns of novel polypeptides generated by mutants in one segment of the *cob-box* region of yeast mitochondrial DNA. Submitted.

Claisse, M. L., A. Spyridakis, M. L. Wambier-Kluppel, P. Pajot, and P. P. Slonimski. 1978. Mosaic organization and expression of the mitochondrial DNA region controlling cytochrome *c* reductase and oxidase. II. Analysis of proteins translated from the *box* region. *In* M. Bacila, B. L. Horecker, and A. O. M. Stoppani (eds.), Biochemistry and genetics of yeast, pp. 369–90. Academic Press, New York.

Cleveland, D. W., G. S. Fischer, M. W. Kirschner, and U. K. Laemmli. 1977. Peptide mapping by limited proteolysis in sodium dodecyl sulfate and analysis by gel electrophoresis. J. Biol. Chem. 252:1102–6.

Cobon, G. S., D. J. Groot Obbink, R. M. Hall, R. Maxwell, M. Murphy, J. Rytka, and A. W. Linnane. 1976. Mitochondrial genes determining cytochrome *b* (complex III) and cytochrome oxidase function. *In* Th. Bucher, W. Neupert, W. Sebald, S. Werner (eds.), Genetics and biogenesis of chloroplasts and mitochondria, pp. 453–60. Elsevier/North-Holland Biomedical Press, Amsterdam.

Colson, A. M., The Van Luu, B. Convent, M. Briquet, and A. Goffeau. 1977. Mitochondrial heredity of resistance to 3-(3,4-dichlorophenyl)-1, 1-dimethylurea, an inhibitor of cytochrome *b* oxidation, in *Saccharomyces cerevisiae*. Europ. J. Biochem. 74:521–6.

Colson, A. M., and P. P. Slonimski. 1979. Genetic localization of diuron- and mucidin-resistant mutants relative to a group of loci of the mitochondrial DNA controlling coenzyme QH_2-cytochrome *c* reductase in *Saccharomyces cerevisiae*. Molec. Gen. Genet. 167:287–98.

Davidson, E. H., and R. J. Britten. 1979. Regulation of gene expression: possible role of repetitive sequences. Science 204:1052–59.

Dhawale, S. 1980. Regulatory interactions between mitochondrial genes. M.Sc. Thesis, Ohio State University, Columbus, Ohio.

Din, N., J. Engberg, W. Kaffenberger, and W. A. Eckert. 1979. The intervening sequence in the 26S rRNA coding region of *T. thermophila* is transcribed within the largest stable precursor for rRNA. Cell 18:525–32.

Douglas, M. G., and R. A. Butow. 1975. Variant forms of mitochondrial translation products in yeast: evidence for location of determinants on mitochondrial DNA. Proc. Nat. Acad. Sci. USA 73:1083–86.

Douglas, M. G., K. Finkelstein, and R. A. Butow. 1979. Analysis of products of mitochondrial protein synthesis in yeast: genetic and biochemical aspects. *In* S. Fleischer and L. Packer (eds.), Methods in enzymology LVI, pp. 58–66. Academic Press, New York.

Dujon, B., P. P. Slonimski, and L. Weill. 1974. Mitochondrial genetics. IX. A model for recombination and segregation of mitochondrial genomes in *Saccharomyces cerevisiae*. Genetics 78:415–37.

Ephrussi, B., H. Hottinguer, and A. M. Chimenes. 1949. Action de l'acriflavine sur les levures. I. La mutation "petite colonie." Ann. Inst. Pasteur (Paris) 76:351–64.

Faye, G., N. Dennebouy, C. Kujawa, and C. Jacq. 1979. Inserted sequence in the mitochondrial 23S ribosomal RNA gene of the yeast *Saccharomyces cerevisiae*. Molec. Gen. Genet. 168:101–9.

Foury, F., and A. Tzagoloff. 1976a. Localization on mitochondrial DNA of mutations leading to a loss of rutamycin-sensitive adenosine triphosphatase. Eur. J. Biochem. 68:113–19.

Foury, F., and A. Tzagoloff. 1976b. Assembly of the mitochondrial membrane system. XIX. Genetic characterization of *mit⁻* mutants with deficiencies in cytochrome oxidase and coenzyme QH_2-cytochrome *c* reductase. Molec. Gen. Genet. 149:43–50.

Foury, F., and A. Tzagoloff. 1978. Assembly of the mitochondrial membrane system: genetic complementation of mit⁻ mutations in mitochondrial DNA of *Saccharomyces cerevisiae*. J. Biol. Chem. 253:3792–97.

Fox, T. D. 1979. Genetic and physical analysis of the mitochondrial gene for subunit II of yeast cytochrome *c* oxidase. J. Mol. Biol. 130:63–82.

Gillham, N. W. 1978. Organelle heredity. Raven Press, New York. 602 pp.

Glover, D., and Hogness, D. 1977. A novel arrangement of the 18S and 28S sequences in a repeating unit of *Drosophila melanogaster* rDNA. Cell 10:167–76.

Grivell, L. A., A. C. Arnberg, P. H. Boer, P. Borst, J. L. Bos, E. F. J. Van Bruggen, G. S. P. Groot, N. B. Hecht, L. A. M. Hensgens, G. J. B. Van Ommen, and H. F. Tabak. 1979. Transcripts of yeast mitochondrial DNA and their processing. *In* D. Cummings, P. Borst, I, Dawid, S. Weissman and C. F. Fox (eds.), Extrachromosomal DNA. ICN-UCLA Symp. on Molecular and Cellular Biology, 15. Academic Press, New York. (In press.)

Haid, A., R. J. Schweyen, H. Bechman, F. Kaudewitz, M. Solioz, and G. Schatz. 1979. The mitochondrial *COB* region in yeast codes for apocytochrome *b* and is mosaic. Eur. J. Biochem. 94:451–64.

Hanson, D. K., D. H. Miller, H. R. Mahler, N. J. Alexander, and P. S. Perlman. 1979. Regulatory interaction between mitochondrial genes. II. Detailed characterization of novel mutants mapping within one cluster in the *cob2* region. J. Biol. Chem. 254:2480–90.

Hensgens, L. A. M., L. A. Grivell, P. Borst, and J. L. Bos. 1979. Nucleotide sequence of the mitochondrial structural gene for subunit 9 of yeast ATPase complex. Proc. Nat. Acad. Sci. USA 76:1663–67.

Heyting, C., and H. H. Menke. 1979. Fine structure of the 21S ribosomal RNA region on yeast mitochondrial DNA. III. Physical location of mitochondrial genetic markers and the molecular nature of ω. Molec. Gen. Genet. 168:279–91.

Heyting, C., J.-L. Talen, P. J. Weijers, and P. Borst. 1979. Fine structure of the 21S ribosomal RNA region on yeast mitochondrial DNA. II. The organization of sequences in petite mitochondrial DNAs carrying genetic markers from the 21S region. Molec. Gen. Genet. 168:251–77.

Jacq, C., C. Kujawa, C. Grandchamp, and P. Netter. 1977. Physical characterization of the difference between yeast mtDNA alleles ω⁺ and ω⁻. *In* W. Bandlow, R. J. Schweyen, K. Wolf, F. Kaudewitz (eds.), Genetics and biogenesis of mitochondria, pp. 255-70. de Gruyter, Berlin.

Johnson, L. 1980. Regulation of dihydrofolate reductase gene expression. *In* this volume.

Kotylak, Z., and P. P. Slonimski. 1976. Joint control of cytochromes *a* and *b* by a unique mitochondrial DNA region comprising four genetic loci. *In* C. Saccone and A. M. Kroon, (eds.), The genetic function of mitochondrial DNA, pp. 143–54. Elsevier/North-Holland Biomedical Press, Amsterdam.

Lancashire, W. E. 1976. Mitochondrial mutations conferring heat or cold sensitivity in *Saccharomyces cerevisiae*. *In* Th. Bucher, W. Neupert, W. Sebald, and S. Werner (eds.), Genetics and biogenesis of chloroplasts and mitochondria, pp. 481–90. Elsevier/North-Holland Biomedical Press, Amsterdam.

Lewin, A., R. Morimoto, M. Rabinowitz, and H. Fukuhara. 1978. Restriction enzyme analysis of mitochondrial DNAs of petite mutants of yeast: classification of petites, and deletion mapping of mitochondrial genes. Molec. Gen. Genet. 163:257–75.

Locker, J., A. Lewin, and M. Rabinowitz. 1979. Review: The structure and organization of mitochondrial DNA from petite yeast. Plasmid 2:155-81.

Locker, J., M. Rabinowitz, and G. S. Getz. 1974. Tandem inverted repeats in mitochondrial DNA of petite mutants of *Saccharomyces cerevisiae*. Proc. Nat. Acad. Sci. USA 71:1366-70.

Macino, G., and A. Tzagoloff. 1979. Assembly of the mitochondrial membrane system: the DNA sequence of a mitochondrial ATPase gene in *Saccharomyces cerevisiae*. J. Biol. Chem 254:4617-23.

Mahler, H. R., D. Hanson, D. Miller, C. C. Lin, N. J. Alexander, R. D. Vincent, and P. S. Perlman. 1978. Regulatory aspects of mitochondrial biogenesis. *In* M. Bacila, B. L. Horecker, and A. O. M. Stoppani (eds.), Biochemistry and genetics of yeast, pp. 513-47. Academic Press, New York.

Mahler, H. R., and J. Johnson. 1979. Specific effects of nalidixic acid on mitochondrial gene expression in *Saccharomyces cerevisiae*. Molec. Gen. Genet. 176:25-31.

Mahler, H. R., and P. S. Perlman. 1979. Mitochondrial biogenesis: evolution and regulation. *In* D. Cummings, P. Borst, I. Dawid, S. Weissman, and C. F. Fox (eds.), Extrachromosomal DNA. ICN-UCLA Symp. on Molecular and Cellular Biology, 15. Academic Press, New York. (In press.)

Mannella, C., R. A. Collins, M. R. Green, and A. M. Lambowitz. 1979. Defective splicing of mitochondrial rRNA in cytochrome-deficient nuclear mutants of *Neurospora crassa*. Proc. Nat. Acad. Sci. USA 76:2635-39.

Martin, N. C., M. Rabinowitz, and H. Fukuhara. 1977. Yeast mitochondrial DNA specifies tRNA for 19 amino acids: deletion mapping of the tRNA genes. Biochemistry 16:4672-77.

Michaelis, G., and E. Pratje. 1977. Mapping of the two mitochondrial antimycin A resistance loci in *Saccharomyces cerevisiae*. Molec. Gen. Genet. 156:79-85.

Mizuuchi, K., M. H. O'Dea, and M. Gellert. 1978. DNA gyrase: subunit structure and ATPase activity of the purified enzyme. Proc. Nat. Acad. Sci. USA 75:5960-63.

Morimoto, R., and M. Rabinowitz. 1979. Physical mapping of the yeast mitochondrial genome: derivation of the fine structure and gene map of strain D273-10B and comparison with a strain (MH41-7B) differing in genome size. Molec. Gen. Genet. 170:25-48.

Pajot, P., M. L. Wambier-Kluppel, Z. Kotylak, and P. P. Slonimski. 1976. Regulation of cytochrome oxidase formation by mutations in a mitochondrial gene for cytochrome *b*. pp. 443-51. *In* Th. Bucher, W. Neupert, W. Sebald, S. Werner (eds.). Genetics and biogenesis of chrloroplasts and mitochondria, pp. 443-51. Elsevier/North-Holland Biomedical Press, Amsterdam.

Perlman, P. S. and C. W. Birky, Jr. 1974. Mitochondrial genetics in baker's yeast: a molecular mechanism for recombinational polarity and suppressiveness. Proc. Nat. Acad. Sci. USA 71:4612-16.

Prunell, A., and G. Bernardi. 1977. The mitochondrial genome of wild-type yeast cells. VI. Genome organization. J. Mol. Biol. 110:53-74.

Rochaix, J. D., and P. Malnoe. 1978. Anatomy of the choroplast ribosomal DNA of *Chlamydomonas reinhardii*. Cell 15:661-70.

Sanders, J. P. M., C. Heyting, M. Ph. Verbeet, F. C. P. W. Meijlink, and P. Borst. 1977. The organization of genes in yeast mitochondrial DNA. III. Comparison of the physical maps of the mitochondrial DNAs from three wild-type *Saccharomyces* strains. Molec. Gen. Genet. 157:239–61.

Sanzey, B. 1979. Modulation of gene expression by drugs affecting deoxyribonucleic acid gyrase. J. Bact. 138:40–47.

Slonimski, P. P. 1979. Organization and expression of interspersed gene segments in mitochondrial DNA coding for cytochrome *b* and controlling cytochrome oxidase. *In* D. Cummings, P. Borst, I. Dawid. S. Weissman, and C. F. Fox (eds.), Extrachromosomal DNA. ICN-UCLA Symp. on Molecular and Cellular Biology, 15. Academic Press, New York. (In press.)

Slonimski, P. P., M. L. Claisse, M. Foucher, C. Jacq, A. Kochko, A. Lamouroux, P. Pajot, G. Perrodin, A. Spyridakis, and M. L. Wambier-Kluppel. 1978a. Mosaic organization and expression of the mitochondrial DNA region controlling cytochome *c* reductase and oxidase. III. A model of structure and function. *In* M. Bacila, B. L. Horecker and A. O. M. Stoppani (eds.). Biochemistry and genetics of yeast, pp. 391–401. Academic Press, New York.

Slonimski, P. P., P. Pajot, C. Jacq, M. Foucher, G. Perrodin, A, Kochko, and A. Lamouroux. 1978b. Mosaic organization and expression of the mitochondrial DNA region controlling cytochrome *c* reductase and oxidase. I. Genetic, physical, and complementation maps of the *box* region. *In* M. Bacila, B. L. Horecker, and A. O. M. Stoppani (eds.), Biochemistry and genetics of yeast, pp. 339–68. Academic Press, New York.

Slonimski, P. P., and A. Tzagoloff. 1976. Localization in yeast mitochondrial DNA of mutations expressed in a deficiency of cytochrome oxidase and/or coenzyme QH_2-cytochrome *c* reductase. Eur. J. Biochem. 61:27–41.

Smith, C. L., M. Kubo, and S. Imamoto. 1978. Promoter-specific inhibition of transcription by antibiotics which act on DNA gyrase. Nature 275:420–23.

Steinkeler, J., and H. R. Mahler. 1980. Regulatory interactions between mitochondrial genes: exon and intron phenotypes observed *in vivo* can be expressed *in vitro*. Plasmid (submitted).

Subik, J., V. Kovacova, and G. Takacsova. 1977. Mucidin resistance in yeast: isolation, characterization, and genetic analysis of nuclear and mitochondrial mucidin-resistant mutants of *Saccharomyces cerevisiae*. Eur. J. Biochem. 73:275–86.

Tzagoloff, A., A. Akai, R. B. Needleman, and G. Zulch. 1975. Assembly of the mitochondrial membrane system: cytoplasmic mutants of *Saccharomyces cerevisiae* with lesions in enzymes of the respiratory chain and in the mitochondrial ATPase. J. Biol. Chem. 250:8236–42.

Tzagoloff, A., F. Foury, and A. Akai. 1976. Assembly of the mitochondrial membrane system. XVIII. Genetic loci on mitochondrial DNA involved in cytochrome *b* biosynthesis. Molec. Gen. Genet. 149:33–42.

Tzagoloff, A., G. Macino, M. P. Nobrega, and M. Li. 1979. Organization of mitochondrial DNA in yeast. *In* D. Cumming, P. Borst, I. Dawid, S. Weissman, and C. F. Fox (eds.), Extrachromosomal DNA. ICN-UCLA Symp. on Molecular and Cellular Biology, 15. Academic Press, New York. (In press.)

Van Ommen, G.-J. B., G. S. P. Groot, and P. Borst. 1977. Fine structure physical mapping of 4S RNA genes on mitochondrial DNA of *Saccharomyces cerevisiae*. Molec. Gen. Genet. 154:255–62.

Vincent, R. D., P. S. Perlman, R. L. Strausberg, and R. A. Butow. 1980. Physical mapping of genetic determinants on yeast mitochondrial DNA affecting the size of the var1 polypeptide. Submitted.

Yang, H. L., K. Heller, M. Gellert, and G. Zubay. 1979. Differential sensitivity of gene expression *in vitro* to inhibitors of DNA gyrase. Proc. Nat. Acad. Sci. USA 76:3304-8.

JERRY B. LINGREL, KY LOWENHAUPT, KATE SMITH,
JOEL HAYNES, PAUL ROSTECK, JR., ERIC SCHON,
PATTY GALLAGHER, DOUG BURKS, AND
DONALD VAN LEEUWEN

Globin Genes: Structure and Regulation of Their Expression

11

INTRODUCTION

Considerable progress has been made during the past few years in understanding how mRNAs of eukaryotic cells are synthesized. Many mRNAs are transcribed as larger precursors, and these additional sequences must be removed during processing. Also, modifications such as polyadenylation and capping must occur. Once these processing steps have taken place, the mRNA is transported to the cytoplasm. These events, along with mRNA turnover, represent possible regulatory points in determining the amount of mRNA accumulated in cells. Our studies are concerned with both the pathway of globin mRNA synthesis and the identification of factors that affect the accumulation of this mRNA. Some of our studies in this area are discussed below.

SYNTHESIS OF GLOBIN mRNA

Globin mRNA was the first mammalian mRNA to be purified (Marbaix and Burney, 1964) and translated in a heterologous cell-free system (Lockard and Lingrel, 1969, 1971). The availability of a relatively pure cell population, namely, reticulocytes, that synthesize one predominant protein, and the development of an active cell-free translation system

Jerry B. Lingrel, University of Cincinnati Medical Center, Department of Biochemistry, 231 Bethesda Avenue, Cincinnati, Ohio 45267

(Adamson et al., 1968) made this possible. Approximately 95% of the proteins synthesized by reticulocytes are α and β globin chains, and a corresponding amount of the total mRNA codes for these proteins. Thus, the presence of two predominant mRNAs in these cells allowed for their easy purification.

In subsequent studies these RNAs, as well as most other mRNAs from higher organisms, were found to be modified at both the 5' and 3' termini. The 3' ends of the mouse α and β globin mRNAs contain a poly(A) region that at steady state exists in two size classes, one 35 nucleotides in length and the other 45 nucleotides long (Gorski et al., 1974). Newly synthesized globin mRNA contains approximately 150 nucleotides in the poly(A) region (Merkel et al. 1975), which is shortened to the steady state size with time (Gorski et al. 1975, Merkel et al., 1975).

Post-transcriptional alteration of the 5' termini of mRNAs also occurs (Rottman et al., 1975). The modified structures present on globin mRNA are $^{-m}$Gppp^{m}AmC and ^{7m}Gppp^{m}AmCm (Muthukrishnan et al., 1975; Perry et al., 1975; Heckle et al., 1977; Cheng and Kazazian, 1977). The A and C residues are transcribed but the G residue, and methyl groups are added post-transcriptionally.

Globin mRNA (Kwan et al., 1977; Curtis and Weissmann, 1976; Ross, 1976) as well as many other mRNAs (Darnell, 1979) are transcribed in the form of precursors. These precursors were identified and in some instances isolated using a complementary DNA copy of globin mRNA (cDNA). The polyacrylamide gel analysis of mature globin mRNA and its precursor isolated from erythroleukemia cells is shown in figure 1. Cells were incubated in the presence of ^{3}H uridine for 10 min and total RNA isolated. Globin RNA sequences were then purified and isolated from the total RNA using globin cDNA cellulose affinity chromatography (Wood and Lingrel, 1977). Analysis of globin RNAs in denaturing gels revealed that, in addition to mature globin mRNA, an RNA of about 1,500 nucleotides was present. This larger RNA also occurs in nucleated erythroid cells of mouse spleens (Kwan et al., 1977) and fetal liver (Ross, 1976). Subsequent studies showed that the larger RNA was a precursor to β globin (Curtis et al., 1977) and that it exhibited a precursor-product relationship to mature β globin mRNA (Ross, 1976).

The extra sequences in the precursor are located not at the ends, as might have been expected, but rather between coding sequences or sequences that appear in mature mRNA. The sequence organization of this RNA was determined in several laboratories (Kinniburgh et al., 1978; Tilghman et al., 1978), including our own (Smith and Lingrel, 1978). Our approach utilized a hybridization-nuclease procedure. Complimentary DNA was

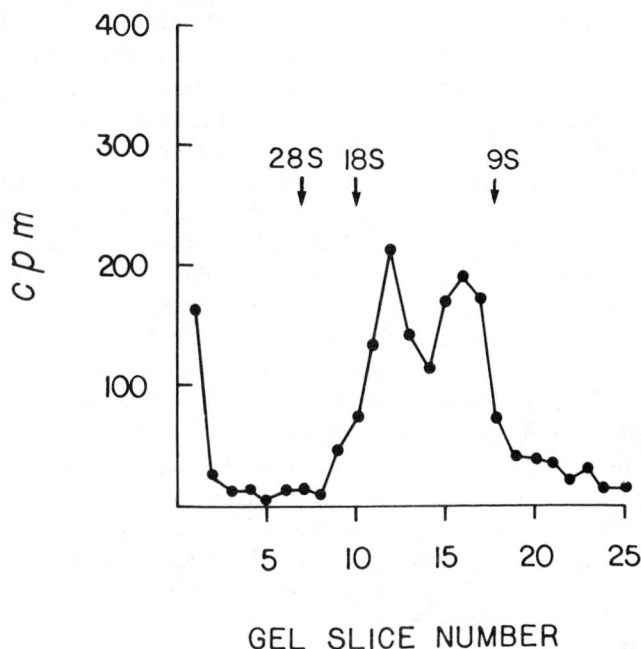

Fig. 1. Analysis of globin mRNA-containing sequences in erythroleukemia cells. Erythroleukemia cells were labeled for 10 min in the presence of ^{3}H uridine, total RNA isolated, and the globin mRNA sequences purified using cDNA cellulose affinity chromatography as previously described (Kwan et al., 1977). The RNA was analyzed by polyacrylamide gel electrophoresis in the presence of 99% formamide (Smith et al., 1978).

synthesized using mature globin mRNA as a template, and this cDNA was hybridized to the ^{3}H β globin mRNA precursor (see fig. 2). The cDNA-precursor RNA hybrid was then digested separately with either RNase A, which degrades unhybridized RNA, or RNase H, which digests only RNA hybridized to DNA, and the resistant precursor RNA fragments analyzed by polyacrylamide gel electrophoresis. If the additional sequences are present between mRNA sequences, they will be looped out in the hybrid and be degraded by RNase A (fig. 2a). Thus, the resistant or hybridized material will be recovered as two or more fragments. If the extra sequences are located on one end or the other, or on both ends, only one fragment will be obtained. When the RNase A-resistant material was analyzed, three fragments were obtained (Smith and Lingrel, 1978), indicating that two intervening sequences do indeed interrupt the mature mRNA sequence. Following RNase H digestion of hybridized RNA (Smith et al., 1978), two fragments were recovered that represent the looped-out intervening sequences. Their sizes were 730 and 100 nucleotides respectively. The digestion pattern of the hybrid formed between cDNA and mature globin mRNA is diagrammed in figure 2b for comparison. Similar results were obtained by Kinniburgh et al. (1978). During the course of these studies, Jeffreys and Flavell (1977) presented evidence that the sequences coding for globin mRNA in the genomic DNA are discontinuous; that is, sequences that eventually end up in mature globin mRNA are interrupted by DNA sequences not present in globin mRNA. Electron microscopic studies of duplexes formed between genomic clones of a β globin gene and mature globin mRNA further supported this finding (Leder et al., 1978). Regions of the DNA that are not present in mature globin mRNA do not hybridize, and exist as loops. These studies were extended to duplexes formed between the DNA of a clone containing the β globin gene and β globin mRNA precursor (Tilghman et al., 1978). A linear hybrid was obtained in electron micrographs demonstrating that the intervening sequences present in the gene are transcribed in the precursor. The entire mouse β globin gene has now been sequenced (Konkel et al., 1978) and the location of the intervening sequences unequivocally established. The sequence organization of the β globin precursor is shown in figure 3. A precursor of a globin mRNA smaller than the β globin precursor has also been observed (Ross, 1976; Curtis et al., 1977).

Thus, there appear to be several levels at which the amount of globin mRNA can be controlled. These are summarized in figure 4 and include transcription, precursor processing (i.e., polyadenylation, methylation, capping, and removal of intervening sequences), transport to the cytoplasm, and, finally, turnover of mRNA in the cytoplasm.

REGULATION OF mRNA ACCUMULATION

General Considerations and Description of the Biological System

It is generally accepted that transcription is the major mechanism for regulating gene expression. Studies based on the accumulation of mature cytoplasmic globin mRNA support this conclusion (Ross et al., 1974; Harrison et al., 1974). However, the rate of transcription is not generally measured directly nor is the contribution of precursor processing and mRNA turnover evaluated. For this reason, we have examined these parameters in erythroid cells.

It would be advantageous to have a pure population of pluripotent stem cells (CFUs) that could be stimulated by a defined signal to differentiate

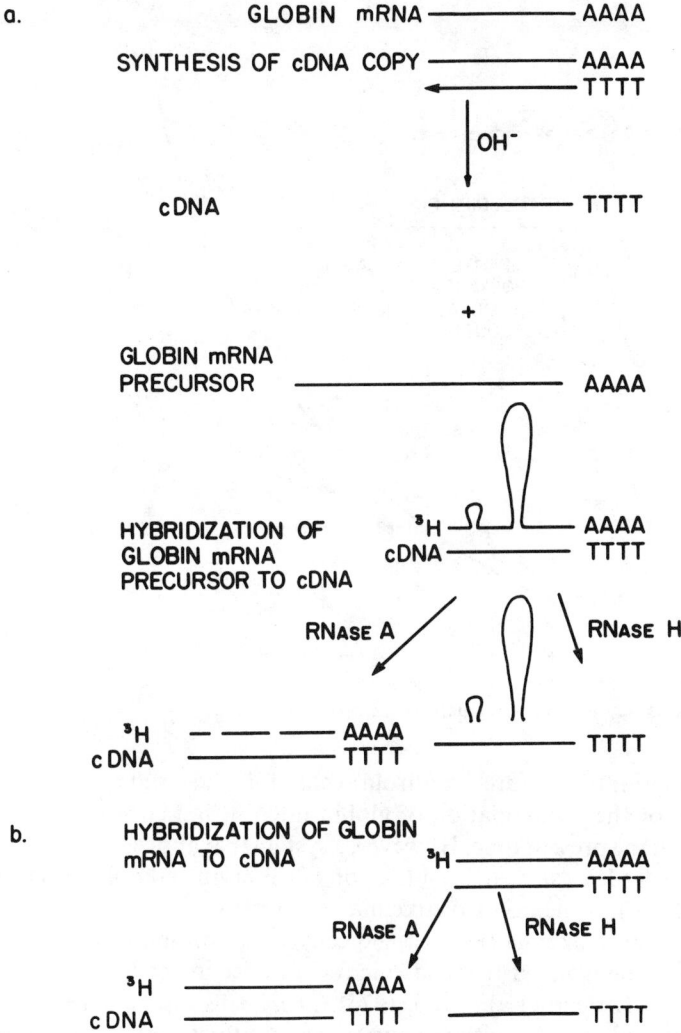

Fig. 2. Schematic representation of the method used for showing that intervening sequences are present in the β-globin mRNA precursor. See text for details.

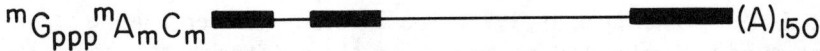

Fig. 3. Sequence organization of the β-globin mRNA precursor of mouse. Heavy lines represent sequences that appear in mature mRNA; thin lines depict intervening sequences.

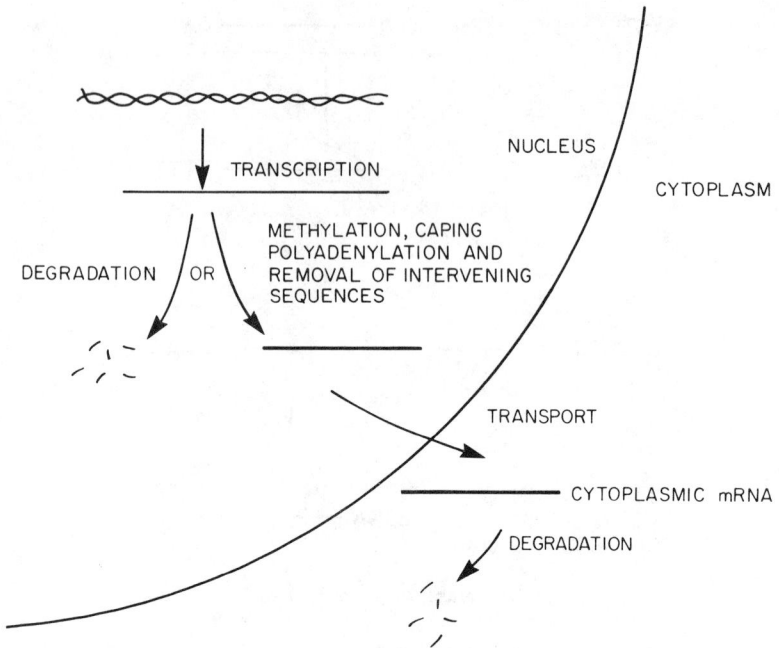

Fig. 4.   Potential levels of gene regulation in eukaryotic cells.

synchronously into mature erythroid cells for the study of factors responsible for the accumulation of globin mRNA, but such a system is unavailable at the present time. However, a system that approximates these requirements, at least over the last portion of the erythroid differentiation program, is the murine erythroleukemia cell system.

Murine erythroleukemia cells (Friend cells) are a continuously growing tissue culture line isolated from splenic foci of mice infected with Friend leukemia virus (Friend et al., 1966, 1974). These cells resemble proerythroblasts morphologically, with the exception that they contain both integrated Friend virus and A- and C-type particles (Sato et al., 1971). Friend cells spontaneously differentiate at a low frequency (probability of less than 0.5%) into hemoglobin-containing cells resembling orthochromatic normoblasts (Friend et al., 1971; Sato et al., 1971) or reticulocytes (Furusawa et al., 1972; Ostertag et al., 1972). The frequency of differentiation can be increased to nearly 100% by the addition of a wide variety of chemical inducers (Leder and Leder, 1975; Tanaka et al, 1975; Bernstein et al., 1976; Gusella and Housman, 1976; Reuben et al., 1976; Gazitt et al., 1978; Scher and Friend, 1978; Terada et al., 1978), of which dimethylsulfoxide (DMSO) is the most widely used (Friend et al., 1971).

A large number of biochemical events have been identified in Friend cell differentiation. These events include the accumulation of globin mRNAs and their respective globins (Boyer et al., 1972; Ross et al., 1974), changes in sensitivity to agglutination (Eisen et al., 1977a), cell surface antigens (Ikawa et al., 1973), and activities of heme biosynthetic enzymes (Sassa, 1976; see also review by Marks and Rifkind, 1978). Differentiation of Friend cells occurs with a moderate degree of synchrony, making it possible to study the temporal order of events in this process. It is not yet possible to compare the program of erythroleukemia cell induction with that of normal erythroid cell differentiation, because it has been difficult to isolate enriched populations of normal erythroid cells at any intermediate stage of differentiation. However, uninduced cells resemble early erythroid cells, whereas cells induced for four or more days resemble both cell populations from the spleens of anemic mice (predominantly orthochromatic normoblasts) and reticulocytes with respect to most of the various parameters studied. Interferon and viral induction are obvious exceptions (Friend et al., 1974).

Whereas uninduced Friend cells are capable of unlimited proliferation, exposure to DMSO or other inducing agents reprograms these cells for terminal erythroid differentiation (Gusella et al., 1976; Friedmann and Schildkraut, 1977). A significant event in this reprogramming has been designated "commitment." Commitment occurs when cells lose the ability to proliferate without limit and initiate the first of four final cell divisions. These four divisions are comparable to the four divisions that occur in normal erythroid differentiation as proerythroblasts develop into red blood cells. The commitment of individual cells can be observed by cloning these cells *in vitro* in semisoft agar (Gusella et al., 1976). If Friend cells exposed to inducer are cloned without the inducer, the proportion of small colonies (16 cells) after four days reflects the proportion of committed cells in the original population. In the presence of most inducers, this proportion increases within the first day of induction and reaches a maximum by three days of induction (Gusella et al., 1976).

Other events in the program of differentiation of Friend cells have been classified with respect to their temporal and causal relationship to commitment (Marks and Rifkind, 1978). Events that precede commitment or that do not require it are designated "early" and include changes in membrane permeability, in cAMP levels, and in the accumulation of spectrin, globin mRNA, and globin protein. Events that occur only after commitment can be called "late"; these include the increase in activities of the enzymes involved in heme biosynthesis, the appearance of IP_{25}, a chromatin protein found in Friend cells induced with DMSO (Keppel et al., 1977), and the accumulation of hemoglobin. Classification of events into early and late programs allows for some elucidation of the relationships that make up the over-all program of erythroid differentiation.

Although most effectors of Friend cell differentiation, including DMSO, initiate the entire program of events, hemin appears to induce only early events. Friends cells exposed to 75-100 μM hemin accumulate spectrin (Rifkind et al., 1979) and synthesize increased amounts of globin mRNA (Ross and Sautner, 1976; Nudel et al., 1977; Rovera et al., 1978) and globin (Rovera et al., 1977); however, they do not cease cell division or become committed (Housman et al., 1978). Cultures induced with hemin give identical results in the commitment assay as to uninduced cultures. When cloned in semisoft agar, less than 1% of the clones contain 16 or fewer cells. This has been established for a number of Friend cell lines including 745, the line used in the following studies (Housman et al., 1978; Marks et al., 1979). Friend cells induced with hemin do not stain positive for hemoglobin with benzidine (Rovera et al., 1977). In addition, hemin does not induce IP_{25}, nor does it affect enzyme activities that have been identified as part of the late program.

Accumulation of Globin mRNA in Erythroleukemia Cells Induced with Hemin

In order to evaluate the contributions of transcription, post-transcriptional processing, and turnover in the accumulation of globin mRNA, we have chosen erythroleukemia cells induced with hemin. These cells divide continuously and do not carry out the later program of differentiation, namely, the termination of cell division. This is advantageous to our studies since one does not have to be concerned with changes in cell division and terminal maturation of the erythroid cells. The accumulation of globin mRNA in erythroleukemia cells induced with hemin is shown in figure 5. Cells accumulate globin mRNA rapidly and reach a steady state level of the RNA following one-day exposure to the inducer. This steady state level is maintained as long as the cells are subcultured. Cells induced with hemin must be subcultured when they reach a density of 1×10^6 cells/ml or they die. In practice, cells are seeded at a concentration of 1×10^5 cells/ml and subcultured when they reach 10^6 cells/ml. The times of subculture are indicated on the figure by arrows.

Evaluation of Transcriptional and Post-transcriptional Processing in the Accumulation of Globin mRNA

In order to evaluate the contributions of transcription and post-transcriptional processing to the accumulation of globin mRNA

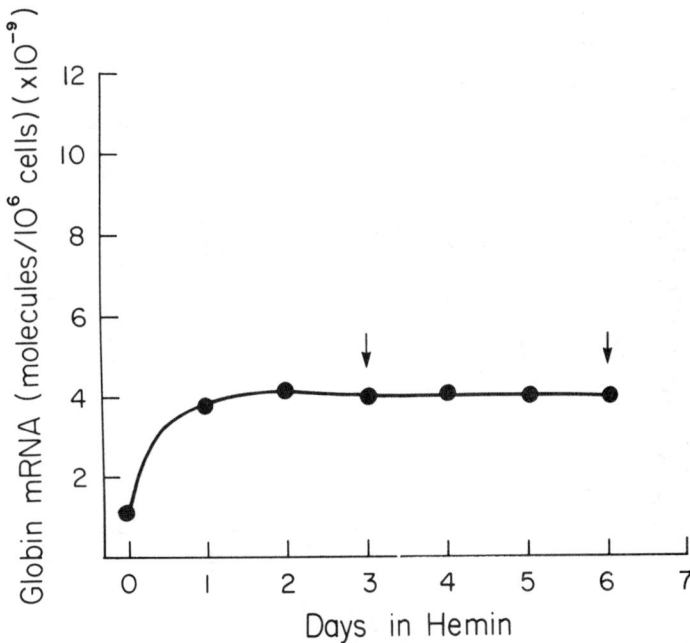

Fig. 5.   Accumulation of globin mRNA in erythroleukemia cells induced with hemin. Cells were innoculated at 1×10^5 cells per ml and 75 μM hemin added. The cells were subcultured to 1×10^5 cells per ml at the times indicated by the arrows on the figure. At various times cells were harvested and total cytoplasmic RNA isolated. The amount of globin mRNA present was determined by hybridization to radioactive cDNA prepared to globin mRNA. For details of the hybridization, see Lowenhaupt et al., 1978.

throughout the differentiation process, RNA was labeled for periods of 10 min and 2 hr at various times during induction; then the globin mRNA sequences were isolated and analyzed. Globin mRNA isolated following 2 hr of label represents the sum of both transcriptional and post-transcriptional events, since it is known that processing events are more rapid than 2 hr, while radioactivity incorporated into the globin mRNA sequences at 10 min is a more accurate reflection of transcription alone. At 10 min most of the label is present in the precursor rather than in mature mRNAs (Lowenhaupt et al., 1978). It is possible that during induction there is no change in the rate of transcription of the globin mRNA, but rather a stabilization of the mRNA or a more efficient processing of the precursor occurs. However, this does not appear to be the case, as shown in figure 6. (Also see Lowenhaupt and Lingrel, 1979). Very little incorporation into either mature globin mRNA or its precursor occurs in uninduced cells, indicating that these cells or those early in differentiation do not

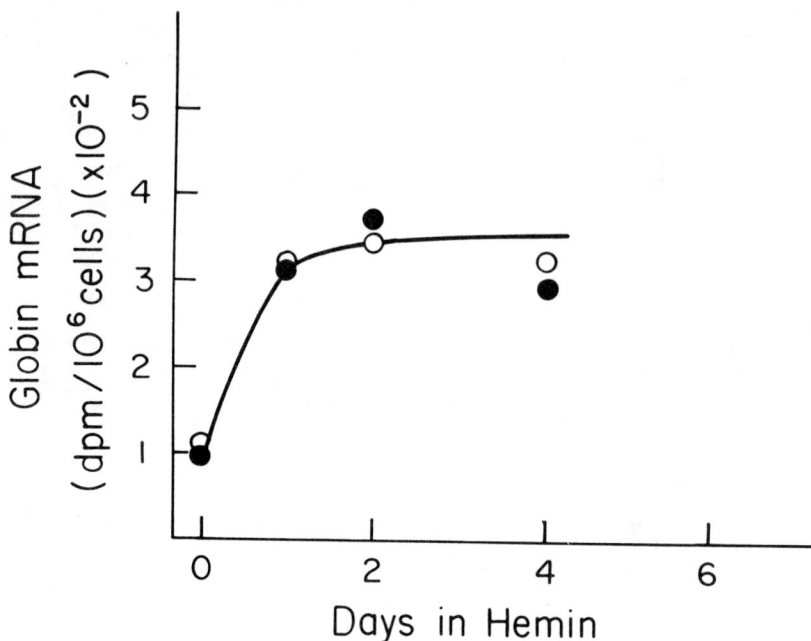

Fig. 6.   Synthesis of globin mRNA in erythroleukemia cells induced with hemin. Uninduced cells and cells at various times following induction with hemin were isolated and incubated in the presence of ^{3}H uridine for either 10 min or 2 hr. Total RNA was isolated from cells labeled for 10 min and the globin mRNA sequences isolated by cDNA chromatography as described by Lowenhaupt et al., 1979. With cells labeled for 2 hr only cytoplasmic RNA was isolated. The amounts of radioactivity binding to globin cDNA cellulose on a second chromatography are given in the figure. Closed circles represent 10 min samples, open circles represent 2 hr samples.

synthesize mature globin mRNA or its precursor to an appreciable extent. Thus, the globin genes are not actively transcribed in uninduced cells, and the accumulation of globin mRNA is not the result of more efficient processing of precursor following induction. The coincidence of the 10 min and 2 hr labeling values is the result of the larger amount of ^{3}H uridine used at the shorter labeling times. However, the fact that these two lines are parallel indicates that there is no change in precursor processing efficiency during hemin-induced globin mRNA synthesis. It has already been determined by Ross (1976) and Curtis and colleagues (1977) that globin mRNA precursors are processed efficiently; therefore, we can conclude that post-transcriptional processing does not play an important role in the accumulation of globin mRNA in erythroleukemia cells induced with hemin. We have obtained similar results in erythroleukemia cells induced with DMSO (Lowenhaupt et al., 1978).

Determination of Globin mRNA Half-life in
Erythroleukemia Cells Induced with Hemin

The half-life of a mRNA is just as important in determining the final amount of that mRNA in the cell as is the rate of transcription (Lowenhaupt and Lingrel, 1978). This parameter was evaluated by directly measuring the half-life of globin mRNA in cells induced with hemin. Cells were labeled for 2 hr with ^3H uridine at various times throughout the induction period, followed by incubation in media containing no uridine. The loss of radioactivity from various RNAs was determined following incubation in media not containing the labeled precursor. These results are summarized in figure 7. As can be seen, ribosomal RNA exhibits a long half-life, greater than 150 hr, as has been observed in other tissue culture cells. The total poly(A)$^+$ RNA decays with a half-life between 2 and 20 hr. In contrast to the total poly(A)$^+$ mRNA, globin mRNA is stable at all times of induction, exhibiting a half-life similar to that of ribosomal RNA. This indicates that the globin mRNA in these cells is essentially immortal and that the steady state level of mRNA is determined by the rate of transcription and cell division.

PREPARATION AND ISOLATION OF CLONES
CARRYING THE GOAT GLOBIN GENES

From the studies above it can be inferred that post-transcriptional processing is not an important factor in regulating the amount of globin mRNA during erythroid cell differentiation, at least in erythroleukemia cells, and that the rate of transcription controls the amount of globin mRNA present. Therefore, our emphasis has shifted to defining the factors that affect the rate of transcription. Our approach to this problem has been to develop a system where the expression of genes at the level of DNA can be studied. To this end we have isolated a set of developmentally regulated genes; namely, the embryonic, fetal, juvenile, and adult globin genes of the goat. We plan to study the transcription of these genes *in vitro* using factors isolated from erythropoietic tissues at various developmental stages in which these genes are expressed. The goat provides an excellent system for such studies. During the first twenty days after fertilization, embryonic hemoglobins are synthesized that are then replaced by fetal hemoglobin. Goat fetal hemoglobin, like that of humans, is composed of two α and two γ chains. Before birth, the γ chains are replaced by β^c chains, giving rise to HbC or juvenile hemoglobin. During the first year of life, this hemoglobin is replaced by adult hemoglobin, HbA, which is composed of α and β^A globin chains. Thus, one can obtain tissue or cells that exclusively express only one of the genes of the β family. For example, fetal cells synthesize α

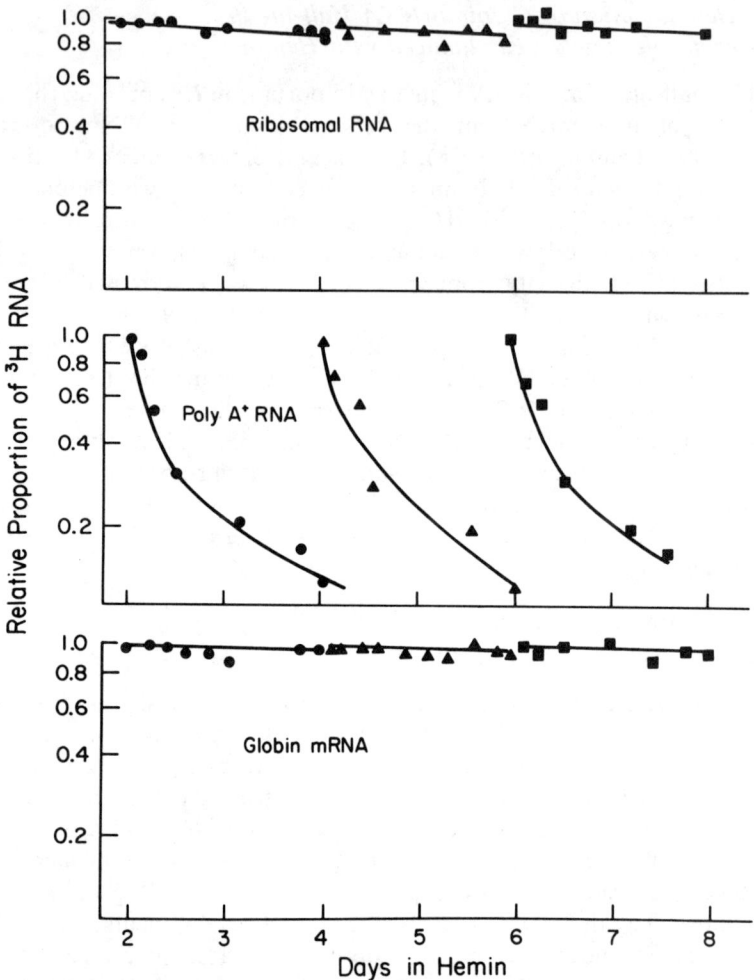

Fig. 7. Turnover of ribosomal, poly(A)$^+$ and globin mRNA in erythroleukemia cells induced with hemin. Erythroleukemia cells induced with hemin for 2 days, 4 days, and 6 days were harvested and incubated for 2 hr in the presence of ^{3}H uridine. Following removal of the radioactive uridine, the culture was continued in the presence of unlabeled media and the loss of radioactivity in the various RNA fractions followed with time. Dots represent RNAs from cells induced for 2 days; triangles represent RNA isolated from cells induced for 4 days; and blocks represent RNA from cells induced for 6 days.

and γ chains, bone marrow of young goats synthesize α and β^C chains, adult bone marrow synthesizes α and β^A globins.

Our approach to isolating these genes has been to prepare a recombinant library of the genomic DNA of the goat and to select those recombinants carrying the goat globin genes (Robbins et al., 1979). The procedure is summarized in figure 8. The cloning vector was the λ bacteriophage derivative Charon 4A developed by Blattner and colleagues (1977). The DNA of this phage contains three Eco RI restriction sites. The two internal fragments produced from Eco RI cleavage are not required for lytic infection of the host bacterium and can be replaced by heterologous DNA. In practice, the cohesive ends of the phage are ligated before Eco RI digestion, thus facilitating separation of the phage "arms" from the two internal fragments. Viable phages are produced when foreign DNA fragments of between 12 and 22 kb are ligated to the 31 kb arms of the vector. This size requirement means that entire genes, including linked genes and/or extensive regions of DNA surrounding genes of interest, will be present in the cloned DNA. The goat DNA to be inserted into the phage was isolated as high molecular weight DNA and partially digested with the restriction endonuclease Eco RI. The digestion conditions were such that

CLONING GOAT GLOBIN GENES

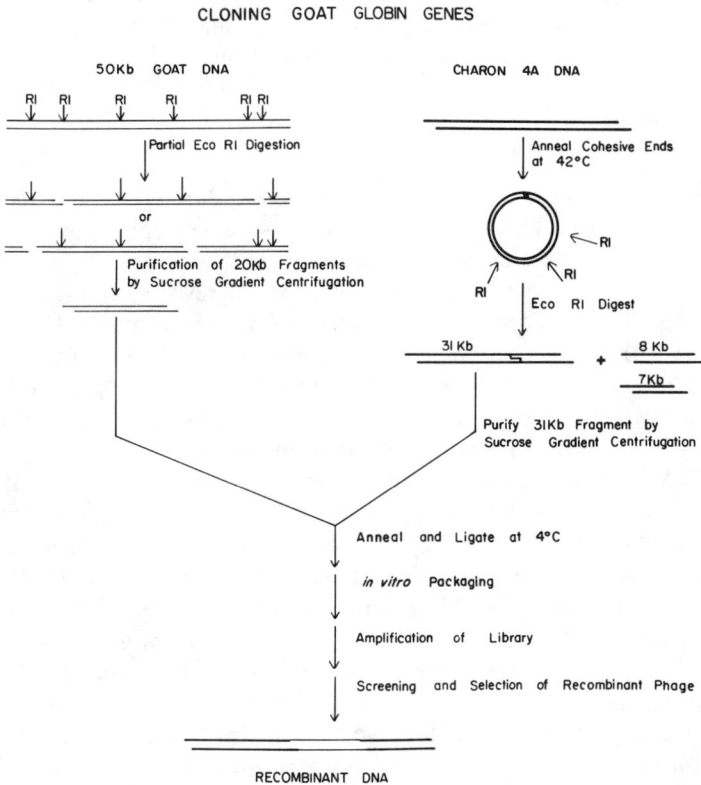

Fig. 8.   Summary of procedures used in cloning goat DNA. See text for description.

DNA fragments of approximately 15 to 20 kb were produced. By performing a partial digestion, a series of random fragments were obtained that should provide overlapping sequences in the collection of recombinants. Thus, it is possible to isolate clones with overlapping sequences, allowing construction of an extensive map of the regions surrounding particular genes.

The goat DNA Eco RI fragments were covalently linked to the Charon 4A tails and the resulting recombinants packaged into infectious particles using λ packaging extracts (Sternberg et al., 1977). The recombinant phages were allowed to infect *E. coli* strain DP50 *sup* F and amplified. This resulted in a collection of recombinants, or genomic library, in which most of the goat DNA sequences are represented. (For further discussion of the preparation of libraries, see Maniatis et al., 1978.) The clones of the library were screened using cDNA prepared to goat globin mRNA. Positive recombinants were mapped with various restriction enzymes and the location of the globin coding sequences determined by hybridization. One of the clones isolated was designated Clone 4, and the construction of its restriction endonuclease map is described below.

Construction of a Restriction Site Map for the Insert of Clone 4

The locations of the cleavage sites of Eco RI, Bam HI, Kpn I, and Hind III within the goat DNA insert of Clone 4 were determined by a series of digestions with these enzymes used alone and in various combinations. Some sites were determined by isolating specific restriction fragments and digesting with a second restriction enzyme. Digestion of the DNA from Clone 4 with the restriction endonuclease Bam HI produces fragments of 25.0, 6.2, and 3.7 kb that contain inserted goat DNA (fig. 9a). The 25.0 kb fragment is the only Bam HI digestion product large enough to contain the 14 kb Bam HI-Eco RI segment found in the left-hand portion of Charon 4A DNA, between the Bam HI site at map position 5,540 and the left-hand Eco RI site of insertion at position 19,608 (refer to fig. 14 for map assignments). The Bam HI, Kpn I, Eco RI, and Hind III restriction sites within Charon 4A DNA have been determined by Blattner and colleagues (personal communication). The remainder of this 25 kb Bam HI fragment is a 12 kb portion of the inserted eukaryotic DNA terminating at position 31,303. The second largest insert-containing Bam HI fragment, 6.2 kb, by the same reasoning, must contain the right-hand Eco RI site of insertion and 5 kb of the right-hand 10.9 kb vector arm from the Eco RI site at position 36,263 to the next Bam HI site at position 41,220. The assignment

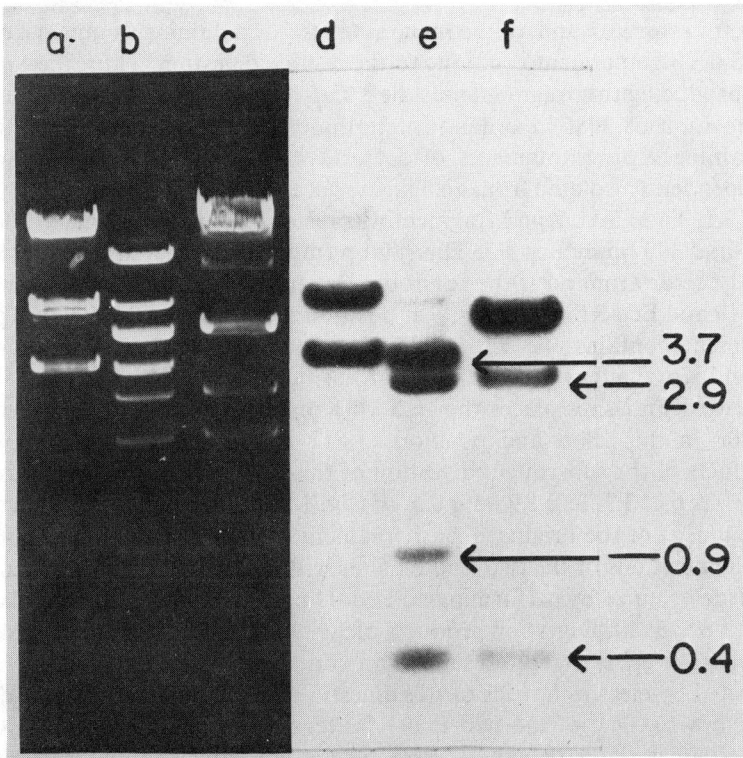

Fig. 9. Restriction endonuclease cleavage of Clone 4 DNA and detection of fragments containing globin sequences. Approximately 1 μg of purified DNA was digested with 2 units of Eco RI, Bam HI, or both enzymes and electrophoresed on a 0.3% agarose slab gel in the presence of ethidium bromide as described previously (Robbins et al., 1979). Fragments were transferred to nitrocellulose paper by the method of Southern (1975) and hybridized to globin [^{32}P] cDNA. Lane (a) is a Bam HI digest of Clone 4, and (d) is the corresponding autoradiograph. Lane (b) is an Eco RI-Bam HI double digest of Clone 4, and (e) is the corresponding autoradiograph. Arrows indicate fragments referred to in the text.

of the 6.2 kb fragment to the right-hand overlap places its Bam HI terminus within the insert at position 35,003. The remaining Bam HI fragment of 3.7 kb must lie between the other two vector-containing Bam HI fragments and consist entirely of inserted genomic DNA.

In order to assign the Kpn I sites within the insert, a double digestion with the enzymes Kpn I and Bam HI was performed (fig. 10). The only insert-containing Bam HI fragment cleaved by Kpn I is the 25 kb fragment. Kpn I digestion of this fragment produced fragments of 11.5, 5.4, 4.3, 3.0, and 1.5 kb. The 11.5 and 1.5 kb fragments are derived entirely from the 19.6

kb left vector arm and, of the remaining insert-containing fragments, only the 3.0 kb component is unique to the double digestion. Thus, the Kpn I site producing this fragment must lie 3 kb from the terminal Bam HI site or at position 28,308. In order to determine the arrangement of the two remaining Kpn I fragments of 5.4 and 4.3 kb, these fragments were independently isolated from an agarose gel and redigested with the enzyme Eco RI. The 4.3 kb Kpn I fragment was recleaved into fragments of 2,900, 989, and 390 bp by Eco RI. The 989 bp fragment is derived from the left-hand vector arm and is the result of cleavage at its Kpn I site at position 18,619 and Eco RI insertion site at position 19,608. Thus, the 4.3 kb Kpn I fragment contains the left-hand vector-insert junction and places the second Kpn I site within the insert at position 22,908. This assignment of the two Kpn I sites places the 15.5 kb Kpn I fragment between position 28,308 in the insert and position 43,612 in the vector right arm. The products of the subsequent digestion of this fragment with Bam HI are of the sizes 6.2, 3.7, 3.0, 1.9, and 0.5 kb (fig. 10) and are consistent with this assignment of the original Kpn I fragment.

The positions of the two Eco RI sites within the 4.3 kb Kpn I fragment were determined by performing an Eco RI partial digestion on the isolated Kpn I fragment. Digestion products of 3.9, 3.3, 2.9, and 0.99 kb were sized by agarose gel electrophoresis and an additional fragment of 390 bp was detected by electrophoresis of this digestion in 4.0% polyacrylamide (data not shown). One of the two Eco RI sites responsible for this digestion pattern, based on the above data, is the left-hand site of insertion at position 19,608. It can be concluded from an analysis of the partial (3.9 and 3.3 kb) and complete (2,900, 990, and 390 hp) digestion products that the second Eco RI site within the 4.3 kb Kpn I fragment lies 2.9 kb from the Eco RI site of insertion and 390 bp from the internal Kpn I terminus, or at position 22,508.

The single Eco RI site within the 5.4 kb Kpn I fragment was positioned by isolating this fragment from an agarose gel and redigesting it with Eco RI. Products of 3.7 and 1.8 kb were obtained. This result indicates that the 1.8 kb fragment and the 390 bp fragment derived from Eco RI digestion of the isolated 4.3 kb Kpn I fragment lie adjacent to each other in the insert and together constitute the 2.2 Eco RI fragment. Thus, the unassigned Eco RI site lies at position 24,708. The remaining 3.7 kb Eco RI-Kpn I fragment could be derived from either the 5.1 or 4.85 kb Eco RI fragments. In order to determine which of these fragments contains a Kpn I site, they were coisolated from an agarose gel, redigested with Kpn I, and the products fractionated on a 0.8% agarose gel. Only the 5.1 kb Eco RI fragment was recleaved by Kpn I and produced fragments of 3.6 and 1.5 kb

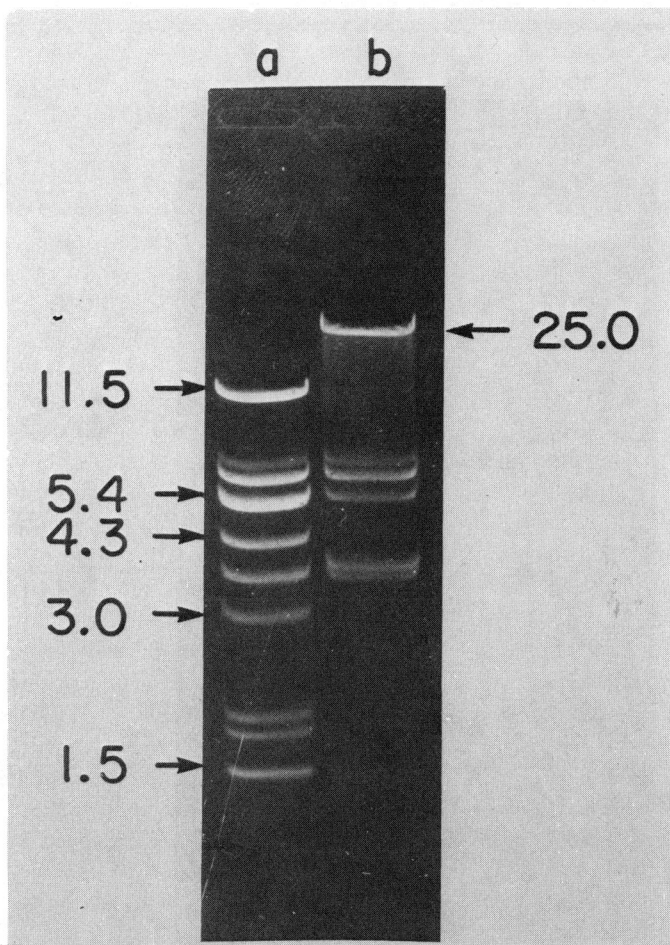

Fig. 10.   Agarose gel analysis of Bam HI-Kpn I double digestion and Bam HI digestion of Clone 4 DNA. Restriction endonuclease digestions and electrophoresis on 0.6% agarose are as described in the legend to figure 9. Lane (a) is a double digestion of Clone 4 DNA with Kpn I and Bam HI. Lane (b) is Clone 4 DNA digested with Bam HI.

(fig. 11c). The 5.1 kb Eco RI fragment, therefore, maps between positions 24,708 and 29,808.

The two remaining Eco RI sites within the genomic insert were assigned based on the results of Eco RI redigestion of the 6.2 kb Bam HI fragment isolated from an agarose gel. Electrophoresis of the resulting digestion products resolved fragments of 4,957 bp, 850 bp, and 405 bp. The 4,957 bp digestion product is derived solely from the λ Charon 4A right vector arm between the Eco RI site of insertion at position 36,263 and the nearest Bam

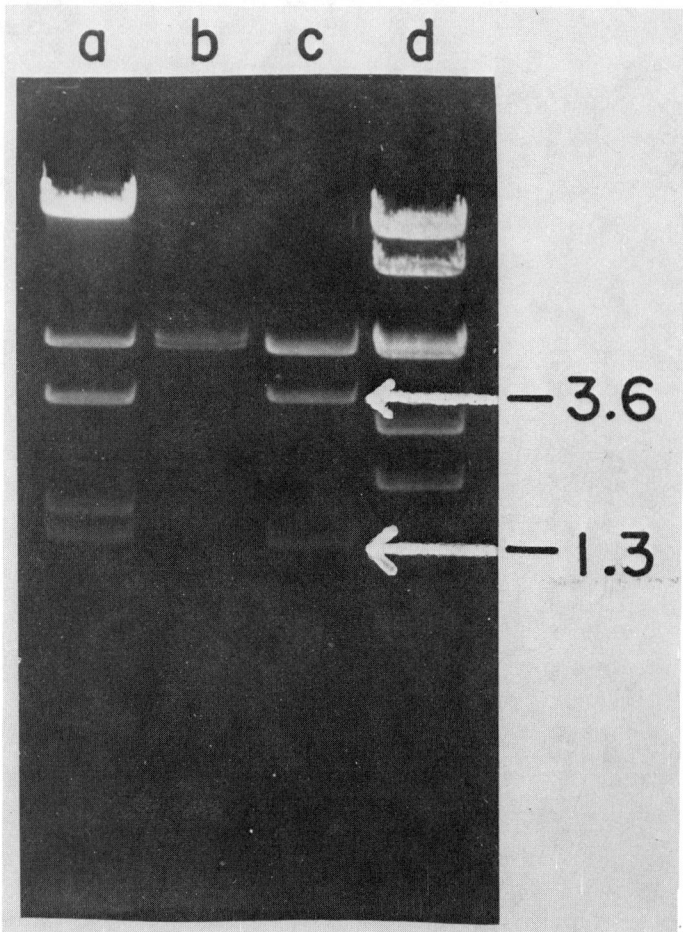

Fig. 11.   Agarose gel analysis of the products of Kpn I digestion of the 4.85 and 5.1 kb Eco RI fragments of Clone 4. The 4.85 and 5.1 kb Eco RI fragments were isolated from an agarose gel by electroelution onto hydroxylapatite by the method of Tabak and Flavell (1978) and redigested with 30–40 units of Kpn I per μg of DNA in the presence of 1 mg/ml nuclease-free bovine serum albumin. Electrophoresis on 0.8% agarose was as described in the legend to fig. 9. Lane (a) is Charon 4A DNA digested with Bgl II as a size marker. Lane (b) is undigested, gel-isolated 4.85 and 5.1 kb Eco RI fragments of Clone 4. Lane (c) is the Kpn I digestion of the 4.85 and 5.1 kb Eco RI fragments of Clone 4. Lane (d) is Clone 4 DNA digested with Eco RI.

HI site at position 41,220. Based on the presence of the 405 bp Eco RI fragment in this digestion, it can be concluded that one of the two Eco RI sites yet to be assigned lies this distance from the right-hand Eco RI site of insertion, or at position 35,858, and that the remaining site, which defines the limits of the 4.85 kb Eco RI fragment, is at position 31,008.

The single Hind III site within the inserted DNA was mapped by performing a double digestion of Clone 4 DNA with Hind III and Eco RI. This digestion resulted in the loss of the 4.85 kb Eco RI fragment and the appearance of new fragments of 2.9 and 2.0 kb (fig. 12). It was concluded that the Hind III site is at position 33,903 within the inserted DNA. Figure 14 represents the composite restriction site map of Clone 4 for the enzymes Eco RI, Bam HI, Kpn I, and Hind III.

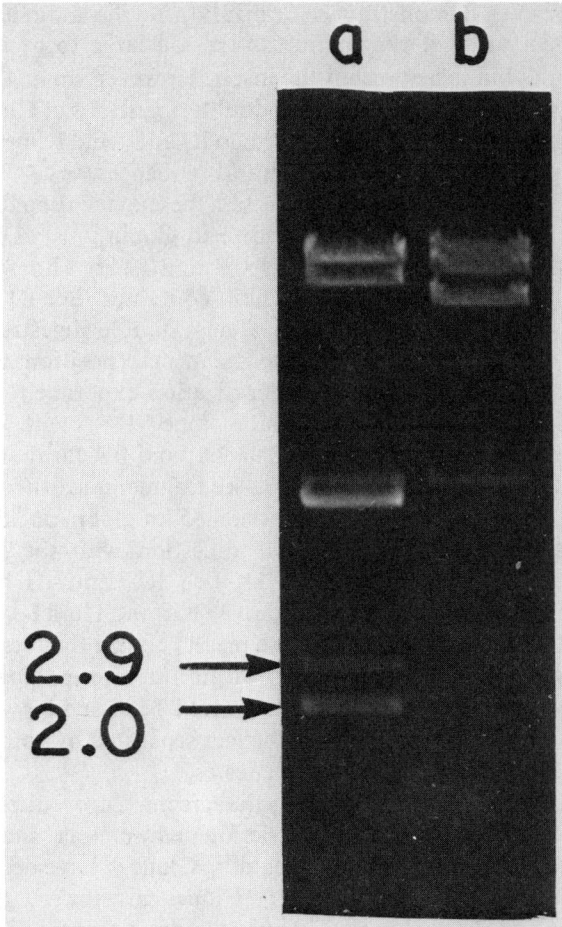

Fig. 12. Agarose gel analysis of Hind III digestion and Eco RI-Hind III double digestion of Clone 4. Restriction digests and electrophoresis on 0.8% agarose were as described in the legend to fig. 9. Lane (a) is Clone 4 DNA digested with Eco RI and Hind III. Lane (b) is Clone 4 DNA digested with Hind III.

Localization of Globin Coding Sequences
within the insert of Clone 4

The positions of "globin hybridizable sequences" within the goat DNA insert of Clone 4 were determined by fractionating the products of various restriction digests on agarose gels, transferring the DNA directly to nitrocellulose paper by the method of Southern (1975), and hybridizing the immobilized fragments to goat globin [^{32}P] cDNA. Although numerous fragments generated by cleavage of the recombinant DNA with single restriction enzymes demonstrate hybridization to the radioactive probe (figs. 9 and 13), most of the fragments are too large to be of value in localizing the globin genes within the insert. However, an analysis of the globin-positive fragments generated by double digestions of Clone 4 DNA with the enzymes Eco RI and Bam HI or Eco RI and Hind III permits more precise localization of the globin structural sequences.

Sequential digestion of Clone 4 DNA with the enzymes Eco RI and Bam HI produces four fragments that hybridize to globin [^{32}P] cDNA (fig. 9, lanes b and e). The fragments are 3.7, 2.9, 0.9, and 0.4 kb. Three of the four, the 3.7, 2.9, and 0.4 kb fragments, are globin-containing Eco RI or Bam HI fragments. The 0.9 kb component is a new double digestion product generated by Eco RI and Bam HI cleavage at map positions 35,858 and 35,003 respectively. A similar filter hybridization experiment performed with the products of a double digestion of Clone 4 DNA with Eco RI and Hind III reveals three fragments, of 2.9, 2.0, and 0.4 kb that anneal to globin [^{32}P] cDNA. Again, the 2.9 and 0.4 kb fragments are those observed in an Eco RI digestion of Clone 4, but the 4.85 kb globin positive Eco RI fragment has been lost in the double digestion with the subsequent appearance of a globin positive 2.0 kb Eco RI-Hind III component resulting from Eco RI cleavage at position 35,858 and Hind III cleavage at position 33,903 (data not shown). These results permit the localization of globin coding regions at two positions within the inserted goat DNA of Clone 4, between map coordinates 19,608 and 22,508 and between 33,903 and 36,263. The presence of globin sequences separated by approximately 11 kb suggest the presence of linked genes.

Several of the other clones selected by the screening procedure have been characterized with respect to their restriction endonuclease sites and their maps constructed. Several of these, including Clone 4, have been partially sequenced to identify the genes they carry. Clones carrying $\epsilon, \gamma, \beta^C, \beta^A$, and α globin genes have been identified. Thus, it appears that we have isolated the complete set of globin genes, which should now allow us to determine the basis of their regulation during development.

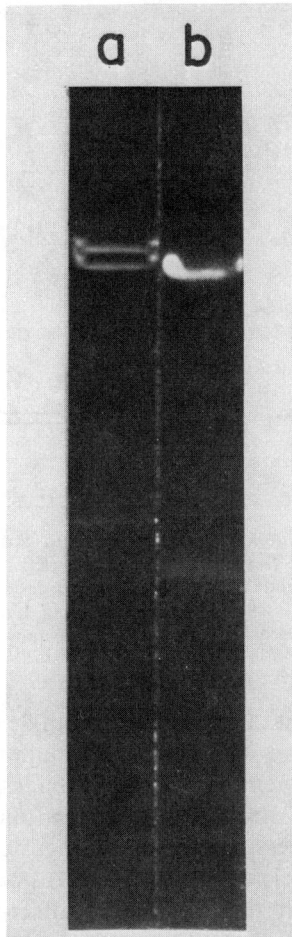

Fig. 13. Restriction endonuclease cleavage of Clone 4 DNA and detection of fragments containing globin sequences. Restriction endonuclease digestion and hybridization of immobilized restriction fragments to globin [^{32}P] cDNA were as described in the legend to fig. 9. Lane (a) is a Kpn I digestion of Clone 4 DNA, and lane (b) is the corresponding autoradiograph.

SUMMARY

The pathway of mRNA synthesis in eukaryotic cells is complex and encompasses not only transcription but also processing of the initial transcript and subsequent transport to the cytoplasm. In addition, mRNA turnover can play an important role in determining the amount of mRNA accumulated in cells. We have investigated these processes in erythroleu-

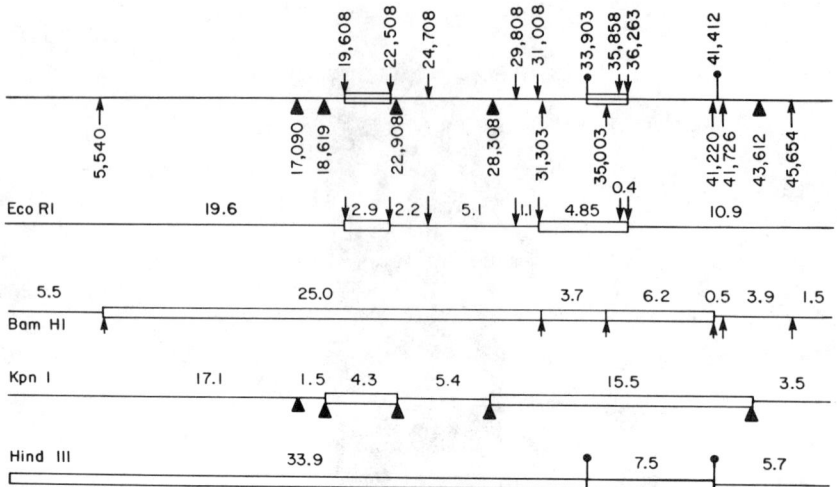

Fig. 14. Location of restriction endonuclease sites and globin hybridizable sequences in Clone 4. The top line represents the composite restriction site map of Clone 4 for the enzymes Eco RI, Bam HI, Kpn I, and Hind III. Succeeding lines represent the individual maps for each restriction endonuclease and include fragment sizes in kilobase pairs. Fragments that hybridize to globin [^{32}P] cDNA are depicted as open rectangles.

kemia cells, a cell line that is arrested early in development (possibly the pronormoblast stage), but can proceed through the developmental program when compounds such as dimethylsulfoxide, butyric acid, and hemin are added to the culture medium. It has been found that the efficiency of post-transcriptional processing does not change during differentiation and that globin mRNA is stable in uninduced cells as well as in cells undergoing differentiation. Although these processes affect the amount of globin mRNA in erythroid cells, the only parameter that changes during differentiation appears to be transcription.

Based on these observations, we have directed our attention toward developing a system where the basis for the differential transcription of genes can be studied. The globin genes of goat have been selected for study because the type of hemoglobin synthesized is under developmental control, i.e., there is a switch from embryonic hemoglobin, $\alpha_2\epsilon_2$, to fetal hemoglobin, $\alpha_2\gamma_2$, to pre-adult or juvenile hemoglobin, $\alpha_2\beta_2{}^C$, to adult hemoglobin, $\alpha_2\beta_2{}^A$, during development. Our approach involved the formation of a genomic recombinant library of goat DNA and the selection of those recombinants that carry globin genes. Recombinants carrying each of these genes have now been isolated, and it is hoped that we will be able to identify and isolate those factors in the respective tissues that allow these genes to be specifically expressed. We hope to show, for example, that factors are present in fetal liver tissue which either allow γ globin genes to be expressed or, conversely, inhibit the expression of the other β-like globin genes.

ACKNOWLEDGMENTS

This work was supported by grants from the American Cancer Society, NP59, National Science Foundation, PCM 76-80222, and the National Institutes of Health, GM 10999 and AM 20119. K. S. is a postdoctoral fellow of the National Institutes of Health. The expert technical assistance of Lon Williamson and Nancy Dischar is gratefully acknowledged.

REFERENCES

Adamson, S. D., E. Herbert, and W. Godchaux, III. 1968. Factors affecting the rate of protein synthesis in lysate systems from reticulocytes. Arch. Biochem. Biophys. 125:671–83.

Bernstein, A., D. S. Hunt, V. Crichley, and T. W. Mak. 1976. Induction by ouabain of hemoglobin synthesis in cultured friend erythroleukemic cells. Cell 9:375–81.

Blattner, F. R., B. G. Williams, A. E. Blechl, K. D. Thompson, H. E. Faber, L. A. Furlong, D. J. Grunwald, D. O. Kiefer, D. D. Moore, J. W. Schumm, E. L. Sheldon, and O. Smithers. 1977. Charon phages: safer derivatives of bacteriophage lambda for DNA cloning. Science 196:161–69.

Boyer, S. H., K. D. Wuu, A. N. Noyes, R. Young, W. Scher, C. Friend, H. D. Priesler, and A. Bank. 1972. Hemoglobin biosynthesis in murine virus-induced leukemic cells *in vitro*: structure and amounts of globin chains produced. Blood 40:823–35.

Cheng, T-C., and H. H. Kazazian, Jr. 1977. The 5'-terminal structures of murine α- and β-globin messenger RNA. J. Biol. Chem. 252:1758–63.

Curtis, P. J., N. Mantei, J. van den Berg, and C. Weissmann. 1977. Presence of a putative 15S precursor to β-globin mRNA but not α-globin mRNA in Friend cells. Proc. Nat. Acad. Sci. USA 74:3184–88.

Curtis, P. J., and C. Weissmann. 1976. Purification of globin messenger RNA from dimethylsulfoxide-induced Friend cells and detection of a putative globin messenger RNA precursor. J. Mol. Biol. 106:1061–75.

Darnell, J. E. 1979. Transcription-units for messenger RNA-production in eukaryotic-cells and their DNA viruses. Prog. in Nucl. Acid Res. & Mol. Biol. 22:327.

Eisen, H., S. Nasi, C. P. Georgopoulos, D. Arndt-Jovin, and W. Ostertag. 1977. Surface changes in differentiating Friend erythroleukemic cells in culture. Cell 10:689–95.

Friedmann, E. A., and C. L. Schildkraut. 1977. Terminal differentiation in cultured Friend erythroleukemia cells. Cell 12:901–13.

Friend, C., M. C. Patuleia, and E. de Harven. 1966. Erythrocytic maturation *in vitro* of murine (Friend) virus-induced leukemic cells. Nat. Cancer Inst. Monograph 22:505–22.

Friend, C., H. D. Preisler, and W. Scher. 1974. Studies of the control of differentiation of murine virus-induced erythroleukemic cells. Curr. Top. Develop. Biol. 8:81–101.

Friend, C., W. Scher, J. G. Holland, and T. Sato. 1971. Hemoglobin synthesis in murine virus-induced leukemic cells *in vitro*. 2. Stimulation of erythroid differentiation by dimethyl sulfoxide. Proc. Nat. Acad. Sci. USA 68:378–82.

Furusawa, M., Y. Ikawa, and H. Sugano. 1972. Development of erythrocyte membrane specific antigen(s) in clonal cultured cells of Friend virus-induced tumor. Proc. Jap. Acad. 47:220–29.

Gazitt, Y., R. C. Reuben, A. D. Deitch, P. A. Marks, and R. A. Rifkind. 1978. Changes in cyclic adenosine 3′:5′-monophosphate levels during induction of differentiation in murine erythroleukemia cells. Cancer Res. 38:3779–83.

Gorski, J., M. R. Morrison, C. G. Merkel, and J. B. Lingrel. 1974. Size heterogeneity of polyadenylate sequences in mouse globin messenger RNA. J. Mol. Biol. 86:363–71.

Gusella, J., R. Geller, B. Clarke, U. Weeks, and D. Housman. 1976. Commitment to erythroid differentiation by Friend erythroleukemia cells: a stochastic analysis. Cell 9:221–29.

Gusella, J., and D. Housman, 1976. Induction of erythroid differentiation *in vitro* by purines and purine analogues. Cell 8:263–69.

Harrison, P. R., D. Conkie, N. Affara, and J. Paul. 1974. *In situ* localization of globin messenger RNA formation. J. Cell Biol. 63:402–13.

Heckle, W. L., R. G. Fenton, T. G. Wood, C. G. Merkel, and J. B. Lingrel. 1977. Methylated nucleotides in globin mRNA from mouse erythroid cells. J. Biol. Chem. 252:1964–70.

Housman, D., J. Gusella, R. Geller, R. Levenson, and S. Weil. 1978. Differentiation of murine erythroleukemia cells: the central role of the commitment event. *In* B. Clarkson, J. E. Till, and P. Marks (eds.), Differentiation of normal and neoplastic hematopoietic cells, pp. 193–207. Cold Spring Harbor Laboratory, Cold Spring Harbor, New York.

Ikawa, Y., M. Furusawa, and H. Sugano. 1973. Erythrocyte membrane specific antigens in Friend's virus-induced leukemia cells. In unifying concepts of leukemia. Bibl. Haemat. 39:955–67.

Jeffreys, A. J., and R. A. Flavell. 1977. The rabbit β-globin gene contains a large insert in the coding sequence. Cell 12:1097–1108.

Keppel, F., B. Allet, and H. Eisen. 1977. Appearance of a chromatin protein during the erythroid differentiation of Friend virus-transformed cells. Proc. Nat. Acad. Sci. USA 74:653–56.

Kinniburgh, A. J., J. E. Mertz, and J. Ross. 1978. The precursor of mouse β-globin messenger RNA contains two intervening RNA sequences. Cell 14:681–93.

Konkel, D. A., S. M. Tilghman, and P. Leder. 1978. The sequence of the chromosomal mouse β-globin major gene: homologies in capping, splicing, and poly (A) sites. Cell 15:1125–32.

Kwan, S.-P., T. G. Wood, and J. B. Lingrel. 1977. Purification of a putative precursor of globin messenger RNA from mouse nucleated erythroid cells. Proc. Nat. Acad. Sci. USA 74:178–82.

Leder, A., and P. Leder. 1975. Butyric acid, a potent inducer of erythroid differentiation in cultured erythroleukemic cells. Cell 5:319–22.

Leder, A., H. I. Miller, D. H. Hamer, J. B. Seidman, B. Norman, M. Sullivan, and P. Leder. 1978. Comparison of cloned mouse α- and β-globin genes: conservation of intervening sequence locations and extragenic homology. Proc. Nat. Acad. Sci. USA 75:6187–91.

Lockard, R. E., and J. B. Lingrel. 1971. Identification of mouse hemoglobin mRNA. Nat. New Biol. 233:204–6.

Lockard, R. E., and J. B. Lingrel. 1969. The synthesis of mouse hemoglobin β-chains in a rabbit reticulocyte cell-free system programmed with mouse reticulocyte 9S RNA. Biochem. Biophys. Res. Commun. 37:204.

Lowenhaupt, K., and J. B. Lingrel. 1978. A change in the stability of globin mRNA during the induction of murine erythroleukemia cells. Cell 14:337–44.

Lowenhaupt, K., and J. B. Lingrel. 1979. Synthesis and turnover of globin mRNA in murine erythroleukemia cells induced with hemin. Proc. Nat. Acad. Sci. USA 76:5173–77.

Lowenhaupt, K., C. Trent, and J. B. Lingrel. 1978. Mechanisms for accumulation of globin mRNA during dimethyl sulfoxide induction of murine erythroleukemia cells: synthesis of precursors and mature mRNA. Developmental Biol. 63:441–54.

Maniatis, T., R. C. Hardison, E. Lacy, J. Lauer, C. O'Connell, and D. Quon. 1978. The isolation of structural genes from libraries of eucaryotic DNA. Cell 15:687–700.

Marbaix, G., and A. Burny. 1964. Separation of the messenger RNA of reticulocyte polyribosomes. Biochem. Biophys. Res. Commun. 16:522–527.

Marks, P. A., and R. A. Rifkind. 1978. Erythroleukemic differentiation. Ann. Rev. Biochem. 47:419–28.

Marks, P. A., R. A. Rifkind, A. Bank, M. Terada, R. Gambari, E. Ribach, G. Maniatis, and R. Reuben. 1979. Cellular and molecular regulation of hemoglobin switching. *In* G. Stamatoyannopoulos and A. W. Nienhius (eds.), Cellular and molecular regulation of hemoglobin switching, pp. 437–55. Grune & Stratton, New York.

Merkel, C. G., S.-P. Kwan, and J. B. Lingrel. 1975. Size of polyadenylic acid region of newly synthesized globin mRNA. J. Biol. Chem. 250:3725–28.

Muthukrishnan, S., G. W. Both, Y. Furuichi, and A. J. Shatkin. 1975. 5'-Terminal 7-methylguanosine in eukaryotic RNA is required for translation. Nature 25:33–37.

Nudel, J., J. Salmon, E. Fiback, N. Terada, R. Rifkind, P. A. Marks, and A. Bank. 1977. Accumulation of α and β-globin messenger RNAs in mouse erythroleukemia cells. Cell 12:463–69.

Ostertag, W. H., G. Melderes, G. Steinherder, N. Klug, and S. Dube. 1973. Synthesis of mouse hemoglobin and globin mRNA in leukemic cell cultures. Nature New Biol. 239:231–34.

Perry, R. P., D. E. Kelley, K. Friderici, and F. Rottman. 1975. The methylated constituents of L cell messenger RNA: evidence for an unusual cluster at the 5' terminus. Cell 4:387–94.

Reuben, R. C., R. L. Wife, R. Breslow, R. A. Rifkind, and P. A. Marks. 1976. A new group of potent inducers of differentiation in murine erythroleukemia cells. Proc. Nat. Acad. Sci. USA 73:862–66.

Rifkind, R. A., E. Fiback, G. Maniatis, R. Gambino, and P. A. Marks. 1979. Commitment to differentiation of normal and transformed erythroid precursors. In G. Stamatoyannopoulos and A. W. Nienhius (eds.), Cellular and molecular regulation of hemoglobin switching, pp. 421–37. Grune & Stratton, New York.

Robbins, J., P. Rosteck, Jr., J. R. Haynes, G. Freyer, M. B. Cleary, H. D. Kalter, K. Smith, and J. B. Lingrel. 1979. The isolation and partial characterization of recombinant DNA containing genomic globin sequences from the goat. J. Biol. Chem. 254:6187–95.

Ross, J. 1976. A precursor of globin messenger RNA. J. Mol. Biol. 106:403–20.

Ross, J., J. Gielen, S. Packman, Y. Ikawa, and P. Leder. 1974. Globin gene expression in cultured erythroleukemic cells. J. Mol. Biol. 87:697–714.

Ross, J., and D. Sautner. 1976. Induction of globin mRNA accumulation by hemin in cultured erythroleukemic cells. Cell 8:513–20.

Rottman, F., A. J. Shatkin, and R. P. Perry. 1974. Sequences containing methylated nucleotides at the 5' termini of messenger RNAs: possible implications for processing. Cell 3:197–99.

Rovera, G., J. Abramczuk and S. Surrey. 1977. The effect of hemin on the expression of β globin genes in Friend cells. FEBS Lett. 81:366–70.

Rovera, G., J. Tortikar, G. R. Connolly, C. Magarian, and T. W. Dolvey. 1978. Hemin controls the expression of the β minor globin gene in Friend erythroleukemic cells at the pretranslational level. J. Biol. Chem. 253:7588–90.

Sassa, S. 1976. Sequential induction of heme pathway enzymes during erythroid differentiation of mouse Friend leukemia virus-infected cells. J. Exp. Med. 143:305–15.

Sato, T., C. Friend, and E. de Haven. 1971. Ultrastructural changes in Friend erythroleukemia cells treated with dimethylsulfoxide. Cancer Res. 31:1402–7.

Scher, W., and C. Friend. 1978. Breakage of DNA and alterations in folded genomes by inducers of differentiation in Friend erythroleukemic cells. Cancer Res. 38:841–49.

Smith, K., and J. B. Lu. 1978. Sequence organization of the β-globin in RNA precursor. Nucl. Acid Res. 5:3295–301.

Smith, K., P. Rosteck, Jr. and J. B. Lingrel. 1978. The location of the globin mRNA sequence within its 16S precursor. Nucl. Acids Res. 5:105–15.

Southern, E. M. 1975. Detection of specific sequences among DNA fragments separated by gel electrophoresis. J. Mol. Biol. 98:503–17.

Sternberg N., D. Tiemeier, and L. Enquist. 1977. *In vitro* packaging of a λ *Dam* vector containing *Eco* RI DNA fragments of *Escherichia coli* and phage P1*. Gene 1:255–80.

Tabak, H. F., R. A. Flavell. 1978. A method for the recovery of DNA from agarose gels. Nucl. Acid Res. 5:2321–32.

Tanaka, M., J. Levy, M. Terada, R. Breslow, R. A. Rifkind, and P. A. Marks. 1975. Induction of erythroid differentiation in murine virus-infected erythroleukemia cells by highly polar compounds. Proc. Nat. Acad. Sci. USA 72:1003–6.

Terada, M., E. Epner, U. Nudel, J. Salmon, E. Fiback, R. Rifkind, and P. Marks. 1978. Induction of murine erythroleukemia cell differentiation by actinomycin D. Proc. Nat. Acad. Sci. USA 75:2795–99.

Tilghman, S. M., P. J. Curtis, D. C. Tiemeier, P. Leder, and C. Weissmann. 1978. Intervening sequence of a mouse β-globin gene is transcribed as the 15S β-globin mRNA precursor. Proc. Nat. Acad. Sci. USA 75:1309–13.

Wood, T. G., and J. B. Lingrel. 1977. Purification of biologically active globin mRNA using cDNA-cellulose affinity chromatography. J. Biol. Chem. 252:457–63.

LEE F. JOHNSON

Regulation of Dihydrofolate Reductase Gene Expression

12

INTRODUCTION

The study of the regulation of gene expression in eukaryotes, and especially in higher eukaryotes, has lagged far behind similar studies in prokaryotes. This is true for a number of reasons, including the complexity of the genome, the difficulty in isolating well-defined mutant cell lines, and the difficulty in maintaining cultured lines of cells from higher organisms. However, recent advances in the techniques of molecular biology, such as the development of methods for cloning and sequencing specific DNA fragments, have enabled a number of laboratories to study in greater detail the complexities of gene structures and expression in higher eukaryotes. Such studies have shown that there are a variety of fundamental differences between prokaryotes and eukaryotes in gene organization and the control of gene expression. A striking example is the existence of intervening sequences in many eukaryotic structural genes.

Investigations of the regulation of gene expression are directed toward answering the following questions. Under what physiological conditions is a particular gene expressed or not expressed? What are the molecular signals (both intracellular and extracellular) that are responsible for bringing about changes in the level of gene expression? Are the changes brought about by regulating the transcription of the structural gene, the processing of the initial transcription product (hnRNA) into mature mRNA, the efficiency of translation of the mRNA or by post-translational regulatory events?

Department of Biochemistry, Ohio State University, Columbus, Ohio 43210

A number of convenient model systems have been studied to answer these questions. Most of the regulated genes of higher eukaryotes that have been studied in detail have been the genes for major proteins. Examples include the genes for globin (Tilghman et al., 1978; Lingrel et al., in this volume), ovalbumin (Breathnach et al., 1977; Dugaiczyk et al., 1978), and immunoglobulin (Tonegawa et al., 1977; Seidman et al., 1978; Schibler et al., 1978). Because these "luxury" proteins are synthesized at extremely high levels, the isolation of the protein and its mRNA and hnRNA, as well as studies of the metabolism of each, are relatively simple matters with technology that is presently available. Once the mRNA is purified, it can be used as a hybridization probe to detect the presence of DNA sequences corresponding to the structural gene of the protein. Using standard recombinant DNA and cloning procedures, the gene can be isolated and its organization and sequence determined.

Detailed studies of the structure and regulation of the expression of genes for enzymes in higher eukaryotes have been particularly difficult. This is because enzymes and other "housekeeping proteins" are normally present at low concentrations in the cell. Since the mRNA for these proteins represents a similarly low proportion of total mRNA, isolation of the message for a specific enzyme is a formidable task. Without pure mRNA to serve as a source of hybridization probe, it is difficult to exploit the power of cloning procedures to isolate and study the gene for the enzyme or to study the metabolism of its mRNA or hnRNA.

The level of expression of many genes for housekeeping enzymes is regulated over a very wide range in response to a variety of stimuli. Furthermore, these genes may have unusual structural features, and may be controlled by mechanisms quite different from genes for luxury proteins. Therefore, it is important that procedures be developed to facilitate the analysis of the structure and expression of genes for enzymes and other proteins found at low levels in the cell.

In this article I describe an approach taken by my laboratory and others toward this goal. Our studies have centered primarily on the genes for dihydrofolate reductase (DHFR) and, more recently, thymidylate synthetase (TS). We have found that the expression of both of these genes is regulated over a wide range and in an interesting manner during the cell cycle. I describe first our studies on the regulation of DHFR gene expression in cultured mouse fibroblasts. Next I describe an approach that we and others have used to obtain a cell line in which the level of DHFR, DHFR mRNA, and the DHFR gene are greatly amplified. Since DHFR gene expression appears to be regulated in the overproducing cells in the same manner as in normal cells, the overproducing cell line should be an excellent model system for studying the molecular mechanism(s) for regulating DHFR gene expression. Finally, I discuss the possibility of generalizing the approach to facilitate the investigation of the structure and expression of genes for other enzymes that cannot be studied by more conventional methods. I describe our recent isolation of a cell line that overproduces TS to illustrate this approach.

DHFR GENE EXPRESSION IN NORMAL CELLS

Dihydrofolate reductase is the enzyme responsible for the reduction of folate and dihydrofolate to tetrahydrofolate, the biochemically active form. Derivatives of tetrahydrofolate are involved in many reactions in single carbon metabolism. In most reactions tetrahydrofolate serves as the donor or acceptor of single carbon groups and does not change its oxidation state during the transfer reaction. Examples of this type of reaction include the transfer of single carbon groups during the biosynthesis of the purine ring and the interconversion of serine and glycine. However, in one important reaction the oxidation state of tetrahydrofolate is changed during the single carbon transfer reaction. The enzyme TS catalyzes the methylation of deoxyuridylic acid to form thymidylic acid. N^5-N^{10} methylene tetrahydrofolate supplies the methyl group in the reaction. Tetrahydrofolate is oxidized to dihydrofolate in the course of the reaction and must be converted back to tetrahydrofolate by the enzyme DHFR (Blakley, 1969). The synthesis of thymidylic acid occurs primarily (if not exclusively) during the S phase of the cell cycle. Therefore, the need for DHFR is greatest in cells that are actively synthesizing DNA.

The activity of DHFR is strongly inhibited by methotrexate and other 4-amino-derivatives of folic acid, which bind extremely tightly at the active site of the enzyme (Blakley, 1969). These drugs are quite toxic to rapidly proliferating cells since they lead to starvation for tetrahydrofolic acid and, therefore, thymidylic acid. This results in the inhibition of DNA synthesis and "thymineless death" (Cohen, 1971). This fact has been exploited by chemotherapists for the treatment of various diseases characterized by improper control of cell proliferation, such as cancer and psoriasis (Bertino, 1979).

The fact that the need for DHFR is greatest during S phase is in line with observations that the cellular level of DHFR is severalfold higher in exponentially growing cells than in quiescent or noncycling cells (Hillcoat et al., 1967; Chello et al., 1977). The level of DHFR is also increased in response to infection of cells with certain DNA viruses (Frearson et al., 1966). These observations led us to consider the possibility that DHFR gene expression is regulated in response to the cellular need for tetrahydrofolate and, in particular, that the DHFR gene is expressed primarily, if not exclusively, during the S phase of the cell cycle.

Our studies have been conducted primarily with the mouse fibroblast cell line 3T6 (Todaro and Green, 1963). These cells rest in the "G_0" state of the cell cycle, upon reaching confluence, when grown in medium containing 0.5% serum. The resting cells can be stimulated to reenter the cell cycle by increasing the serum concentration to 10% (Johnson et al., 1974). The rate of DNA synthesis is extremly low in resting cells but increases sharply 10–12 hr after stimulation and reaches a maximum at about 16–18 hr. Mitotic cells are first observed about 22 hr following stimulation. Thus, stimulated 3T6 cells provide a convenient and simple model system for studying the biochemical events required for the transition from the resting to the growing state and, in particular, for the entry into the S phase of the cell cycle.

Our initial studies showed that exponentially growing 3T6 cells contained 3–5 times as much DHFR as resting 3T6 cells (Johnson et al., 1978a). We next wanted to determine if the rate of synthesis of DHFR was significantly greater in growing than in resting 3T6 cells. Since the enzyme was present at extremely low levels in these cells, we had to resort to rather indirect measurements of this rate. We took advantage of the fact that methotrexate (MTX) binds "essentially irreversibly" to the active site of the enzyme. We first inactivated all cellular DHFR by a brief exposure to a low concentration of MTX. All unbound MTX was then removed, and the recovery of DHFR enzyme activity was determined by measuring either the ability of a cell extract to bind ^{3}H-MTX or to reduce ^{3}H-folate to ^{3}H-tetrahydrofolate. Figure 1 shows that this rate was about 40 times greater in exponentially growing 3T6 cells than in resting 3T6 cells. The recovery of enzyme activity was inhibited by cycloheximide, suggesting that the recovery was due to *de novo* synthesis of DHFR rather than activation of a stored enzyme precursor. Finally we found that recovery of DHFR activity was not blocked by 5 μg/ml actinomycin D, which is sufficient to block all cellular transcription. This indicates that the translation of DHFR mRNA is not inhibited by the drug and suggests that DHFR mRNA is fairly stable (Johnson et al., 1978b).

The above results indicate that the gene for DHFR is not expressed in resting cells but is expressed at a high level in growing cells. To determine if DHFR gene expression occurred in a particular portion of the cell cycle, we studied the accumulation of DHFR in resting cells that had been serum-stimulated to reenter the cell cycle. Figure 2 shows that the rate of accumulation of DHFR increases sharply at about 10 hr following

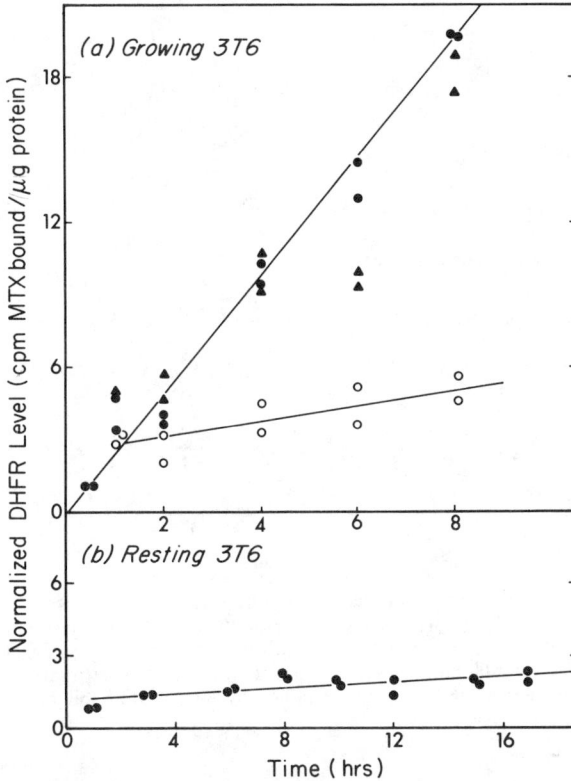

Fig. 1. Rate of accumulation of dihydrofolate reductase in resting and growing cells. Cultures of resting or growing 3T6 cells were pretreated with 10^{-6}M MTX to inactivate preexisting DHFR, then rinsed extensively with serum-free medium and fed at time = 0 with appropriate medium. Resting cultures were fed with "conditioned medium" containing 0.5% serum, taken from sister cultures of resting 3T6 cells. Growing cultures were fed with fresh medium containing 10% serum (solid circles). Some growing cultures were fed at time = 0 with medium containing either 5 μg/ml cycloheximide (open circles) or 5 μg/ml actinomycin (triangles). The rate of accumulation of DHFR was determined by harvesting cultures at various times and measuring the level of active (newly synthesized) reductase by determining the ability of a cell extract to bind ^{3}H-MTX (^{3}H-MTX binding assay). The level of DHFR was normalized to the amount of protein present at time = 0 so that the rates of accumulation of DHFR in resting and growing cells could be compared. (From Johnson et al., 1978.)

stimulation. The enzyme level increases in a linear manner until at least 20 hr following stimulation. We also found that this increase was blocked by cycloheximide, again indicating that the increase in the DHFR level was the result of an increase in *de novo* synthesis (Johnson et al., 1978b).

As mentioned earlier, 3T6 cells begin DNA synthesis about 10 hr following serum stimulation. The temporal coordination between the

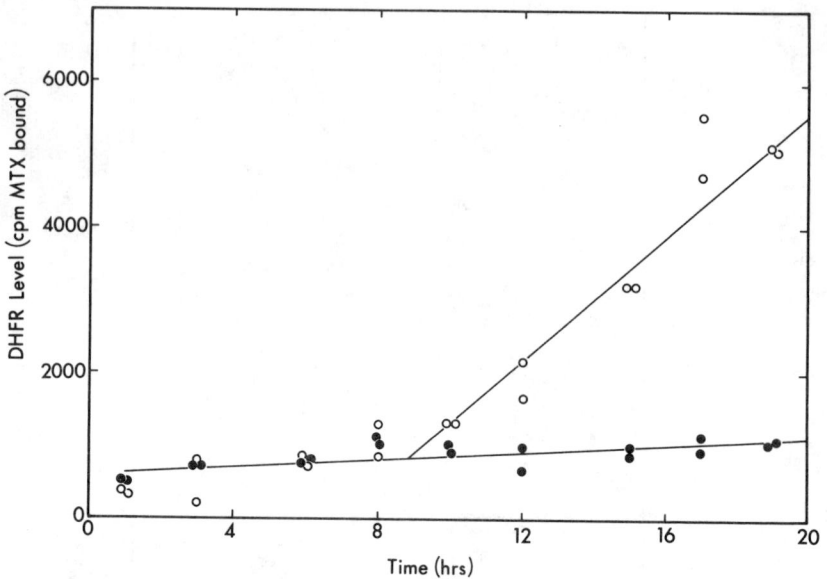

Fig. 2.   Increase in dihydrofolate reductase following serum stimulation of resting cells. Cultures of resting 3T6 cells were pretreated with 10^{-6}M MTX to inactivate preexisting reductase, then fed at time $= 0$ with conditioned medium containing 0.5% serum (solid circles) or with fresh medium containing 10% serum (open circles). Cultures were harvested at various times; the level of active DHFR was determined by the ^{3}H-MTX binding assay. (From Johnson et al., 1978.)

increase in DHFR gene expression and entry into S phase is not surprising, considering the role of the enzyme in the biosynthesis of thymidylic acid. To determine if there was a direct coupling between these two events, we studied the effect of inhibiting DNA synthesis on DHFR gene expression in stimulated cells. DNA synthesis was blocked by either cytosine arabinoside or hydroxyurea, which were added to the culture medium at the time of stimulation. We found that neither inhibitor had any effect on the increase in DHFR gene expression even though DNA synthesis was blocked by greater than 98%. Therefore, active DNA synthesis is not required for the increase in DHFR gene expression (Johnson et al., 1978b). This is quite different from the coordination between DNA synthesis and histone gene expression. Histones are also synthesized primarily during the S phase of the cell cycle. However, blocking of DNA synthesis leads to an immediate inhibition of histone synthesis (Borun et al., 1967).

It should be emphasized that the lack of direct coupling between the increase in DHFR accumulation and DNA synthesis does not imply that the events are not coordinated by a common biochemical "signal." It has been suggested that at some point in G_1 phase the cell becomes committed to enter S phase and complete one round of cell division (Pardee et al., 1978). This commitment process may involve the synthesis of one or more control molecules (e.g., regulatory proteins) that act at a variety of specific DNA recognition sites to turn on the expression of the spectrum of genes necessary for entry into the S phase. Other genes that are expressed at high levels at the beginning of S phase include the genes for thymidine kinase, thymidylate synthetase, and ribonucleotide reductase (Prescott, 1976). Of course, the expression of each of these genes might also be regulated by other processes, but at least the primary signal responsible for coordinating the turning on of these genes might be the same.

To determine if the increase in DHFR accumulation was regulated at the level of transcription, we examined the sensitivity of DHFR gene expression to actinomycin D. Figure 3 shows that when transcription was inhibited 7.5 hr following stimulation, there was no subsequent increase in DHFR accumulation. This is consistent with the idea that there is no accumulation of untranslated DHFR mRNA in resting or stimulated cells prior to this time. The accumulation of reductase was only partly inhibited when actinomycin was added 12.5 hr following stimulation, and was not inhibited at all if added 15 hr after stimulation (Johnson et al., 1978b). These observations are consistent with the idea that DHFR gene expression is regulated at the transcriptional level and that DHFR mRNA begins to accumulate just prior to the increase in DHFR accumulation. However, since actinomycin is known to affect a number of processes besides transcription, these conclusions must be regarded as tentative until verified by more direct measurements of DHFR mRNA content and transcription.

In summary, we have found that the expression of the gene for DHFR is regulated over a very wide range during the G_0 to S transition. The expression appears to be controlled at the level of transcription and is linked temporally, but not directly, with DNA replication. The DHFR gene appears to be a member of the family of genes involved in the synthesis of DNA precursors that are expressed in a coordinated fashion at the beginning of S phase.

The next logical step in this investigation would be to quantitate the amount and rate of synthesis of DHFR mRNA. Unfortunately DHFR is present at extremely low levels even in exponentially growing cells. Even the isolation of the enzyme in pure form is an involved process. Therefore, isolation and quantitation of DHFR mRNA, to say nothing of its precursor or its gene, by procedures presently available would be extremely

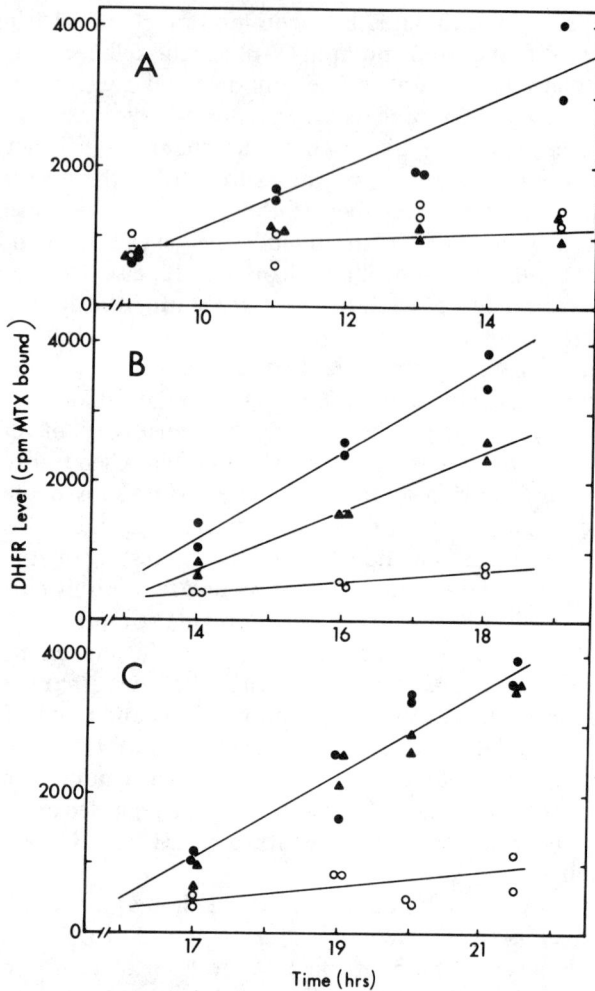

Fig. 3.   Effect of inhibition of RNA synthesis on dihydrofolate reductase gene expression. Cultures of resting 3T6 cells were serum-stimulated at time = 0. At time = 7.5 hr (A), 12.5 hr (B), or 15 hr (C) experimental cultures (triangles) were exposed to 10^{-6} M MTX to inactivate preexisting DHFR and 5 μg/ml actinomycin to inhibit RNA synthesis. After 30 min the cultures were rinsed extensively to remove excess MTX, then fed with fresh medium containing 10% serum and 5 μg/ml actinomycin. Control stimulated cultures (solid circles) were treated the same way except that actinomycin was omitted. Control resting cultures (open circles) were pretreated with MTX, then fed with conditioned medium containing 0.5% serum. Cultures were harvested at later times and the level of active DHFR was determined for each by the ^{3}H-MTX binding assay. (From Johnson et al., 1978.)

difficult. Fortunately, a remarkable series of discoveries in a number of laboratories has eliminated this potential obstacle.

ISOLATION OF 3T6 CELLS THAT OVERPRODUCE DHFR

When cancer patients are treated for long periods with MTX, it has been observed that in some cases the tumor cells become resistant to the toxic effects of the drug (Bertino et al., 1963). The same phenomenon occurs when cultured cells are exposed to the drug for many generations (Hakala et al., 1961). This process has been studied recently in great detail and has led to some interesting and quite unexpected observations.

In general, cells develop resistance to a toxic drug by lowering the rate of transport of the drug into the cell, by altering the structure of the target enzyme (by missense mutation) so that it has a lower affinity for the drug, or by accelerating the breakdown of the toxic drug. Several of these mechanisms have been found to explain the resistance of cells to low levels of MTX (Flintoff et al., 1976). However, it has been found that cells develop resistance to high levels of the drug by massive (several hundred-fold) overproduction of DHFR, the target enzyme of the drug (Hakala et al., 1961; Littlefield, 1969). Apparently the cell uses the excess enzyme to titrate the MTX transported into the cell, allowing a sufficient amount of active enzyme for cell survival. Resistance to this high level of drug is achieved by growing cells for prolonged periods in gradually increasing levels of the drug. Normal growing cells are killed quite readily, by 10^{-8} M MTX. However, it is possible to obtain cells that will grow normally in 10^{-4} M MTX using this selection procedure. Dihydrofolate reductase becomes the major translation product of the cell, accounting for up to 5% of total cell protein (Alt et al., 1976; Kellems et al., 1979; Wiedemann and Johnson, 1979). The mechanism of overproduction of the enzyme has been studied extensively in the laboratories of Littlefield, Schimke, and others. They found that the overproduction of reductase was directly attributable to the overproduction of the mRNA for the reductase (Chang and Littlefield, 1976; Kellems et al., 1976). Recent studies in Schimke's group have shown that the increase in DHFR mRNA is due to an increase in the number of copies of the DHFR structural gene (Alt et al., 1978). The gene appears to be localized either to an unusual "homogeneously staining" chromosomal region observed in MTX-resistant hamster cells (Biedler and Spengler, 1976) or to so-called double minute chromosomes in mouse cells. The physical location of the DHFR gene appears to be related to the genetic stability of the overproduction trait. In mouse cells, where the trait is lost in the absence of selective pressure (Alt et al., 1976; Kellems et al., 1979; Wiedemann and Johnson, 1979), the gene appears to be localized to the double minute chromosomes (Kaufman et al., 1979). However, in hamster cells, where the trait is maintained even when the cells are grown for many months in the absence of MTX, the genes are found in the homogenously staining region (Nunberg et al., 1978).

The actual mechanism of gene amplification is not known at present. The process, which appears to occur in a stepwise manner, may involve tandem gene duplication as a first step. The duplicated genes may then be amplified by unequal crossing over or rolling circle replication (Schimke et al., 1978). Comparison of the sequences of the DHFR gene in normal and overproducing cells may shed light on the mechanism of overproduction.

The presence of high levels of DHFR mRNA sequences in the overproducing cells have greatly facilitated the quantitation of DHFR mRNA in such cells. We reasoned that if DHFR gene expression were regulated in overproducing cells in the same manner as in normal cells, the overproducing cells would be an excellent model system for studying DHFR gene expression. In particular, the content and metabolism of DHFR mRNA and its precursor could be studied by standard DNA or RNA-excess hybridization procedures. Unfortunately, none of the MTX-resistant cell lines that had been isolated appeared to be particularly susceptible to growth control. Because this property was critical as far as our studies of the regulation of DHFR gene expression were concerned, we decided that we would attempt to isolate a MTX-resistant 3T6 cell line. Our objective was to isolate a line that would overproduce DHFR (and its mRNA) several hundredfold, that would retain the ability to rest in the G_0 state when kept in medium containing 0.5% serum, and finally, that would regulate DHFR gene expression in the same manner as the parental 3T6 cells.

After exposing 3T6 cells for many months to gradually increasing levels of MTX in the culture medium, we were able to isolate a clone (designated M50L3) that was able to grow normally in medium containing 50 μM MTX and that retained the ability to rest in the G_0 state in medium containing 0.5% serum. The cells overproduced DHFR by a factor of 300 compared with normal 3T6 cells (Wiedemann and Johnson, 1979). Comparison of the cellular proteins of 3T6 and M50L3 by one-dimensional SDS polyacrylamide gel electrophoresis (fig. 4) revealed that the only obvious difference was a protein with a molecular weight of 21,000 that was prominent in the M50L3 pattern but undetectable in 3T6. This protein was shown to be the overproduced DHFR following its purification by affinity chromatography on folate-sepharose. Dihydrofolate reductase represented about 4% of soluble cytoplasmic protein.

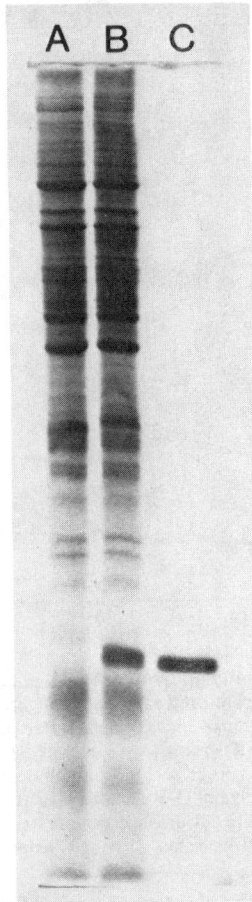

Fig. 4. Comparison of the cytoplasmic proteins of 3T6 and M50L3 cells. Cultures of exponentially growing 3T6 or M50L3 cells were harvested, and cytoplasmic extracts were fractionated by SDS polyacrylamide slab gel electrophoresis. The gels were then stained with Coomassie Blue. Slot A: 3T6 extract; Slot B: M50L3 extract; Slot C: pure DHFR isolated from M50L3 cells by affinity chromatography on folate-Sepharose. (From Wiedemann and Johnson, 1979.)

REGULATION OF DHFR GENE EXPRESSION
IN OVERPRODUCING CELLS

We next showed that the expression of the DHFR gene was regulated in the same manner in the overproducing cells as in normal 3T6 cells. We found that the rate of accumulation of DHFR was about 25 times greater in the exponentially growing M50L3 cells than in resting M50L3 cells. When resting M50L3 cells were serum-stimulated, the level of DHFR began to increase about 10 hr later (fig. 5). The increase in accumulation was

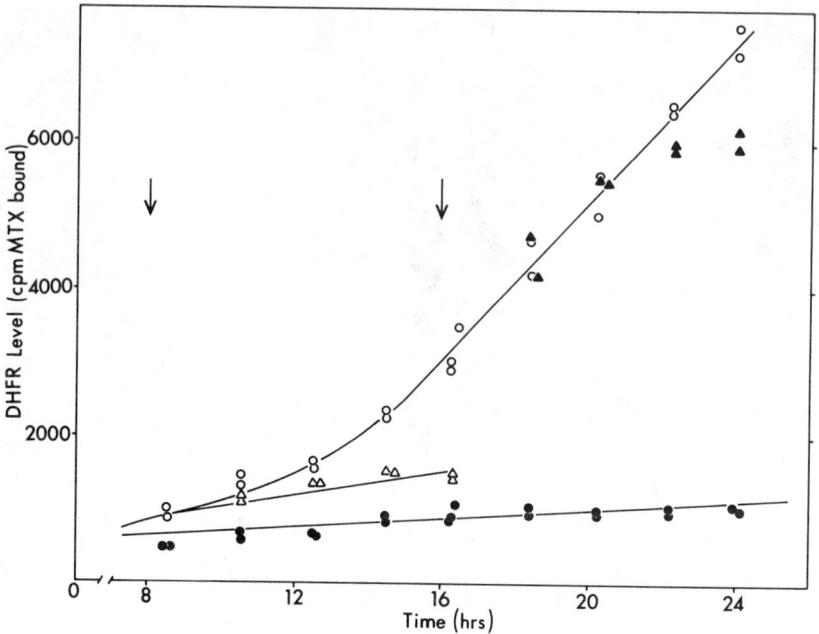

Fig. 5. Increase in rate of accumulation of DHFR in serum stimulated M50L3 cells. Cultures of resting M50L3 cells were pretreated with 50 μM MTX for 2 days to inactivate preexisting DHFR, then rinsed extensively with serum-free medium and fed at time = 0 with fresh medium containing 10% calf serum (open circles). Control cultures were fed with conditioned medium containing 0.5% calf serum (solid circles). At time = 8 hr (open triangles) or 16 hr (solid triangles) actinomycin D was added to the stimulated cultures at a final concentration of 5 μg/ml. Duplicate cultures were harvested at various times and assayed for DHFR level by the ^{3}H-MTX binding assay. (From Wiedemann and Johnson, 1979.)

sensitive to actinomycin if the drug was added 8 hr following stimulation, but insensitive if the drug was added 16 hr following stimulation (fig. 5). Finally, the increase in DHFR was not inhibited when the cells were stimulated in the presence of inhibitors of DNA synthesis (Wiedemann and Johnson, 1979). From these studies it appeared that DHFR gene expression was regulated in the same manner in the overproducing cells as in normal 3T6 cells. Therefore, the M50L3 cells were an excellent model system for studying the molecular details involved in controlling DHFR gene expression.

We extended our previous studies by measuring directly the rate of synthesis of DHFR in stimulated M50L3 cells. Cultures of M50L3 cells were pulse labeled at various times following stimulation and the rate of labeling of DHFR was determined. We found that the rate of synthesis

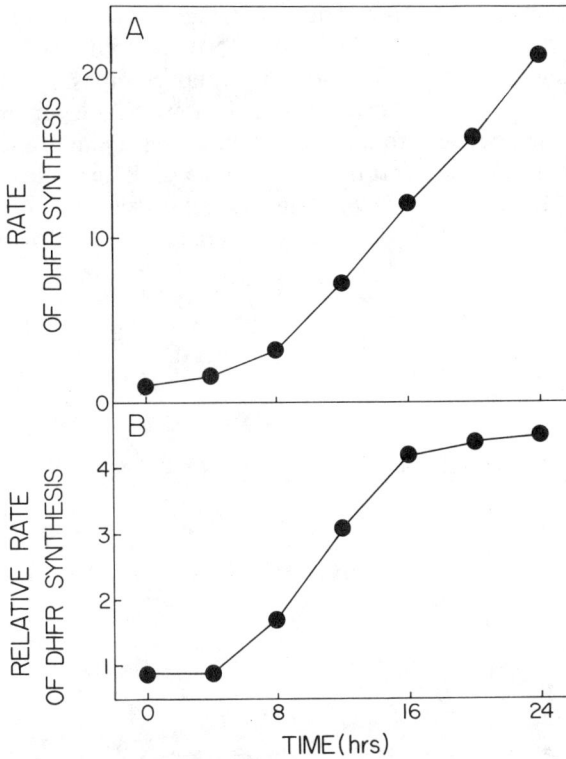

Fig. 6.   Rate of synthesis of DHFR following serum stimulation of M50L3 cells. Cultures of resting M50L3 cells were serum-stimulated at time = 0. At various times the culture medium was replaced with fresh "labeling medium" containing 10% calf serum, 0.5% of the normal amount of leucine, and 25 μCi/ml of ^{3}H-leucine (final specific activity = 6 Ci/m-mole). The cultures were incubated at 37° for 1 hr, then harvested. Labeled proteins were fractionated by SDS slab gel electrophoresis, and the dried gels were fluorographed. The region corresponding to labeled DHFR was cut from the dried slab gel. Radioactivity was quantitated, normalized to the amount of radioactivity found in "high molecular weight proteins," and plotted as the relative rate of DHFR synthesis (lower panel). The absolute rate of DHFR synthesis (upper panel) was the product of the relative rate of DHFR synthesis and the rate of total cellular protein synthesis. Ordinates are expressed in arbitrary units. (Adapted from data published by Wiedemann and Johnson, 1979.)

increased about 10- to 20-fold by 20 hrs following stimulation (fig. 6). Therefore, the increase in the rate of accumulation of DHFR activity was due primarily to an increase in the rate of synthesis rather than an increase in enzyme stability (Wiedemann and Johnson, 1979).

The increase in the rate of synthesis of DHFR could be the result of an increase in the total amount of cytoplasmic DHFR mRNA or to an increase in the efficiency of translation of DHFR mRNA or possibly both. We determined the amount of DHFR mRNA in M50L3 cells at various times following stimulation by isolating total cytoplasmic RNA from cultures of cells and using it as a template in a mRNA-dependent *in vitro* translation system derived from rabbit reticulocytes (Pelham and Jackson, 1976). Our results indicate that there is a good correlation between the rate of synthesis of DHFR in the *in vitro* translating system (fig. 7) and the rate of synthesis of DHFR *in vivo*, indicating that the rate of DHFR synthesis is

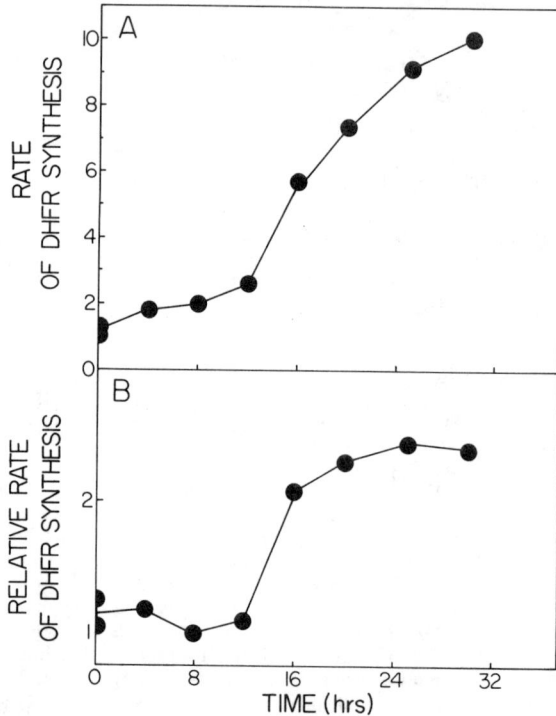

Fig. 7.   DHFR mRNA levels in serum stimulated M50L3 cells. Roller bottles of resting M50L3 cells were serum-stimulated at time = 0. At various times cultures were harvested and cytoplasmic RNA purified (Palmiter, 1974). The RNA was added to an *in vitro* mRNA-dependent translating system derived from rabbit reticulocytes (Pelham and Jackson, 1976). The amount of DHFR synthesized (proportional to the amount of translatable DHFR mRNA added to the system) was determined by immunoprecipitation (Kellems et al., 1976) and normalized to the total amount of labeled protein synthesized in each reaction (proportional to the total amount of translatable mRNA added to the system) to give the relative rate of DHFR synthesis shown in panel B. This value was multiplied by the total amount of mRNA per cell at various times following stimulation to give the total amount of DHFR mRNA in the cell (panel A). Ordinates are expressed in arbitrary units.

controlled in serum-stimulated cells by regulating the amount of DHFR mRNA present in the cytoplasm (Wu, Wiedemann, Pratt, and Johnson, in preparation).

To thoroughly understand the mechanism(s) used to regulate the expression of a particular gene, one must study not only the turning on, but also the turning off, of gene expression. We have recently found that if the serum stimulus is withdrawn from M50L3 cells, the rate of synthesis of DHFR quickly decreases to about the same level as found in resting cells, as shown in figure 8. The half-life of the decrease is approximately 6–8 hr, which is close to the half-life of total mRNA in 3T6 cells (Abelson et al., 1974). This suggests that the production of DHFR mRNA ceases shortly

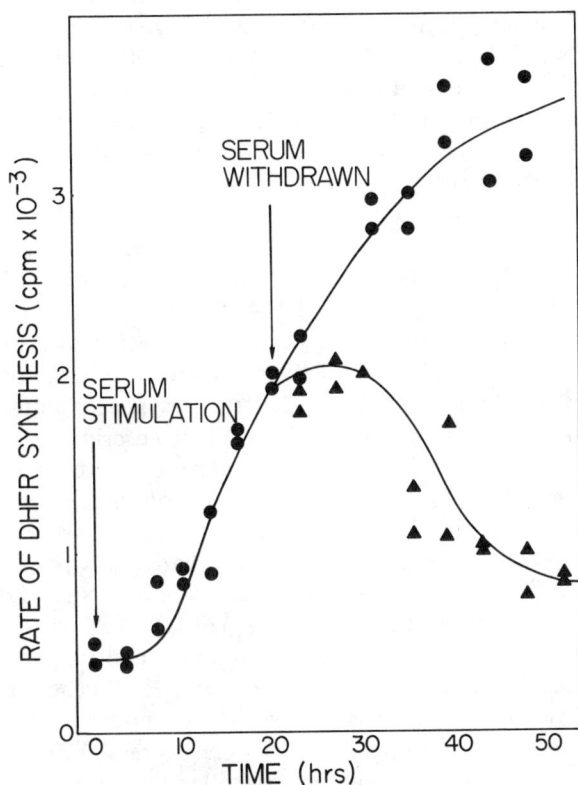

Fig. 8.   Rate of DHFR synthesis following withdrawal of the serum stimulus. Cultures of resting M50L3 cells were serum stimulated at time = 0 (solid circles). 20 hrs following stimulation some of the cultures were washed with serum-free medium and fed with medium containing 0.5% serum (solid triangles). Cultures were labeled for 1 hr with ^{3}H-leucine at various times and the amount of labeled DHFR was determined by immunoprecipitation.

after the serum stimulus is removed, and that DHFR synthesis decreases directly in parallel to the decrease in DHFR mRNA content (Wu et al., in preparation). Direct quantitation of DHFR mRNA levels at various times following serum withdrawal are being made to verify this hypothesis.

Recent studies with another MTX-resistant 3T6 cell line isolated in Schimke's laboratory have shown that infection of quiescent cells with polyoma virus resulted in a 4- to 5-fold increase in the relative rate of DHFR synthesis and a corresponding increase in DHFR mRNA abundance. This increase is analogous to that observed following infection of normal cells with polyoma virus (Kellems et al., 1979). They also observed that addition of fresh serum to stationary phase 3T6 cells resulted in a 2-fold increase in DHFR synthesis, and that inhibitors of DNA synthesis did not block the increase in DHFR gene expression.

Our results and those of Kellems and colleagues (1979) show that DHFR gene expression is regulated in the same manner in overproducing cells as in normal cells. Amplification of the DHFR gene does not appear to affect the ability of the cells to control the level of expression of the gene. Therefore, if specific regulatory molecules are responsible for controlling the level of DHFR gene expression by controlling the level of transcription of the gene, processing of the initial transcription product into mature DHFR mRNA, export of the mRNA from the nucleus to the cytoplasm, or the translation of the cytoplasmic mRNA, they must either be amplified to the same extent as the structural gene or be present in great excess in normal cells. Alternatively if DHFR gene expression is controlled at the level of transcription, and if the DHFR structural gene is amplified but the regions responsible for controlling DHFR transcription (presumably upstream from the amplified structural genes) are not, the control of DHFR gene expression would be retained without overproduction of the control proteins. Comparison of the amplified DHFR sequences and the normal gene (plus adjacent sequences) may clarify this point.

Kellems and colleagues (1979) have also shown that DHFR gene expression appears to be affected by the levels of cellular cAMP. They found that when cells were serum-stimulated under conditions that caused an elevated level of cyclic AMP, the increase in DHFR gene expression did not occur. However, infection with polyoma virus under similar conditions resulted in the normal increase in DHFR gene expression. They concluded that DHFR gene expression may be controlled by at least two regulatory pathways: one involving serum that is blocked by high levels of cAMP and another involving polyoma induction that is not inhibited by cyclic AMP. We have also found that high levels of cAMP block the increase in DHFR synthesis in serum-stimulated M50L3 cells. However, we observed that in normal 3T6 cells DHFR gene expression increased normally under the same high cyclic AMP conditions (Wiedemann and Johnson, in preparation). Therefore, the cAMP effect may represent an artifact of the gene amplification process and not a fundamental aspect of DHFR gene regulation.

DHFR gene expression appears to be regulated primarily by controlling the level of DHFR mRNA. Similar observations have been made for a variety of other systems, although control of mRNA translation may also play an important role in regulating the expression of some genes. A more fundamental question may now be asked. What controls the level of cytoplasmic DHFR mRNA? The most efficient control would be the regulation of gene transcription. Our actinomycin results suggest (but do not prove) that this is the case. However, one must also consider the possibility that the DHFR gene is transcribed throughout the cell cycle, but that the cell regulates the processing and/or export of the initial transcription product, DHFR hnRNA, into the mature cytoplasmic species.

The rate of synthesis of DHFR hnRNA as well as cytoplasmic DHFR mRNA can be determined by standard DNA-excess hybridization assays if large amounts of DNA sequences corresponding to DHFR mRNA are available. The recent cloning of full-length DHFR cDNA in *Escherichia coli* has provided such a source of DHFR DNA sequences (Chang et al., 1978). The availability of these cloned DNA sequences will greatly facilitate these and other studies, such as the isolation and analysis of the structure of the DHFR gene and the DHFR hnRNA processing scheme. Schimke's laboratory has recently found that the DHFR gene is about 40 kb in length and that the initial transcription product appears to be of similar length (R. Schimke, personal communication). The amount of noncoding sequences in the DHFR gene appears to be much larger than that found in other eukaryotic genes. The length of the DHFR mRNA is about 1.5 kb, which means that about 95% of the initial transcription product must be "spliced out" or degraded to produce the mature mRNA molecule.

OVERPRODUCTION OF GENES FOR OTHER ENZYMES

The results described above have demonstrated a new approach that can be taken to permit detailed studies of the structure and expression of the gene for DHFR, one of the so-called housekeeping genes. Recent studies have also shown that the same type of approach can be used to facilitate the study of genes coding for other proteins or enzymes found at low levels in normal cells. For example, Stark's laboratory has found that cells resistant to N-(phosphonacetyl)-L-aspartate overproduce the target enzyme (aspartate transcarbamylase), as well as the mRNA and the gene for the enzyme, about 100-fold (Kempe et al., 1976; Padgett et al., 1979; Wahl et al., 1979). Cells resistant to 25-hydroxycholesterol also exhibit elevated levels of the enzyme hydroxymethyl glutaryl CoA reductase (Sinensky, 1977). In fact, it is possible that overproduction of the target protein as a means of developing resistance to a toxic agent may very well be a general mechanism, if proper selective conditions are used. These conditions are likely to involve exposure of cultured cells to gradually increasing levels of a toxic agent that exhibits a high degree of specificity for the target enzyme (or protein) to be amplified. The target enzyme must be essential for cell survival, at least under the selective conditions. During the selection process, the vast majority of cells in the population are killed. However, those cells able to grow in high levels of the toxic agent may have achieved their resistance by virtue of target enzyme overproduction.

ISOLATION OF CELLS THAT OVERPRODUCE
THYMIDYLATE SYNTHETASE

As mentioned earlier, TS is the enzyme responsible for the conversion of deoxyuridylic acid to thymidylic acid. Previous studies have shown that the level of TS is very low in stationary phase cells, but increases dramatically when the cells are replated at low density (Conrad, 1971). We have found that resting 3T6 cells (as well as resting M50L3 cells) also contain very low levels of the enzyme. When the resting cells are serum-stimulated, the level of the enzyme begins to increase about 10–12 hrs following stimulation (fig. 9). We also found that the increase occurs when the cells are stimulated in the presence of inhibitors of DNA synthesis (Navalgund et al., 1980). Thus, the expression of the genes for TS and DHFR appear to be controlled in a similar manner. In fact, the expression of the two genes may very well be regulated by the same cell cycle specific control signal(s).

To study TS gene expression in detail will also require a cell line that overproduces the mRNA for the enzyme. We have used the approach discussed above to devise a selective system for isolating a 3T6 cell line that overproduces TS. The enzyme is strongly inhibited by 5-fluo-rodeoxyuridylic acid, a structural analogue of the normal substrate (Heidelberger et al., 1960). Although the nucleotide is not taken up by the cell, the nucleoside form of the drug is. The nucleoside is converted to the nucleotide by the enzyme thymidine kinase. Our strategy has been to isolate 3T6 cells that are resistant to high levels of 5-fluorodeoxyuridine in

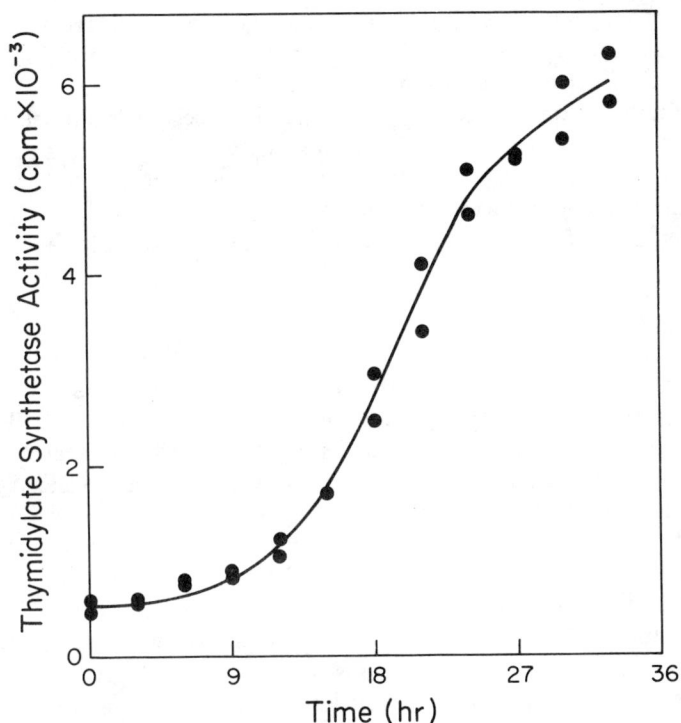

Fig. 9. Thymidylate synthetase activity in serum stimulated 3T6 cells. Cultures of resting 3T6 cells were serum stimulated at time = 0 and harvested at later times. Cytoplasmic extracts were assayed for thymidylate synthetase activity by the procedure of Roberts (1966).

hopes of obtaining cells that overproduce TS. We are aware that drug resistance can also be achieved by a number of other mechanisms, the most obvious being the elimination of thymidine kinase. However, if necessary, we can select for the presence of thymidine kinase using HAT medium (Littlefield, 1964). Unfortunately, both selective pressures cannot be applied simultaneously since the thymidine of HAT medium is able to provide the thymidylic acid needed for DNA synthesis, eliminating the need for *de novo* biosynthesis of thymidylic acid via the TS pathway.

We have been successful in our attempts to obtain TS overproducers. We find that 3T6 cells overproduce the enzyme at least 200-fold even when they are resistant to relatively low (3 μM) drug levels (Johnson and Rossana, in preparation). We are presently attempting to obtain higher degrees of overproduction by selecting for cells able to grow at higher levels of fluorodeoxyuridine. Of course, this cell line will greatly facilitate our studies of TS gene expression. Our ultimate goal will be to determine if the biochemical signals responsible for controlling TS gene expression are the same as those used to control the expression of the DHFR gene, and other genes that are expressed at the same time in the cell cycle. At present, we speculate that these control signals may be specific nonhistone chromosomal proteins, or enzymes that modify specific chromosomal proteins, thereby allowing the coordinated transcription of a class of unlinked genes for the enzymes needed for entry into S phase of the cell cycle.

SUMMARY

In this article I review recent studies from my laboratory and others on the regulation of DHFR gene expression in mammalian cells. The expression of the gene is very low in resting (G_0) cells, but increases sharply about 10–12 hrs following stimulation to reenter the cell cycle. We have found that the increase in DHFR gene expression appears to be regulated at the level of transcription and occurs about the same time the cell starts DNA replication but is not tightly linked to DNA synthesis.

To facilitate our studies, we have isolated a MTX-resistant cell line that overproduces DHFR, as well as the mRNA and the gene for the enzyme. Dihydrofolate reductase gene expression appears to be regulated in the same manner in the overproducing cell line as in the normal cell line. Using this cell line we have shown that the increase in DHFR synthesis is due to an increase in DHFR mRNA content. Studies are in progress to determine directly if the cell regulates the level of DHFR mRNA at the transcriptional or post-transcriptional level.

I also discuss the possibility of isolating other cell lines that overproduce the target enzymes of various toxic drugs. Such cell lines may be of great value in facilitating the investigation of the structure and regulation of the expression of a variety of genes that could not be studied by more conventional procedures. In particular I describe our recent isolation of a cell line that is resistant to 5-fluorodeoxyuridine and that overproduces the target enzyme TS.

ACKNOWLEDGMENTS

Research in the author's laboratory was supported by grants from the American Cancer Society (Ohio Division), the National Cancer Institute (CA 16058-05 and CA 26470-01), and by a Basil O'Connor Starter Research Grant from The National Foundation–March of Dimes.

REFERENCES

Abelson, H. T., L. F. Johnson, S. Penman, and H. Green. 1974. Changes in RNA in relation to growth of the fibroblast. II. The lifetime of mRNA, rRNA, and tRNA in resting and growing cells. Cell 1:161–65.

Alt, F. W., R. E. Kellems, J. R. Bertino, and R. T. Schimke. 1978. Selective multiplication of dihydrofolate reductase genes in methotrexate-resistant variants of cultured murine cells. J. Biol. Chem. 253:1357–70.

Alt, F. W., R. E. Kellems, and R. T. Schimke. 1976. Synthesis and degradation of folate reductase in sensitive and methotrexate-resistant lines of S-180 cells. J. Biol. Chem. 251:3063–74.

Bertino, J. R. 1979. Toward improved selectivity in cancer chemotherapy: the Richard and Hinda Rosenthal Foundation Award Lecture. Cancer Res. 39:293–304.

Bertino, J. R., D. M. Donohue, B. Simmons, B. W., Gabrio, R. Silber, and F. M. Huennekens. 1963. The "induction" of dihydrofolate reductase activity in leukocytes and erthyrocytes of patients treated with amethopterin. J. Clin. Invest. 42:466–75.

Biedler, J. L., and B. A. Spengler. 1976. Metaphase chromosome anomaly: association with drug resistance and cell-specific products. Science 191:185–87.

Blakley, R. L. 1969. The biochemistry of folic acid and related pteridines. North-Holland Publishing Company, Amsterdam.

Borun, T. W., M. D. Scharff, and E. Robbins. 1967. Rapidly labeled polyribosome-associated RNA having the properties of histone messenger. Proc. Nat. Acad. Sci. USA 58:1977–82.

Breathnach, R., J. L. Mandel and P. Chambon. 1977. Ovalbumin gene is split in chicken DNA. Nature 270:314–19.

Chang, A. C. Y., J. H. Nunberg, R. J. Kaufman, H. A. Erlich, R. T. Schimke, and S. N. Cohen. 1978. Phenotypic expression in *E. coli* of a DNA sequence coding for mouse dihydrofolate reductase. Nature 275:617–24.

Chang, S. E., and J. W. Littlefield. 1976. Elevated dihydrofolate reductase messenger RNA levels in methotrexate-resistant BHK cells. Cell 7:391–96.

Chello, P. L., C. A. McQueen, L. M. DeAngelis, and J. R. Bertino. 1977. Comparative effects of folate antagonists versus enzymatic folate depletion on folate and thymidine enzymes in cultured mammalian cells. Cancer Treat. Rep. 61:539–48.

Cohen, S. S. 1971. On the nature of thymineless death. Ann. N. Y. Acad. Sci. 186:292–301.

Conrad, A. H. 1971. Thymidylate synthetase activity in cultured mammalian cells. J. Biol. Chem. 246:1318–23.

Dugaiczyk, A., S. L. C. Woo, E. C. Lai, M. L. Mace, Jr., L. McReynolds, and B. W. O'Malley. 1978. The natural ovalbumin gene contains seven intervening sequences. Nature 274:328–33.

Flintoff, W. E., S. V. Davidson, and L. Siminovitch. 1976. Isolation and partial characterization of three methotrexate-resistant phenotypes from Chinese hamster ovary cells. Somatic Cell Genet. 2:245–61.

Frearson, P. M., S. Kit, and D. R. Dubbs. 1966. Induction of dihydrofolate reductase activity by SV40 and polyoma virus. Cancer Res. 26:1653–60.

Hakala, M. T., S. F. Zakrzewski, and C. A. Nichol. 1961. Relation of folic acid reductase to amethopterin resistance in cultured mammalian cells. J. Biol. Chem. 236:952–58.

Heidelberger, C., G. Kaldor, K. L. Mukherjee, and P. B. Danneberg. 1960. Studies on fluorinated pyrimidines. XI. *In vitro* studies on tumor resistance. Cancer Res. 20:903–9.

Hillcoat, B. L., V. Swett, and J. R. Bertino. 1967. Increase of dihydrofolate reductase activity in cultured mammalian cells after exposure to methotrexate. Proc. Nat. Acad. Sci. USA 58:1632–37.

Johnson, L. F., H. T. Abelson, H. Green, and S. Penman. 1974. Changes in RNA in relation to growth of the fibroblast. I. Amounts of mRNA, rRNA, and tRNA in resting and growing cells. Cell 1:95–100.

Johnson, L. F., C. L. Fuhrman, and H. T. Abelson. 1978a. Resistance of resting 3T6 mouse fibroblasts to methotrexate cytotoxicity. Cancer Res. 38:2408–12.

Johnson, L. F., C. L. Fuhrman, and L. M. Wiedemann. 1978b. Regulation of dihydrofolate reductase gene expression in mouse fibroblasts during the transition from the resting to growing state. J. Cell. Physiol. 97:397–406.

Kaufman, R. J., P. C. Brown, and R. T. Schimke. 1979. Amplified dihydrofolate reductase genes in unstable methotrexate-resistant cells are associated with double minute chromosomes. Proc. Nat. Acad. Sci. USA 76:5669–73.

Kellems, R. E., F. W. Alt, and R. T. Schimke. 1976. Regulation of folate reductase synthesis in sensitive and methotrexate-resistant sarcoma 180 cells. J. Biol. Chem. 251:6987–93.

Kellems, R. E., V. B. Morhenn, E. A. Pfendt, F. W. Alt, and R. T. Schimke. 1979. Polyoma virus and cyclic AMP-mediated control of dihydrofolate reductase mRNA abundance in methotrexate-resistant mouse fibroblasts. J. Biol. Chem. 254:309–18.

Kempe, T. D., E. A. Swyryd, M. Bruist, and G. R. Stark. 1976. Stable mutants of mammalian cells that overproduce the first three enzymes of pyrimidine nucleotide biosynthesis. Cell 9:541–50.

Littlefield, J. W. 1964. Selection of hybrids from matings of fibroblasts *in vitro* and their presumed recombinants. Science 145:709–10.

Littlefield, J. W. 1969. Hybridization of hamster cells with high and low folate reductase activity. Proc. Nat. Acad. Sci. USA 62:88–95.

Navalgund, L. G., C. Rossana, A. Muench, and L. F. Johnson. 1980. Cell cycle regulation of thymidylate synthetase gene expression in cultured mouse fibroblasts. J. Biol. Chem. (in press).

Nunberg, J. H., R. J. Kaufman, R. T. Schimke, G. Urlaub, and L. A. Chasin. 1978. Amplified dihydrofolate reductase genes are localized to a homogeneously staining region of a single chromosome in a methotrexate-resistant Chinese hamster ovary cell line. Proc. Nat. Acad. Sci. USA 75:5553–56.

Padgett, R. A., G. M. Wahl, P. F. Coleman, and G. R. Stark. 1979. N-(Phosphonacetyl)-L-aspartate-resistant hamster cells overaccumulate a single mRNA coding for the multifunctional protein that catalyzes the first steps of UMP synthesis. J. Biol. Chem. 254:974–80.

Palmiter, R. D. 1974. Magnesium precipitation of ribonucleoprotein complexes: expedient techniques for the isolation of undegraded polysomes and messenger ribonucleic acid. Biochemistry 13:3606–15.

Pardee, A. B., R. Dubrow, J. Hamlin and R. F. Kletzien. 1978. Animal cell cycle. Ann. Rev. Biochem. 47:716–50.

Pelham, H. R. B., and R. J. Jackson. 1976. An efficient mRNA-dependent translation system from rabbit reticulocytes. Eur. J. Biochem. 67:247–56.

Prescott, D. M. 1976. Reproduction of eucaryotic cells. Academic Press, New York.

Roberts, D. 1966. An isotopic assay for thymidylate synthetase. Biochemistry 5:3546–48.

Schibler, U., K. B. Marcu, and R. P. Perry. 1978. The synthesis and processing of the messenger RNAs specifying heavy and light chain immuno-globulins in MPC-11 cells. Cell 15:1495–1509.

Schimke, R. T., R. J. Kaufman, F. W. Alt, and R. F. Kellems. 1978. Gene amplification and drug resistance in cultured murine cells. Science 202:1051–55.

Seidman, J. G., A. Leder, M. Nau, B. Norman, and P. Leder. 1978. Antibody diversity. Science 202:11–17.

Sinensky, M. 1977. Isolation of a mammalian cell mutant resistant to 25-hydroxy cholesterol. Biochem. Biophys. Res. Comm. 78:863–67.

Tilghman, S. M., P. J. Curtis, D. C. Tiemeier, P. Leder, and C. Weissmann. 1978. The intervening sequence of a mouse β-globin gene is transcribed within the 15 S β-globin mRNA precursor. Proc. Nat. Acad. Sci. USA 75:1309–13.

Todaro, G. J., and H. Green. 1963. Quantitative studies of the growth of mouse embryo cells in culture and their development into established lines. J. Cell Biol. 17:299–313.

Tonegawa, S., C. Brack, N. Hozumi, G. Matthyssens, and R. Schuller. 1977. Dynamics of immunoglobulin genes. Immunol. Rev. 36:73–94.

Wahl, G. M., R. A. Padgett, and G. R. Stark. 1979. Gene amplification causes overproduction of the first three enzymes of UMP synthesis in N-(Phosphonacetyl)-L-aspartate-resistant hamster cells. J. Biol. Chem. 254:8679–89.

Wiedemann, L. M., and L. F. Johnson. 1979. Regulation of dihydrofolate reductase synthesis in an overproducing 3T6 cell line during transition from resting to growing state. Proc. Nat. Acad. Sci. USA 76:2818–22.

SARAH C. R. ELGIN, SUSAN M. ABMAYR,
IAN L. CARTWRIGHT, GARY C. HOWARD,
MICHAEL A. KEENE, KY LOWENHAUPT,
YUK-CHOR WONG, AND CARL WU

Chromatin Structure in Relation to Gene Activity: A Speculative Essay

13

INTRODUCTION

In essentially all eukaryotes, the DNA of the genome is found in association with the histones and nonhistone chromosomal proteins (NHC proteins) in a stable complex referred to as chromatin. During the last few years, knowledge of the structure of chromatin has increased dramatically. It is now well established that the basic chromatin fiber consists of a chain of repeating subunits, the nu bodies or nucleosomes, each made up of 155–240 base pairs of DNA wrapped around the outside of a core of eight of the small histones, apparently two each of H2A, H2B, H3, and H4. One hundred forty-five base pairs of DNA are firmly associated with the histone core, and the remaining portion forms a "linker" or "spacer" between core particles. It has been suggested that histone H1 is associated with, and stabilizes, the DNA of the linker region. Presumably the NHC proteins interact with the DNA-histone complex by assuming positions within the core, in association with linker DNA, and/or in association with the outer surface of the DNA-histone bead. DNA and histones are present in chromatin in a 1:1 weight ratio; the NHC protein fraction is present at 30–120% of the weight of the histone fraction. Although there are only five major histones, the NHC proteins can be very heterogeneous, with 10^2–10^3

Harvard University, The Biological Laboratories, 16 Divinity Avenue, Cambridge, Massachusetts 02138 different polypeptides typically reported. For a review of the evidence leading to this model of chromatin structure, see the papers of Elgin and Weintraub (1975), Kornberg (1977), and Felsenfeld (1978).

We are interested in investigating the relationship between the structure of the chromatin fiber and the expression of a particular gene. Given the complexity of the eukaryotic genome, it is attractive to suppose that genes are regulated in coordinated sets (Britten and Davidson, 1969). We would like to suggest that certain sets of genes, defined by common functional parameters (such as the state of expression), can also be identified in a given cell type by common features of chromatin structure. Two functional sets of genes will be considered. Those genes that are being transcribed at the moment of assay will be referred to as "immediately active" genes. Those genes that may be active at some time or under some conditions in the cell type being examined will be referred to as "developmentally active" genes. Note that this second set of genes includes the first set. Can these sets of genes be identified by structural characteristics? It appears likely that the nucleosomal pattern of structure is the primary organizational principle for all genomic DNA. Nonetheless, there is now good evidence that the chromatin structure of the immediately active genes is altered in a defined manner relative to inactive genes. This has been demonstrated both in terms of the susceptibility of the chromatin fiber to nuclease digestion and in terms of the pattern of associated NHC proteins, to be discussed below, as well as by other experimental approaches. In contrast, there is little direct evidence to support the hypothesis that the developmentally active loci can be defined by distinctive common features of their chromatin structure. However, the observation that certain NHC proteins are preferentially associated with the developmentally active loci suggests that a particular configuration of the chromatin fiber is necessary but not sufficient for transcription. It is possible that the developmentally active genes possess such a configuration, and perhaps therefore are accessible to immediate regulation by mechanisms analogous to those used in prokaryotes. Such a model makes several predictions that can be tested. Specifically, it predicts that the chromatin structure of a specific gene may be regulated by developmental events, and that shifts in chromatin structure may be correlated with determination as well as with transcription. In the present paper we will review some of the evidence leading us to this viewpoint and speculate on the most profitable direction of our further research. This brief review draws considerably on the work in our own laboratory and is not intended to be comprehensive. In all our work we have used *Drosophila melanogaster* as the model system.

ANALYSIS OF CHROMATIN STRUCTURE USING MICROCOCCAL NUCLEASE

One of the first clues pointing toward a regular repeat as a fundamental feature of chromatin fiber structure was the observation by Clark and Felsenfeld (1971) that when chromatin was dissected into relatively "open" and relatively "covered" DNA segments with micrococcal nuclease, the protected fragments recovered were generally quite small. Subsequent work by Hewish and Burgoyne (1973) using an endogenous nuclease and by Noll (1974), Sahasrabuddhe and Van Holde (1974), and others using micrococcal nuclease demonstrated that these enzymes cleaved the chromatin fiber into a series of oligomers, indicating preferential cleavage of the chromatin fiber between repeating, regular subunits. Experimental results of this type are illustrated in figure 1. Relative protection of the

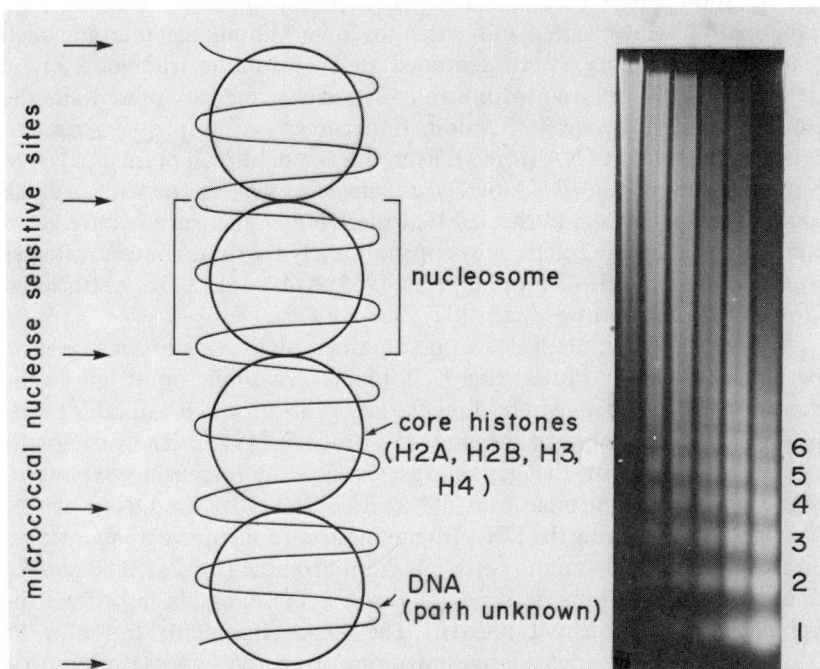

Fig. 1. A schematic representation of the chromatin fiber, indicating micrococcal nuclease-sensitive sites, and gel electrophoresis display of DNA fragments generated by increasing digestion of *Drosophila* chromatin with micrococcal nuclease. Electrophoresis is from top to bottom. Isolated nuclei at a concentration of ~5 × 10⁸/ml were incubated with increasing amounts of the enzyme for 3 min. at 25° C; the DNA has been digested to 0–6% acid solubility by the range of enzyme concentrations used here. The purified DNA fragments were run on a 1% agarose cell and stained with ethidium bromide. See Wu et al., (1979a) for a detailed account of the experimental procedures.

DNA comes from association with the histone cores. The nucleosome model of chromatin fiber structure is now well established on the basis of results from many different kinds of experiments. Consequently, it is now reasonable to take the generation of such a regular cleavage pattern as diagnostic of the regular histone-DNA interaction.

The question of the fundamental mechanisms resulting in limited and specific gene expression in eukaryotic cells is one of long-standing interest. Considerable circumstantial evidence and the results of early studies of transcription by RNA polymerase *in vitro* had suggested that histones might act as general repressors of template activity (reviewed in Elgin et al., 1971). One could now reexamine this hypothesis and probe the structure of the active gene by determining the distribution of such DNA sequences in the digestion pattern produced using micrococcal nuclease. Originally such experiments were carried out either by determining the rate at which transcribed sequences were degraded to acid soluble fragments or by determining the relative proportion of such sequences present in the isolated mononucleosome fraction. Experiments using probes representative of the total mRNA from rat liver, mRNA of human brain, mRNA of human lymphocytes, rRNA of *Tetrahymena*, as well as one using a chick globin gene probe, demonstrated that the DNA sequences of active genes were present in these fractions at approximately the same concentration as in the whole genome (Lacy and Axel, 1975; Kuo et al., 1976; Mathis and Gorovsky, 1976; Brown et al., 1977; Kuo, 1979).

A more definitive analysis can be obtained by looking at the digestion pattern as a whole, rather than by looking at a single point late in the digestion process. This can be done for any gene for which a suitable probe is available by using the method of Southern (1975). The nucleolytic cleavage pattern for the genome as a whole at increasing extents of digestion of total chromatin (0–6% acid solubility of the DNA) can be observed by separating the DNA fragments according to size by agarose gel electrophoresis and staining with ethidium bromide (fig. 2a). The pattern observed is an oligomeric series of discrete DNA bands, up to ca. the octamer, as previously discussed. The DNA fragments can now be transferred in the same relative positions from the agarose gel onto a nitrocellulose sheet by the blotting technique (Southern, 1975). Filter hybridization with radioactive, specific DNA sequences (generally made available through recombinant DNA technology) allows visualization by autoradiography of those fragments on the nitrocellulose filter that contain (wholly or in part) sequences homologous to those of the radioactive probes. This provides a selective visualization of the cleavage pattern of the chromatin fiber of a specific region of the genome. Figure 2b illustrates

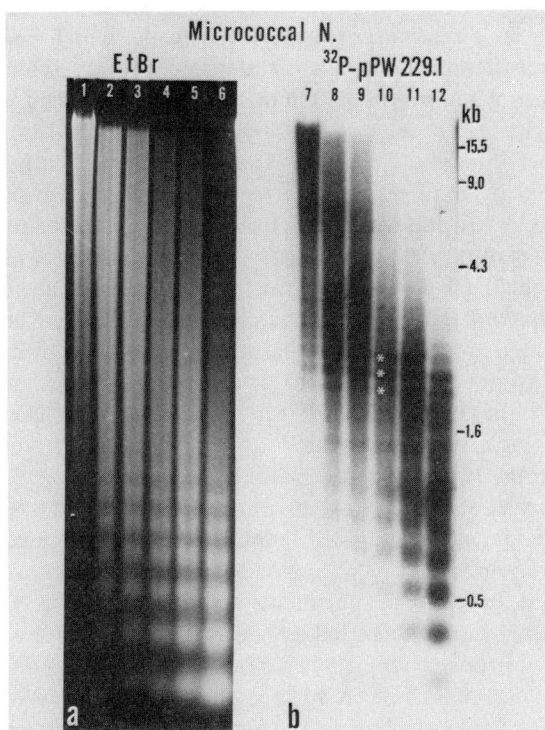

Fig. 2. Comparison of the general micrococcal nuclease digestion pattern with the pattern from sequences homologous to pPW 229.1 in *Drosophila* tissue culture cells. (a) 1% agarose gel stained with ethidium bromide. Digestions were carried out using isolated nuclei as above with enzyme concentrations from 9 to 71 units/ml. (b) Southern blot of gel (a) probed with ^{32}P-pPW 229.1. The nucleosome monomer fragments are hardly visible on the autoradiogram because DNA fragments of this size range do not stick very well to nitrocellulose. In this and subsequent figures, marker restriction fragments sharing sequence homology to the labeled plasmid have been included on flanking slots of the gel. From Wu et al., 1979a. Copyright © 1979 by The MIT Press; reprinted by permission of Cell.

such a result. We have used this approach to look at the structure of the gene encoding the major heat shock protein of *Drosophila melanogaster* in the inactive and active states.

The heat shock response in *Drosophila* is extremely useful as a model system in studying the process of gene activation. Briefly, on raising the culture temperature from 25° to 35° C one observes that a set of nine chromosomal loci puff in the salivary gland polytene chromosomes of *D. melanogaster*, and preexisting puffs regress (Ritossa, 1962; Ashburner, 1970). In parallel, the synthesis of most preexisting proteins ceases, while several specific proteins are induced and synthesized at high rates (Tissieres et al., 1974). In *Drosophila* tissue culture cells, which also exhibit this response, preexisting polysomes disintegrate and are replaced by a new population using newly synthesized mRNA within a few minutes. This new RNA labels the heat shock puff sites by *in situ* hybridization (McKenzie et al., 1975; Spradling et al., 1975, 1977) and directs the synthesis of the heat shock-induced proteins by an *in vitro* translation system (McKenzie and Meselson, 1977; Mirault et al., 1978). Several of the heat shock genes have now been cloned in recombinant DNA plasmids (Lis et al., 1978; Livak et al., 1978; Schedl et al., 1978; Artavanis-Tsakonis et al., 1979). In our analysis we have used primarily the clone pPW 229.1, which contains a *Drosophila* DNA sequence that falls entirely within the transcribed region for the gene encoding the major heat shock protein, one of 70,000 molecular weight (Livak et al., 1978). There are at least five copies of this gene in *D. melanogaster*, located at positions 87A7 and 87C1 (Ish-Horowicz et al., 1979). Several, probably all, are activated in heat shock (Ish-Horowicz et al., 1977; Caggese et al., 1979). The plasmid pPW 244.1 containing sequences entirely within the unique heat shock gene encoded at 63BC (R. Holmgren and M. S. Meselson, personal communication) has also been used. In control experiments we have used plasmids homologous to loci known not to be activated by heat shock. The sequence of pPW 100 is present twice, once each at 57BC and 83C; that of pPW 112 is present once at 78A; and that of pDm 4 is present once at 62F (Wu et al., 1979a).

The results of an investigation of oligonucleosome structure at the major heat shock locus before and after activation are shown in figure 3. Nuclei were isolated quickly from *D. melanogaster* tissue culture cells (Schneider's line 2) that had been cultured at 25° or heat shocked at 35° for 5, 15, or 30 minutes. Aliquots of nuclei were then digested with increasing amounts of micrococcal nuclease, the DNA isolated, and the fragments separated by agarose gel electrophoresis. The same amount of DNA (10 μg) was loaded on the gel for each sample. Samples from each cell population were shown to be digested to an equivalent degree by comparison of the extents of digestion of the DNA of the genome as a whole as shown by ethidium bromide staining of the gel. Visualization of the DNA digestion products from the major heat shock locus using plasmid pPW 229.1 reveals two interesting features. First, the oligonucleosomes of the active gene are preferentially attacked by micrococcal nuclease (compare the amount of hybridizable material in lanes 4, 8, 12, and 16). Second, the nucleosome pattern per se has become considerably "smeared"; distinct oligomers can no longer be detected in the most extreme cases. Essentially similar results are obtained using pPW 244.1 as a probe. Results of the analogous experiment carried out using the three different control probes show no

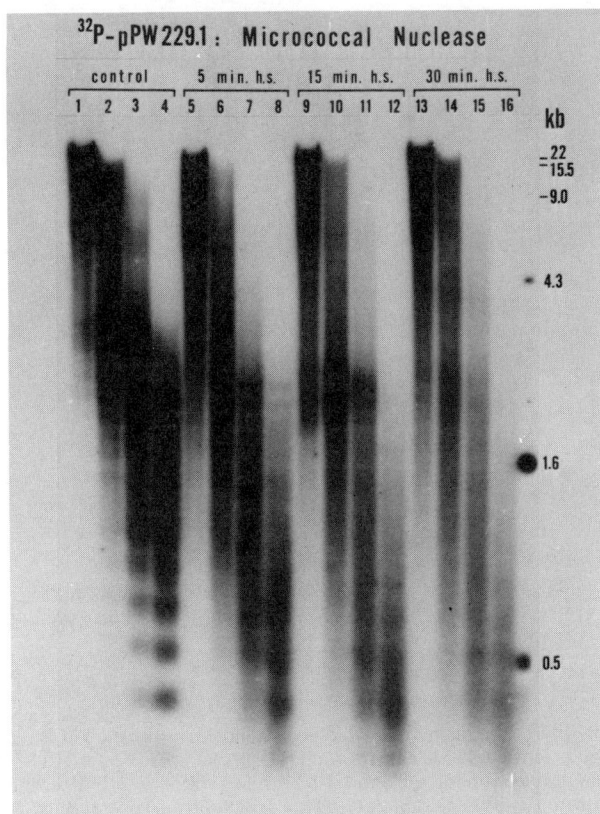

Fig. 3. Pattern of DNA fragments from micrococcal nuclease digestion of chromosomal regions homologous to pPW 229.1 in control and heat-shocked tissue culture cells. Electrophoresis is on a 1.1% agarose gel; the sample load is 10 μg DNA per slot. A Southern blot of the gel was carried out, probed with ^{32}P-pPW 229.1 and autoradiographed. Digestions were carried out at the following enzyme concentrations: slots (1, 5, 9, and 13) 12 U/ml; (2, 6, 10, and 14) 23 U/ml; (3, 7, 11, and 15) 47 U/ml; (4, 8, 12, and 16) 94 U/ml. Slots (1–4) are samples from control cells; (5–8), (9–12), and (13–16) are from cells subjected to 35° C heat shock for 5, 15, and 30 min., respectively. From Wu et al., 1979b. Copyright © 1979 by The MIT Press; reprinted by permission of Cell.

such perturbation of nucleosome structure, demonstrating that this is a specific effect of heat shock, correlated with gene activation rather than with the cell perturbation in general (fig. 4) (Wu et al., 1979b).

Using somewhat more laborious techniques, others have shown that the DNA of highly active genes (rRNA genes in several systems and the ovalbumin gene in chick oviduct) is more rapidly degraded by micrococcal nuclease to small oligomers or mononucleosomes, although there is no

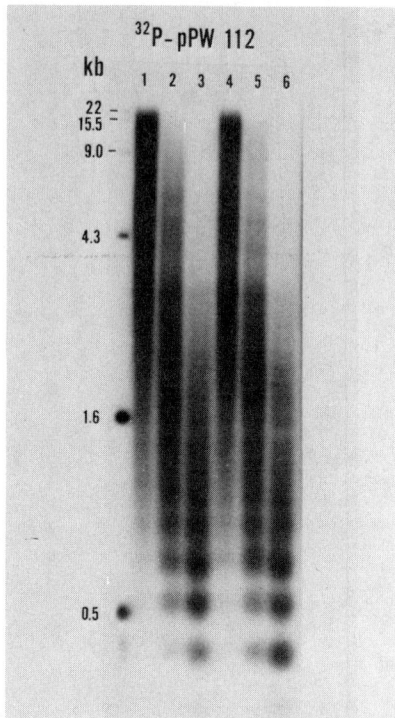

Fig. 4.   Pattern of DNA fragments from micrococcal nuclease digestion of the chromosomal region homologous to pPW 112 in control and heat-shocked tissue culture cells. A Southern blot of DNA samples electrophoresed on a 1.1% agarose gel at 10 μg DNA per slot was carried out and probed with ^{32}P-pPW 112. Slots (1–3) are samples from control cells; (4–6) are from cells heat-shocked for 15 min. Enzyme concentrations are for slots (1 and 4) 23 U/ml; (2 and 5) 47 U/ml; (3 and 6) 94 U/ml. From Wu et al., 1979b. Copyright © 1979 by The MIT Press; reprinted by permission of Cell.

preferential degradation to acid-soluble fragments (Reeves, 1978; Johnson et al., 1978a, 1978b; Bellard et al., 1978; Bloom and Anderson, 1978). One might anticipate that the monomer fragment should be prominent in the heat shock cases in figure 3; unfortunately the nucleosome monomer band scarcely registers on the autoradiogram because DNA fragments of this size range stick to nitrocellulose with very low efficiency during the blotting/hybridization procedure. The smearing of the oligonucleosome pattern is most dramatic, indicating an altered association between the histones and DNA of the active gene. Similar results have subsequently been obtained for the rRNA genes of *Physarum* (Stalder et al., 1979), although extensive smearing was not observed using as a probe cDNA from total polysomal RNA of rat liver (Gottesfield and Melton, 1978). As the latter probe consists primarily of sequences transcribed at a low frequency, the composite results suggest that the perturbation of histone-DNA interaction that destroys the regular micrococcal nuclease cleavage pattern may be quite localized, perhaps limited to the actual sequences associated with RNA polymerase, perhaps including a few hundred base pairs of DNA to either side. This interpretation is supported by many (but not all) results obtained using electron microscopy to observe chromatin fibers. Genes that are being transcribed at a high rate are packed with RNA polymerase, and no nucleosomes are observed. In some cases an "unbeaded" state is observed for rDNA prior to transcription (Foe, 1978; Franke et al., 1978). However, for genes transcribed at a low rate, nucleosome-like structures are observed on the fiber between transcription complexes, suggesting a high proportion of "regular" histone-DNA associations (McKnight and Miller, 1976; McKnight et al., 1978; Scheer, 1978). Whether or not genes transcribed at a low level are preferentially susceptible to digestion to small oligonucleosomes and monomers has not yet been adequately tested. It is clear, however, that the chromatin structure of a highly active gene differs considerably from that of an inactive gene in one, perhaps two, ways: smearing of the nucleosome pattern indicates that the immediate histone-DNA interaction is perturbed, while the more rapid digestion of oligonucleosomes might (but need not) indicate an additional structural perturbation resulting in a more general accessibility of the DNA to the nuclease.

Two structural mechanisms may be suggested that would explain the smearing of the nucleosome pattern. "Sliding" of the nucleosomes relative to the DNA sequence could produce such an effect. Sliding, or lateral displacement of the histone bead along the DNA, has been reported to occur, but at very low rates (Beard, 1978). One could of course postulate that modification of the DNA, the histones, or the binding of some NHC protein might accelerate this process. Note, however, that transcription is obtained from histone-DNA templates *in vitro* under conditions that do not favor sliding (Williamson and Felsenfeld, 1978; Mathis et al., 1978; Gariglio et al., 1979). Alternatively, alteration of the histone-DNA interaction such that even one cleavage may occur at random within the core-associated DNA as frequently as cleavage occurs in the linker DNA would be sufficient to destroy the pattern of oligonucleosome fragments. Such a change might be relatively subtle in structural terms. One way to investigate the situation will be to determine the boundaries of the region of perturbation at an active gene. In the first case (sliding), one might expect the region of perturbation to be large, perhaps the size of a lampbrush chromosome loop. In the second case (unfolding), one might expect the region of perturbation to be limited to the region of transcription. We are currently trying to obtain some answers to these questions by repeating the above experiment using as a plasmid probe a small fragment immediately downstream from the 3' end of one copy of a gene for the major heat shock protein. If the micrococcal nuclease digestion patterns for this region of the genome are the same for chromatin of control and heat shocked cells, it will indicate that the structural perturbations associated with gene activation detected by this nuclease are localized. Further work will be needed to determine accurately the boundaries of the region of structural perturbation (M. Keene, and S. C. R. Elgin, work in progress).

ANALYSIS OF CHROMATIN STRUCTURE USING DNASE I

When chromatin in isolated nuclei is digested with DNase I and the purified double-stranded DNA fragments electrophoresed on agarose gels and displayed by ethidium bromide staining, there is no indication of bands, i.e., no indication of regular structures (fig. 5a). The cleavage of chromatin in general by DNase I produces a continuum of fragments. As the extent of digestion is increased (from 0 to 4% acid solubility of the DNA), a hint of nucleosomal structures at the low molecular weight range of the gel can be discerned. However, as originally observed by Weintraub and Groudin (1976) using the chick hemoglobin gene, DNase I has the interesting property of preferentially attacking genes that are being actively transcribed. Digestion of chromatin to 10–15% acid solubility of the DNA with this enzyme results in almost complete digestion to acid solubility of the DNA sequences encoding immediately active genes. This result has now been confirmed using solution hybridization techniques to monitor the attack on specific active genes in a number of cases, including other globin genes, the ovalbumin gene, protamine genes, murine leukemia proviral DNA, integrated adenovirus, and integrated SV40 (Garel and Axel, 1976; Flint and Weintraub, 1977; Levy and Dixon, 1977; Panet and Cedar, 1977; D. M. Miller et al., 1978; Palmiter et al., 1978; Chae et al., 1978; Frolova et al., 1978; Varshavsky et al., 1978; Breindl and Jaenisch, 1979). For this reason we decided to repeat the above analysis of changes in chromatin structure on activation at the heat shock locus using this enzyme with the Southern blot technique.

We first examined with plasmid 229.1 the digestion pattern at this locus from the chromatin of control cells (gene *not* active), and were surprised to obtain the result shown in figure 5b. The difference between the general and specific digestion patterns is dramatic. On the autoradiogram one observes at least seven discrete bands ranging in size from ca. 2.5 to greater than 20

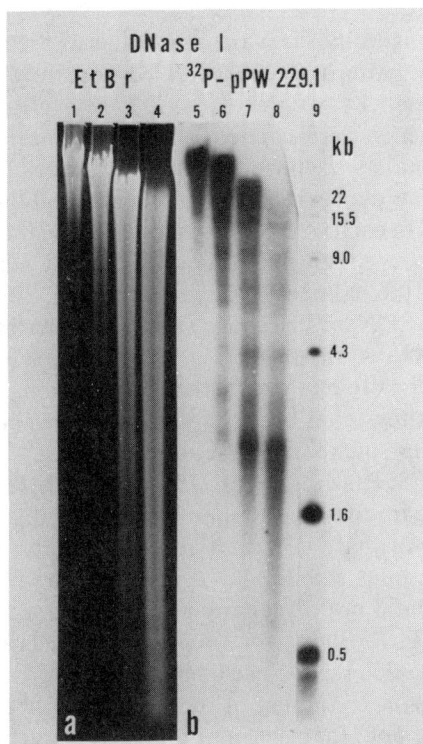

Fig. 5.   Comparison of the general DNase I digestion pattern with the DNase I digestion pattern of sequences homologous to pPW 229.1 in *Drosophila* tissue culture cells. (a) 0.8% agarose gel stained with ethidium bromide. The sample load is 10 μg DNA per slot. Digestions were carried out at 6–16 units/ml. (b) Southern blot of gel (a) probed with ^{32}P-pPW 229.1. The sizes of the bands observed in slot (7) are, in increasing order, approximately 2.4 (doublet, upper band), 3.2, 4.2, 5.8, 6.7, 9.0 (doublet), and 14.4 kb, respectively. From Wu et al., 1979a. Copyright © 1979 by The MIT Press; reprinted by permission of Cell.

kb. The bands are not as sharp as restriction fragments but have a width on the order of a few hundred base pairs, suggesting that they are generated by preferential cleavage at sites of approximately that size. Control experiments carried out using "naked" (purified) DNA as a substrate for the enzyme generated random fragments with no evidence of any band pattern. The specificity of cleavage by DNase I therefore cannot be due in this case to the primary or secondary structure of the DNA alone but must reflect the structure of the chromatin complex (Wu et al., 1979a).

This experiment has been carried out using several other plasmid probes. In almost all cases a unique set of fragments has been obtained. This indicates that the lack of bands in the general digestion pattern is not the consequence of nonspecific cleavage, but rather is the result of the sum of all specific cleavage patterns (5,000?) derived from the *Drosophila* genome. To verify the specificity of the early DNase I cleavage sites, double digestion experiments were carried out. DNA samples obtained after digestion of chromatin were purified, cut with a restriction enzyme known to make one or a few cuts in the probe sequence, and then analyzed by the Southern method. In each of three cases tested, new smaller fragments, not entirely bounded by either the original DNase I cleavage sites or by restriction sites on the naked DNA, were obtained. Since these fragments are well defined, the DNase I cleavage sites in chromatin must be relatively position specific (Wu et al., 1979a).

It will be of interest to explore further the functional significance of the DNase I cleavage sites. It has been suggested that the nucleosome fiber may be further folded in solenoids or "superbeads" (e.g., Hozier et al., 1977; Stratling et al., 1978; F. Miller et al., 1978; Renz, 1979; Butt et al., 1979). However, the large fragments obtained on DNase I digestion do not occur as integer repeats. An analysis of this type on chromatin with DNase I using a probe for the complex satellite DNA (Carlson and Brutlag, 1977; Hsieh and Brutlag, 1979) did not reveal any pattern of large regular bands (I. L. Cartwright and S. C. R. Elgin, work in progress). Regardless of the validity of the superbead model, the DNase I-sensitive sites do not appear to reflect any regular higher-order folding of the chromatin fiber in general.

It is intersting to note that the smallest fragment generated by DNase I digestion of chromatin of the heat shock locus is 2.5 kb, essentially the size of the transcribed region. An analysis with DNase I of the embryo chromatin using a probe for the 5S genes (Artavanis-Tsakonas et al., 1977) indicates two cleavage sites per repeat, producing a regular array of fragments (I. L. Cartwright and S. C. R. Elgin, work in progress). In this instance the result may be a consequence of preferential cleavage of the transcribed region; more detailed experiments are needed to clarify this point. Nonetheless, the data at hand suggest that the DNase I-sensitive sites may be organized in relation to regions of potential transcription. The regulatory implications are intriguing. The matter will be clarified by mapping the position of the DNase I-sensitive sites for several types of genes; such work is in progress. It should be noted that we do not know yet the extent to which these observations will be duplicated using other organisms. *Drosophila* has a relatively small genome, simplifying the technical requirements for looking at specific genes, and could have other peculiarities making it particularly favorable for these experiments. However, recent studies by several groups using SV40 chromatin have demonstrated preferential nucleolytic cleavage of the viral chromatin at specific sites (Scott and Wigmore, 1978; Waldeck et al., 1978; Varshavsky et al., 1978, 1979); similar results have been reported for polyoma viral chromatin (Waldeck et al., 1978).

Let us now return to the question of alteration of chromatin structure on gene activation. An analysis of the chromatin structure using DNase I digestion at the major heat shock loci before and after activation is shown in figure 6. The Southern technique with the plasmid probe 229.1 has been used as previously described. Each set of samples (from control or heat

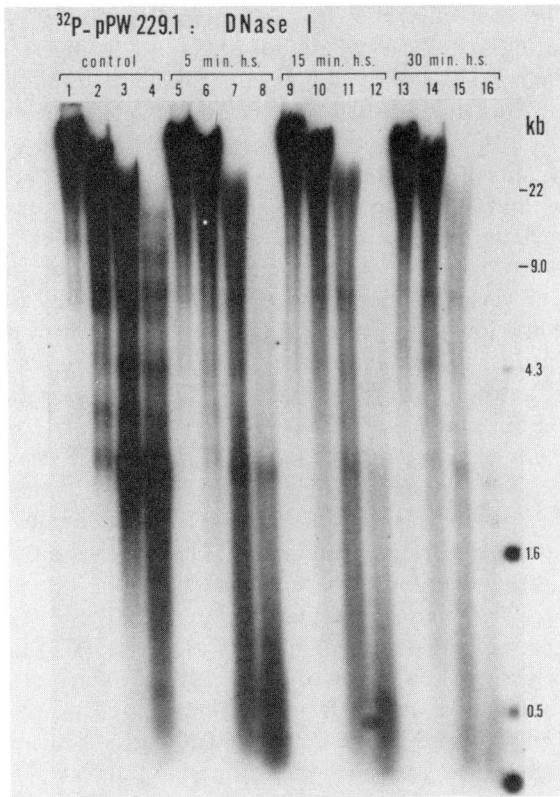

Fig. 6.   Pattern of DNA fragments from DNase I digestion of chromosomal regions homologous to pPW 229.1 in control and heat-shocked tissue culture cells. Electrophoresis is on a 0.7% gel; the sample load is 10 μg DNA per slot. A Southern blot of the gel was carried out, probed with ^{32}P-pPW 229.1 and autoradiographed. Enzyme concentrations are for slots (1, 5, 9, and 13) 6 U/ml; (2, 6, 10, and 14) 8 U/ml; (3, 7, 11, and 15) 12 U/ml; (4, 8, 12, and 16) 16 U/ml. Slots (1–4) are samples from control cells; (5–8), (9–12), and (13–16) are from cells subjected to 35° C heat shock for 5, 15, and 30 min. respectively. From Wu et al., 1979b. Copyright © 1979 by The MIT Press; reprinted by permission of Cell.

shocked cells) has been matched for equivalent extents of digestion (0–4% acid solubility of the DNA) and an equivalent amount of DNA (10 μg) has been loaded on each gel slot. It is immediately observed that the DNA of the active gene is relatively susceptible to digestion by DNase I (compare the extent of hybridization to the DNA of channels 4, 8, 12, and 16), in agreement with the previously reported results. Similar results have been observed for the activated heat shock locus 63BC (plasmid probe pPW 244.1). It also appears that the higher order bands are reduced to low levels of intensity and are much less distinct. In contrast, probes for loci not activated by heat shock (pPW 100 and pDm4) produce hybridization patterns indicating that these loci are not preferentially digested by DNase I and that there is no loss of the higher order band pattern following heat shock (fig. 7). Thus, the effect is specific to activated genes (Wu et al., 1979b).

The results indicate a dramatic alteration in the structure of the chromatin fiber and suggest that a fairly large region (a chromomere?) has undergone a packaging alteration. Certainly, the visual appearance of puffs in polytene chromosomes suggests an "unfolding" of at least the chromomere; however, caution is necessary. A locus may be transcriptionally active at a low level without any visual indication of puffing (e.g., Bonner and Pardue, 1977), yet Garel et al. (1977) have demonstrated that loci transcribed at low levels are as sensitive to DNase I as those transcribed at high levels. Flint and Weintraub (1977) have demonstrated that in the case of the integrated adenovirus genome, the region of DNase I sensitivity maps fairly closely (within a few nucleosomes) to the region of transcription. If, as we suspect, the DNase I-sensitive sites in inactive chromatin bracket the transcribed region, the experiment presented in figure 6 does not give any information concerning the structural state of regions outside the transcribed region when the gene is active; the experiment must be repeated using a probe that is outside the DNase I-sensitive sites closest to the region of transcription. Such an experiment is being carried out using a unique fragment immediately downstream from the 3′ end of one copy of the gene for the 70,000 dalton heat shock protein (M. Keene and S. C. R. Elgin, work in progress). Again, it will be of interest to map the boundaries of the region of structural perturbation on gene activation.

The available evidence indicates that the DNase I-sensitivity of the chromatin of active genes is not dependent on transcription per se. As noted above, genes transcribed at a low level are as DNase I-sensitive as those transcribed at a high level (Garel et al., 1977). Further, it has been reported both for globin genes and for the ovalbumin gene that once active

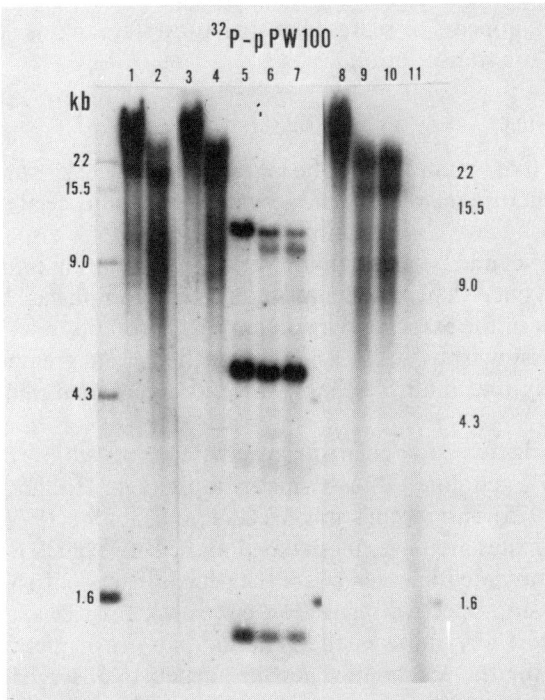

Fig. 7. Pattern of DNA fragments from DNase I digestion and subsequent restriction enzyme analysis of chromosomal regions homologous to pPW 100 in control and heat-shocked tissue culture cells. Southern blots of DNA samples electrophoresed on a 0.6% agarose gel (slots 1–7) and 0.5% agarose gel (slots 8–10) at 10 μg DNA per slot were carried out and probed with ^{32}P-pPW 100. Slots (1–2) and (10) are DNA samples from cells heat-shocked for 20 min., digested with 3, 6, and 3 U/ml DNase I, respectively. Slots (3 and 4) are from two different digestion experiments using control cells; enzyme concentrations are 4 and 3 U/ml, respectively. Slot (5) is Sal 1 digest of naked tissue culture cell DNA. Slots (6 and 7) are Sal 1 digests of DNA samples from DNase I digests identical to those in slots (4) (control) and (2) (heat-shocked), respectively. Slots (8 and 9) are DNA samples from a separate digestion experiment using control cells; enzyme concentrations are 1.5 and 3 U/ml, respectively. From Wu et al., 1979b. Copyright © 1979 by The MIT Press; reprinted by permission of Cell.

these sequences remain sensitive to DNase I even when transcription has ceased (Weintraub and Groudine, 1976; Palmiter et al., 1978; Young et al., 1978). (In contrast, the chromatin structure of the heat shock loci appears to return to the previous state when *Drosophila* tissue culture cells are allowed to recover at 25° C [Wu et al., 1979b; M. Keene, K. Zinn, and S. C. R. Elgin, unpublished observations]. An analysis of the reversibility of chromatin structure alterations for different types of genes in different types of cells may reveal interesting features of gene regulation in developing organisms.) The chromatin structure detected as DNase I-sensitive, then, appears to represent a structural state that is necessary but not sufficient for transcription.

THE MACROMOLECULES OF ACTIVE GENES

One would like to understand the macromolecular basis for the shifts in chromatin structure related to gene activity detected in the above analysis. The evidence suggests an alteration in the histone–DNA association at the nucleosome level and perhaps an alteration in the folding or packaging of the chromatin fiber. A priori, changes could be accomplished by alteration of the DNA or of the associated histones, NHC proteins, or RNA. We will consider the first two possibilities briefly and the third at greater length. It is of course likely that multiple changes occur, some separated in developmental time.

Many investigators have been intrigued with the possibility of systematic modification of genomic DNA by methylation (e.g., Holliday and Pugh, 1975; Riggs, 1975). In a recent study McGhee and Ginder (1979) report that certain CCGG sites are never methylated in the active chick β-globin gene but can be methylated in cases where the gene is inactive. In a similar type of analysis (using selective restriction enzymes), Bird et al. (1979) were unable to detect any shifts in the genomic pattern of methylation over development for the sea urchin, and in particular observed the histone genes to be unmethylated both in tissues where they are believed to be transcribed (embryo) and in those where they are not (sperm). It is possible that a lack of methylation is necessary but not sufficient for gene activity in these cases. However, one should not forget that in *Xenopus laevis* the somatic rDNA and oocyte 5S DNA are both methylated (Bird and Southern, 1978; J. R. Miller et al., 1978), making generalization difficult.

It is well known that despite their very conservative primary structure, the histones are subject to considerable modification. There is a substantial body of circumstantial evidence suggesting that acetylation of the histones (primarily H3 and H4) is an early event in gene activation (reviewed by Allfrey, 1977). Definitive evidence, however, is again lacking. Most intriguing is the observation that the hyperacetylation of histones *in vivo* produced by n-butyrate treatment of cells (which blocks the histone deacetylase) results in chromatin with the nuclease sensitivity normally found only for active genes (Simpson, 1978; Vidalli et al., 1978; Sealy and Chalkley, 1978; Nelson et al., 1978, 1979). However, butyrate-treated HTC cells do not show dramatic changes in their patterns of protein synthesis (Rubenstein et al., 1979). Also, reconstituted templates of SV40 DNA and acetylated histones are not transcribed with increased efficiency *in vitro*, although they are nuclease-sensitive (Mathis et al., 1978).

During the last few years, we have concentrated our efforts on studying the NHC proteins. This fraction is complex, probably including structural proteins of chromatin, enzymes of chromosomal metabolism, and perhaps specific activators and repressors (see Elgin and Weintraub, 1975, for a review). Since functional assays are lacking for most of these proteins, we devised an immunofluorescence technique to examine the distribution of specific proteins on the genomic DNA using antibodies to "stain" the *Drosophila* polytene chromosomes (Silver and Elgin, 1976, 1978a). We and others have now studied many NHC proteins using this technique. The method has allowed demonstration of four different classes of NHC proteins (there are certainly many more) relevant to the present discussion.

1. Those NHC proteins associated with all DNA sequences. Antiserum against a *Drosophila* NHC protein of 21,000 molecular weight, pI = 5.2 (purified by 2-dimensional gel electrophoresis) stains the polytene chromosomes extensively, roughly in proportion to the phase density of the material. This indicates that the protein plays some fundamental role in the chromatin fiber, analogous perhaps in necessity to that of the histones (Silver and Elgin, 1978b).

2. Those NHC proteins preferentially associated with heterochromatin (satellite DNA). Will and Bautz (1979) have obtained an antiserum against a *Drosophila* NHC protein fraction purified by hydrox-yapatite chromatography that gives immunofluorescence staining almost exclusively at the chromocenter. Logically one might expect there to be NHC proteins exclusively associated with the total euchromatin, but none have been identified as yet.

3. Those NHC proteins preferentially associated with the developmentally active loci. Two different antisera reflecting the presence of such proteins have been obtained. One, ρ, was prepared using a 100,000 D molecular weight subfraction of *Drosophila* NHC proteins as immunogen; the second, Band 2, was prepared using a 60,000 D molecular weight subfraction of NHC proteins from those released following limited DNase I digestion of *Drosophila* nuclei. A careful analysis of the pattern of prominent and consistent staining obtained using the ρ antiserum on polytene chromosome arms 3R and 3L indicates a 90% correlation with those loci known to puff at some time during this developmental stage (Silver and Elgin, 1977). The staining pattern obtained with the Band 2 antiserum is very similar, but differs in some details (Mayfield et al., 1978). This pattern is illustrated in figure 8. Recently we have obtained

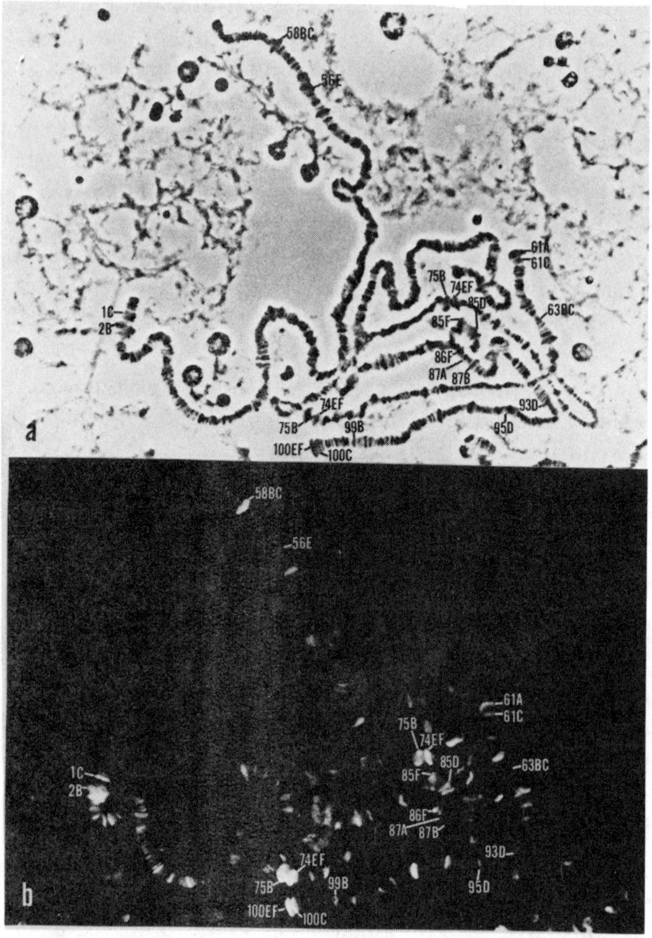

Fig. 8.   Staining pattern obtained using anti-band 2 serum. Glands were obtained from a late third instar larva grown at 25° C and processed through the formaldehyde fixation technique. Anti-band 2 serum was used for staining at a 1:10 dilution. (a) Phase-contrast; (b) fluorescence micrographs. Note that in this particular squash chromosome arm 3L is split from band 64C to the chromocenter. From Mayfield et al., 1978. Copyright © 1978 by The MIT Press; reprinted by permission of Cell.

antibodies from hybridoma cell lines using the Band 2 immunogen that demonstrate a similar staining pattern (fig. 9) (G. C. Howard, S. M. Abmayr, S. C. R. Elgin, work in progress).

4.   Those NHC proteins preferentially associated with the immediately

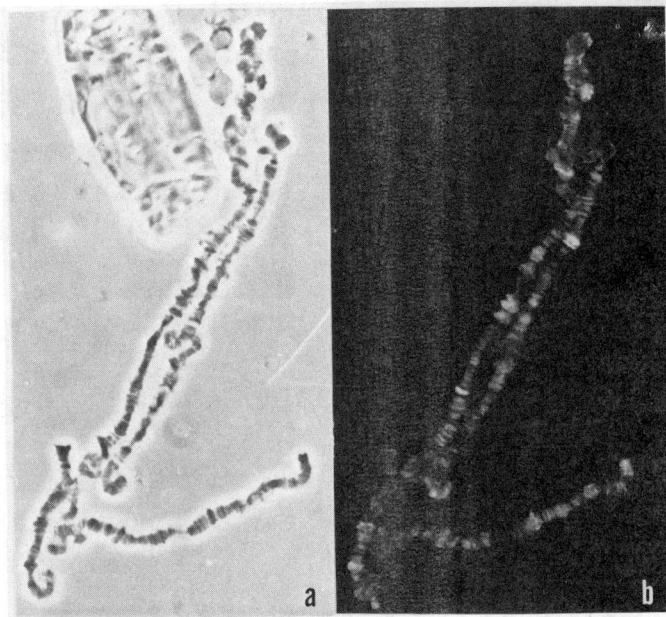

Fig. 9. Staining pattern obtained using antibodies from clone 28 with the formaldehyde fixation technique. (a) Phase contrast; (b) fluorescence micrographs. These antibodies are the IgM type, so staining was achieved by a 3-layer sandwich technique using rabbit IgG directed against mouse IgM and fluorescein-conjugated goat IgG directed against rabbit IgG.

active loci. Antibodies directed against *Drosophila* RNA polymerase II show prominent association with puff sites (Plagens et al., 1976); in particular, only the heat shock puffs are brightly stained in chromosomes from larvae subjected to heat shock (Greenleaf et al., 1978; Elgin et al., 1978). Similar results have been obtained in a preliminary study using an antiserum that reacts with an RNA packaging protein of HeLa cells (Christenson et al., 1978).

The results tabulated above suggest, but do not establish, a hierarchy of chromatin structure related to the transcriptional state of the locus. To further clarify the role of Class 3 proteins, experiments studying the heat shock loci 87A and 87B-C1 have been carried out. These loci are not normally active in the salivary gland of *Drosophila* and are not prominently stained in the chromosomes from animals maintained at 25° (see figure 8). However, using chromosomes from larvae subjected to a 35° heat shock for 15 minutes, one observes prominent puffing at these loci and prominent staining with antisera directed against ρ or Band 2 (e.g., fig. 10) (Silver and Elgin, 1977; Mayfield et al., 1978). This indicates that the

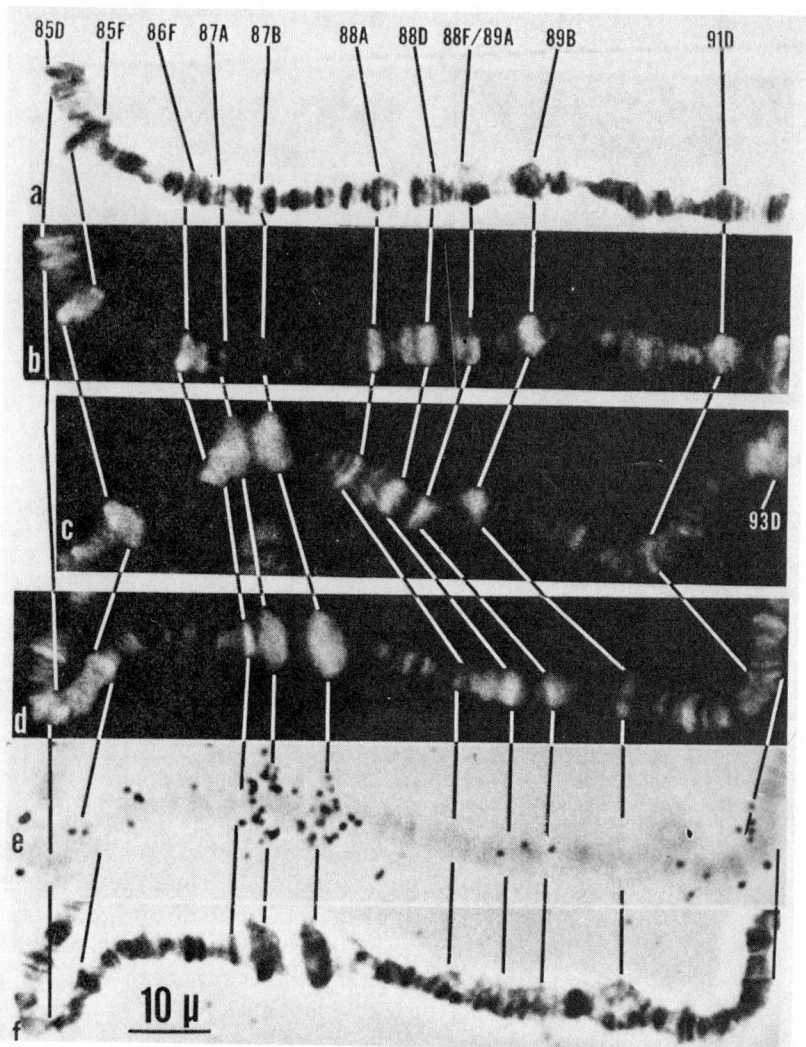

Fig. 10. ρ staining and transcriptional activity after heat shock induction. Glands were obtained from wild-type larvae and processed through the formaldehyde fixation technique. The chromosomes in (c) and (d-f) were obtained from heat-shocked larvae and were stained using anti-ρ serum at a 1:5 dilution. The chromosome in (b) was obtained from a larva grown and dissected at 25° C; this chromosome was stained using anti-ρ serum at a 1:10 dilution. A middle section of chromosome 3R is compared in this figure. 87A, 87B-C1, and 93D are heat shock loci. (a) and (f) are phase-contrast views of the chromosomes in (b) and (d-e), respectively. The chromosome shown in (d-f) was exposed to tritiated uridine for 3 min., subsequent to a 20 min. heat shock treatment. After autoradiography and light Giemsa staining, this chromosome was observed with bright field optics as shown in (e). Actual magnification of each chromosome is ± 15% of the value indicated by the bar. From Silver and Elgin, 1977. Copyright © 1977 by The MIT Press; reprinted by permission of Cell.

antigens in question are not limited to developmentally regulated loci, but are present at other sites of intense transcription. Note that staining is maintained at the developmentally active loci, even though these loci are not active under heat shock conditions. The results imply that a difference in chromatin structure as indicated by ρ or Band 2 antisera staining is a necessary but not sufficient characteristic of the active gene configuration indicated by puffing. One caveat, of course, is that the cytological approach alone cannot absolutely distinguish between conformational shifts (making an antigenic site newly available at a locus) and alterations in the distribution (binding) of an NHC protein (making an antigen newly present at a locus). We have emphasized the latter interpretation since (a) the histones at the major heat shock loci 87A and 87C1 are accessible to antibody probes, and (b) the ρ and Band 2 antigens at loci that are not puffed at the moment of observation are accessible to antibody probes.

Others have supplied biochemical evidence for a role for particular NHC proteins in the chromatin structure of active genes. The HMG proteins (1, 2, 14, 17), small lysine-rich NHC proteins, are of particular interest. Levy and colleagues (1979) have found such proteins to be preferentially associated with an "active" fraction of trout testis chromatin and have suggested that HMG 6 (similar to HMG 14 and 17 of calf) is a component of the nucleosome core while HMG-T (similar to HMG 1 and 2 of calf) replaces H1 to some degree. More convincing are the selective extraction experiments of Weisbrod and Weintraub (1979), which indicate that in the chick red blood cell chromatin, HMG 14 and 17 are necessary to maintain the DNase I-sensitivity of the globin gene.

It is of particular interest that the distribution pattern of Class 3 proteins shows two features reminiscent of the characteristics of loci which are sensitive to digestion by DNase I. First, the condition of these loci appears to be necessary but not sufficient for gene transcription. Second, the continuation of this state is not dependent on transcription per se. The difference between the distribution patterns of Class 3 (ρ) and Class 4 (RNA polymerase II) proteins is illustrated in figure 11. The pattern of association with the set of loci that will be active (puff) at some time during this developmental stage in this tissue suggests a pattern of chromatin structure established at an earlier point in development, perhaps at some determinative stage. One is tempted to predict that genes that will be transcribed at some later time in a committed cell might be DNase I-sensitive or show other indications of a structural alteration prior to the transcriptional event. At present there is no convincing experimental evidence in support of this hypothesis. In general, it is difficult to obtain

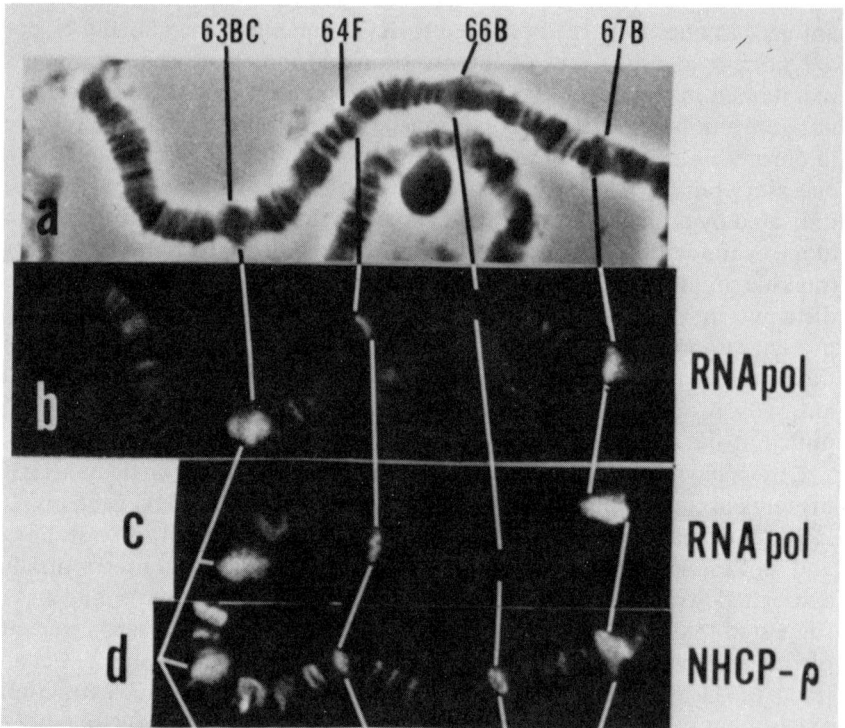

Fig. 11.   Staining in a section of chromosome 3L after heat-shock induction. (a) Phase-contrast view of the chromosome shown in (b); all others are UV dark-field views. The chromosomes were prepared with or without formaldehyde fixation and stained using antisera as follows: (b) formaldehyde fixation, RNA polymerase II antiserum; (c) no formaldehyde fixation, RNA polymerase II antiserum; (d) formaldehyde fixation, ρ antiserum. Loci 63BC, 64F, and 67B are sites of heat-shock puffs; 66B is the site of a developmental puff. (Adapted from data presented in Elgin et al., 1978.)

sufficient quantities of specific precursor cells. One may wish to focus on tissue culture systems where cells can be induced to "differentiate" under appropriate stimuli.

SUMMARY

It is clear that the use of recombinant plasmid probes to analyze specific genes and the use of antibody probes to study specific chromosomal proteins makes it feasible to study the role of chromatin structure in the control of gene expression. Analysis by nuclease digestion indicates alterations of the chromatin structure of the active gene at both a local (nucleosome) and higher order (chromomere?) level. The distribution patterns of

NHC proteins suggest alterations not only at the time of activation but preceding that time, although it must be emphasized that at present there is no conclusive evidence showing that a change in chromatin fiber structure *must* precede transcription. The observation that some alterations, both in the chromatin fiber sensitivity to nucleases and in the distribution pattern of certain NHC proteins, are correlated with, but not sufficient for, transcription indicates that the "activation" process requires two regulated steps at a minimum. Given the power of the techniques at hand, it should be possible to resolve many of the questions raised in this brief review within the next few years.

ACKNOWLEDGMENTS

Research in this laboratory is supported by grants from NIH and NSF to S. C. R. E.; G. C. H. and K. L. are fellows of the Damon Runyon-Walter Winchell Cancer Fund; S. C. R. E. is supported by a Research Career Development Award from N. I. G. M. S. Figures 2–8 and 10 are reprinted with the permission of the MIT Press.

REFERENCES

Allfrey, V. G. 1977. Post-synthetic modifications of histone structure: a mechanism for the control of chromosome structure by the modulation of histone-DNA interactions. *In* H. Li and R. Eckhardt (eds.), Chromatin and chromosome structure, pp. 167–91. Academic Press, New York.

Artavanis-Tsakonas, S., P. Schedl, C. Tschudi, V. Pirrotta, R. Steward, and W. J. Gehring. 1977. The 5S genes of *Drosophila melanogaster*. Cell 12:1057–67.

Artavanis-Tsakonis, S., P. Schedl, M.-E. Mirault, L. Moran, and J. Lis. 1979. Genes for the 70,000 dalton heat shock protein in two cloned *D. melanogaster* DNA segments. Cell 17: 9–18

Ashburner, M. 1970. Patterns of puffing activity in the salivary glands of *Drosophila*. V. Responses to environmental treatments. Chromosoma 31:356–76.

Beard, P. 1978. Mobility of histones on the chromosome of simian virus 40. Cell 15:955–68.

Bellard, M., F. Gannon, and P. Chambon. 1978. Nucleosome structure. III. The structure and transcriptional activity of the chromatin containing the ovalbumin and globin genes in chick oviduct nuclei. Cold Spring Harbor Symp. Quant. Biol. 42:779–92.

Bird, A. P., and E. M. Southern. 1978. Use of restriction enzymes to study eukaryotic DNA methylation. J. Mol. Biol. 118:27–47.

Bird, A. P., M. H. Taggart, and B. A. Smith. 1979. Methylated and unmethylated DNA compartments in the sea urchin genome. Cell 17:889–907.

Bloom, K. S., and J. N. Anderson. 1978. Fractionation of hen oviduct chromatin into transcriptionally active and inactive regions after selective micrococcal nuclease digestion. Cell 15:141–50.

Bonner, J. J., and M. L. Pardue. 1977. Polytene chromosome puffing and *in situ* hybridization measure different aspects of RNA metabolism. Cell 12:227–34.

Breindl, M., and R. Jaemisch. 1979. Conformation of moloney murine leukaemia proviral sequences in chromatin from leukaemic and nonleukaemic cells. Nature 227:320–23.

Britten, R. J., and E. H. Davidson. 1969. Gene regulation for higher cells: a theory. Science 165:349–57.

Brown, I., J. Heikkila, J. Silver, and N. Straus. 1977. Organization and transcriptional activity of brain chromatin subunits. Biochim. Biophys. Acta 447:288–94.

Butt, T. R., D. B. Jump, and M. E. Smulson. 1979. Nucleosome periodicity in HeLa cell chromatin as probed by micrococcal nuclease. Proc. Nat. Acad. Sci. USA 76:1628–32.

Caggese, C., R. Caizzi, M. Morea, F. Scalenghe, and F. Ritossa. 1979. Mutation generating a fragment of the major heat shock-inducible polypeptide in *Drosophila melanogaster*. Proc. Nat. Acad. Sci. USA 76:2385–89.

Carlson, M., and D. Brutlag. 1977. Cloning and characterization of a complex satellite DNA from *Drosophila melanogaster*. Cell 11:371–81.

Chae, C. B., T. K. Wong, and R. A. Gadski. 1978. Transcription, processing, and structure of chromatin in SV40-transformed cell. Biochem. Biophys. Res. Commun. 83:1518–24.

Christenson, M. E., W. M. LeStourgeon, M. Jamrich, and S. C. R. Elgin. 1978. Immunofluorescent localization of heterogeneous nuclear ribonucleoprotein particles in *Drosophila* polytene chromosomes. J. Cell Biol. 79:351a.

Clark, R. J., and G. Felsenfeld. 1971. Structure of chromatin. Nature New Biol. 229:101–6.

Elgin, S. C. R., and H. Weintraub. 1975. Chromosomal proteins and chromatin structure. Ann. Rev. Biochem. 44:725–74.

Elgin, S. C. R., S. C. Froehner, J. E. Smart, and J. Bonner. 1971. The biology and chemistry of chromosomal proteins. Advances in Cell and Molec. Biol. 1:1–57.

Elgin, S. C. R., L. A. Serunian, and L. M. Silver. 1978. Distribution patterns of *Drosophila* nonhistone chromosomal proteins. Cold Spring Harbor Symp. Quant. Biol. 42:839–50.

Felsenfeld, G. 1978. Chromatin. Nature 271:115–22.

Flint, S. J., and H. M. Weintraub. 1977. An altered subunit configuration associated with the actively transcribed DNA of integrated adenovirus genes. Cell 12:783–94.

Foe, V. E. 1978. Modulation of ribosomal RNA synthesis in *Oncopeltus fasciatus*: an electron microscopic study of the relationship between changes in chromatin structure and transcriptional activity. Cold Spring Harbor Symp. Quant. Biol. 42:723–40.

Franke, W. W., V. Scheer, M. Trendelenburg, H. Zentgraf, and H. Spring. 1978. Morphology of transcriptionally active chromatin. Cold Spring Harbor Symp. Quant. Biol. 42:755–72.

Frolova, E. I., E. S. Zalmanzon, E. M. Lukamidin, and G. P. Georgiev. 1978. Studies of the transcription of viral genome in adenovirus 5 transformed cells. Nucl. Acids Res. 5:1–11.

Garel, A., and R. Axel. 1976. Selective digestion of transcriptionally active ovalbumin genes from oviduct nuclei. Proc. Nat. Acad. Sci. USA 73:3966–70.

Garel, A., M. Zolan, and R. Axel. 1977. Genes transcribed at diverse rates have a similar conformation in chromatin. Proc. Nat. Acad. Sci. USA 74:4867–71.

Gariglio, P., R. Llopis, P. Oudet, and P. Chambon. 1979. The template of the isolated native simian virus 40 transcriptional complexes is a minichromosome. J. Mol. Biol. 131:75–105.

Gottesfeld, J. M., and D. A. Melton. 1978. The length of nucleosome-associated DNA is the same in both transcribed and nontranscribed regions of chromatin. Nature 273:317–19.

Greenleaf, A. L., U. Plagens, M. Jamrich, and E. K. F. Bautz. 1978. RNA polymerase B (or II) in heat induced puffs of *Drosophila* polytene chromosomes. Chromosoma 65:127–36.

Hewisch, D., and L. Burgoyne. 1973. Chromatin sub-structure. The digestion of chromatin DNA at regularly spaced sites by a nuclear deoxyribonuclease. Biochem. Biophys. Res. Commun. 52:504–10.

Holliday, R., and J. E. Pugh. 1975. DNA modification mechanisms and gene activity during development. Science 187:226–32.

Hozier, J., M. Renz, and P. Nehls. 1977. The chromosome fiber: evidence for an ordered super structure of nucleosomes. Chromosoma 62:301–17.

Hsieh, T. S., and D. L. Brutlag. 1979. A protein that preferentially binds *Drosophila* satellite DNA. Proc. Nat. Acad. Sci. USA 76:726–30.

Ish-Horowicz, D., J. J. Holden, and W. J. Gehring. 1977. Deletions of two heat-activated loci in *Drosophila melanogaster* and their effects on heat-induced protein synthesis. Cell 12:643–52.

Ish-Horowicz, D., S. M. Pinchin, J. Gausz, H. Gyurkovics, G. Bencze, M. Goldschmidt-Clemont, and J. J. Holden. 1979. Deletion mapping of two *D. melanogaster* loci that code for the 70,000 Dalton heat-induced protein. Cell 17:565–71.

Johnson, E. M., V. G. Allfrey, E. M. Bradbury, and H. R. Matthews. 1978. Altered nucleosome structure containing DNA sequences complementary to 19S and 26S ribosomal RNA in *Physarum polycephalum*. Proc. Nat. Acad. Sci. USA 75:1116–20.

Johnson, E. M., H. R. Matthews, V. C. Littau, L. Lothstein, E. M. Bradbury, and V. G. Allfrey. 1978. The structure of chromatin containing DNA complementary to 19S and 26S ribosomal RNA in active and inactive stages of *Physarum polycephalum*. Arch. Biochem. Biophys. 191:537–50.

Kornberg, R. D. 1977. Structure of chromatin. Ann. Rev. Biochem. 46:931–54.

Kuo, M. T. 1979. Studies in heterochromatin DNA: accessibility of late replicating heterochromatin DNA in chromatin to micrococcal nuclease digestion. Chromosoma 70:183–94.

Kuo, M. T., C. G. Sahasrabuddhe, and G. F. Saunders. 1976. Presence of messenger specifying sequences in the DNA of chromatin subunits. Proc. Nat. Acad. Sci. USA 73:1572–75.

Lacy, E., and R. Axel. 1975. Analysis of DNA of isolated chromatin subunits. Proc. Nat. Acad. Sci. USA 72:3978–82.

Levy W. B., and G. H. Dixon. 1977. Renaturation kinetics of cDNA complementary to cytoplasmic polyadenylated RNA from rainbow trout testis. Accessibility of transcribed genes to pancreatic DNase. Nucl. Acids Res. 4:883–98.

Levy W. B., W. Connor, and G. H. Dixon. 1979. A subset of trout testis nucleosomes enriched in transcribed DNA sequences contains high mobility group proteins as major structural components. J. Biol. Chem. 254:609–20.

Lis, J. T., L. Prestidge, and D. S. Hogness. 1978. A novel arrangement of tandemly repeated genes at a major heat shock site in *D. melanogaster*. Cell 14:901–19.

Livak, K. J., R. Freund, M. Schweber, P. C. Wensink, and M. Messelson. 1978. Sequence organization and transcription at two heat shock loci in *Drosophila*. Proc. Nat. Acad. Sci. USA 75:5613–17.

McGhee, J. D., and G. D. Ginder. 1979. Specific DNA methylation sites in the vicinity of the chicken B globin genes. Nature 280:419–20.

McKenzie, S. L., and M. Meselson. 1977. Translation *in vitro* of *Drosophila* heat-shock messages. J. Mol. Biol. 117:279–83.

McKenzie, S. L., S. Henikoff, and M. Meselson. 1975. Localization of RNA from heat-induced polysomes at puff sites in *Drosophila melanogaster*. Proc. Nat. Acad. Sci. USA 72:1117–21.

McKnight, S. L., and O. L. Miller, Jr. 1976. Ultrastructural patterns of RNA synthesis during early embryogenesis of *Drosophila melanogaster*. Cell 8:305–19.

McKnight, S. L., M. Bustin, and O. L. Miller, Jr. 1978. Electron microscope analysis of chromosome metabolism in the *Drosophila melanogaster* embryo. Cold Springs Harbor Symp. Quant. Biol. 42:741–54.

Mathis, D. J., and M. A. Gorovsky. 1976. Subunit structure of DNA-containing chromatin. Biochem. 15:750–55.

Mathis, D. J., P. Oudet, B. Wasylyk, and P. Chambon. 1978. Effect of histone acetylation on structure and *in vitro* transcription of chromatin. Nucl. Acids Res. 5:3523–48.

Mayfield, J. E., L. A. Serunian, L. M. Silver, and S. C. R. Elgin. 1978. A protein released by DNase I digestion of *Drosophila* nuclei is preferentially associated with puffs. Cell 14:539–44.

Miller, D. M., P. Turner, A. W. Nienhuis, D. E. Axelrod, and T. V. Gopalakrishnan. 1978. Active conformation of the globin genes in uninduced and induced mouse erythroleukemia cells. Cell 14:511–21.

Miller, F., T. Igo-Kemenes, and H. G. Zachau. 1978. Characterization of restriction nuclease prepared chromatin by electron microscopy. Chromosoma 68:327–36.

Miller, J. R., E. M. Cartwright, G. G. Brownlee, N. V. Fedoroff, and D. D. Brown. 1978. The nucleotide sequence of oocyte 5S DNA in *Xenopus laevis*. Cell 13:717–25.

Mirault, M. E., M. Goldschmidt-Clermont, L. Moran, A. P. Arrigo, and A. Tissieres. 1978. The effect of heat shock on gene expression in *Drosophila melanogaster*. Cold Spring Harbor Symp. Quant. Biol. 42:819–28.

Nelson, D. A., M. Perry, L. Sealy, and R. Chalkley. 1978. DNase I preferentially digests chromatin containing hyperacetylated histones. Biochem. Biophys. Res. Commun. 82:1346–53.

Nelson, D., M. E. Perry, and R. Chalkley. 1979. A correlation between nucleosome spacer region susceptibility to DNase I and histone acetylation. Nucl. Acids Res. 6:561–74.

Noll, M. 1974. Subunit structure of chromatin. Nature 251:249–52.

Palmiter, R. D., E. R. Mulvihill, G. S. McKnight, and A. W. Senear. 1978. Regulation of gene expression in the chick oviduct by steroid hormones. Cold Spring Harbor Symp. Quant. Biol. 42:639–48.

Panet, A., and H. Cedar. 1977. Selective degradation of integrated murine leukemia proviral DNA by deoxyribonucleases. Cell 11:933–40.

Plagens, U., A. L. Greenleaf, and E. K. F. Bautz. 1976. Distribution of RNA polymerase on *Drosophila* polytene chromosomes as studied by indirect immunofluorescence. Chromosoma 59:157–65.

Reeves, R. 1978. Nucleosome structure of *Xenopus* oocyte amplified ribosomal genes. Biochemistry 17:4908–16.

Renz, M. 1979. Heterogeneity of the chromosome fiber. Nucl. Acids Res. 6:2761–67.

Riggs, A. D. 1975. X-inactivation, differentiation, and DNA methylation. Cytogenet. Cell Genet. 14:9–25.

Ritossa, F. 1962. A new puffing pattern induced by temperature shock and DNP in *Drosophila*. Experientia 18:571–73.

Rubenstein, P., L. Sealy, S. Marshall, and R. Chalkley. 1979. Cellular protein synthesis and inhibition of cell division are independent of butyrate-induced histone hyperacetylation. Nature 280:692–93.

Sahasrabuddhe, C. G., and K. E. Van Holde. 1974. The effect of trypsin on nuclease-resistant chromatin fragments. J. Biol. Chem. 249:152–56.

Schedl, P., S. Artavanis-Tsakonas, R. Steward, W. J. Gehring, M. E. Mirault, M. Goldschmidt-Clermont, L. Moran, and A. Tissieres. 1978. Two hybrid plasmids with *D. melanogaster* DNA sequences complementary to mRNA coding for the major heat shock protein. Cell 14:921–29.

Scheer, U. 1978. Changes of nucleosome frequency in nucleolar and non-nucleolar chromatin as a function of transcription: an electron microscope study. Cell 13:535–49.

Scott, W. A., and D. J. Wigmore. 1978. Sites in simian virus 40 chromatin which are preferentially cleaved by endonucleases. Cell 15:1511–18.

Sealy, L., and R. Chalkley. 1978. DNA associated with hyperacetylated histone is preferentially digested by DNase I. Nucl. Acids Res. 5:1863–76.

Silver, L. M., and S. C. R. Elgin. 1976. A method for determination of the in situ distribution of chromosomal proteins. Proc. Nat. Acad. Sci. USA 73:423–27.

Silver, L. M., and S. C. R. Elgin. 1977. Distribution patterns of three subfractions of *Drosophila* nonhistone chromosomal proteins: possible correlations with gene activity. Cell 11:971–83.

Silver, L. M., and S. C. R. Elgin. 1978a. Immunological analysis of protein distributions in *Drosophila* polytene chromosomes. *In* H. Busch (ed.), The cell nucleus. *V*: chromatin, part B, pp. 216–63. Academic Press, New York.

Silver, L. M., and S. C. R. Elgin. 1978b. Production and characterization of antisera against three individual NHC proteins: a case of a generally distributed NHC protein. Chromosoma 68:101–14.

Simpson, R. T., 1978. Structure of chromatin containing extensively acetylated H3 and H4. Cell 13:691–99.

Southern, E. M. 1975. Detection of specific sequences among DNA fragments separated by gel electrophoresis. J. Mol. Biol. 98:503–17.

Spradling, A., S. Penman, and M. L. Pardue. 1975. Analysis of *Drosophila* mRNA by in situ hybridization: sequences transcribed in normal and heat shocked cultured cells. Cell 4:395–404.

Spradling, A., M. L. Pardue, and S. Penman. 1977. Messenger RNA in heat-shocked *Drosophila* cells. J. Mol. Biol. 109:559–87.

Stalder, J., T. Seebeck, and R. Braun. 1979. Accessibility of the ribosomal genes to micrococcal nuclease in *Physarum polycephalum*. Biochim. Biophys. Acta 561:452–63.

Stratling, W. H., U. Muller, and H. Zentgraf. 1978. The higher order repeat structure of chromatin is built up of globular particles containing eight nucleosomes. Exp. Cell Res. 117:301–11.

Tissieres, A., H. K. Mitchell, and U. M. Tracy. 1974. Protein synthesis in salivary glands of *Drosophila melanogaster*: relation to chromosome puffs. J. Mol. Biol. 84:389–98.

Varshavsky, A. J., O. H. Sundin, and M. J. Bohn. 1978. SV40 viral minichromosome: preferential exposure of the origin of replication as probed by restriction endonucleases. Nucl. Acids Res. 5:3469–78.

Varshavsky, A. J., O. Sundin, and M. Bohn. 1979. A stretch of "late" SV40 viral DNA about 400bp long which includes the origin of replication is specifically exposed in SV40 minichromosomes. Cell 16:453–66.

Vidalli, G., L. C. Boffa, E. M. Bradbury, and V. G. Allfrey. 1978. Butyrate suppression of histone deacetylation leads to accumulation of multiacetylated forms of histones H3 and H4 and increased DNase I sensitivity of the associated DNA sequences. Proc. Nat. Acad. Sci. USA 75:2239–43.

Waldeck, W., B. Fohring, K. Chowdhury, P. Gruss and G. Saver. 1978. Origin of DNA replication in papovavirus chromatin is recognized by endogenous endonuclease. Proc. Nat. Acad. Sci. USA 75:5964–68.

Weintraub, H., and M. Groudine. 1976. Chromosomal subunits in active genes have an altered conformation. Science 193:848–56.

Weisbrod, S., and H. Weintraub. 1979. Isolation of a subclass of nuclear proteins responsible for conferring a DNase I-sensitive structure on globin chromatin. Proc. Nat. Acad. Sci. USA 76:630–34.

Will, H., and E. K. F. Bautz. 1979. Localization of three nonhistone proteins in polytene nuclei of *Drosophila melanogaster*. J. Supramol. Struc. 53:72.

Williamson, P., and G. Felsenfeld. 1978. Transcription of histone-covered T7 DNA by *Escherichia coli* RNA polymerase. Biochemistry 17:5695–705.

Wu, C., P. M. Bingham, K. J. Livak, R. Holmgren, and S. C. R. Elgin. 1979a. The chromatin structure of specific genes. I. Evidence for higher order domains of defined DNA sequence. Cell 16:797–806.

Wu, C., Y.-C. Wong, and S. C. R. Elgin. 1979b. The chromatin structure of specific genes. II. Disruption of chromatin structure during gene activity. Cell 16:807–14.

Young, N. S., E. J. Benz, Jr., J. A. Kantor, P. Kretschmer, and A. W. Nienhuis. 1978. Hemoglobin switching in sheep: only the γ gene is in the active conformation in fetal liver but all the β and γ genes are in the active conformation in bone marrow. Proc. Nat. Acad. Sci. USA 75:5884–88.

Abstracts of Contributed Papers

E. P. AMANN and J. N. REEVE

The Expression of *Bacillus subtilis*
Phage SPP1 Genes in *Escherichia coli*

Bacteriophage SPP1 is a virulent, double-stranded DNA-containing (M_r-28.6 Mdal) phage that grows on *B. subtilis*. Since SPP1 development is highly host dependent, it is impossible to block host synthesis without interfering with the phage development. We therefore infected *B. subtilis* minicells with SPP1 WT phage, *sus*- and deletion-mutants of SPP1 and analyzed ^{14}Caa-labeled polypeptides by PAA gel electrophoresis. Forty-six phage encoded peptides were detected, which represents the expression of 86% of the phage genome's coding capacity. To assign these polypeptides to specific regions of the phage genome, we used isolated restriction fragments of SPP1 DNA as template in a DNA-dependent *E. coli*-derived cell-free protein-synthesizing system.

In addition we have cloned 13 of the 15 *Eco*RI generated fragments of SPP1 into a λ imm4324 vector. Infection of *E. coli* DS410 minicells with the hybrid phages results in the expression of both λ genes and the SPP1 cloned genes. Infection of *E. coli* DS410 (pGY101) minicells with the hybrid phages allows selective expression of SPP1 specific genes, since plasmid pGY101 codes for the imm434 repressor, which is therefore present in high amount in these minicells and inhibits expression of λ promoters. Polypeptides are only synthesized if the cloned fragment carries a promoter(s) that functions in *E. coli*. The expression of *B. subtilis* phage SPP1 in *E. coli* minicells and *in vitro* appears to be correct in terms of the molecular weights of polypeptides synthesized and in the recognition of suppressor-sensitive mutations in the SPP1 DNA. Peptides resulting from the expression of the DNA at the fusion between the λ and SPP1 DNA are also observed.

Comparison of the *in vitro* and *in vivo* results allowed the assignment of most of the SPP1 polypeptides to specific *Eco*RI fragments.

The potential and problems of using λ vectors and minicell infection to analyze "foreign" gene expression were discussed.

Max-Planck-Institut für molekulare Genetick, Berlin, West Germany

P. J. RIZZO

Histones and Chromatin Structure in Unicellular Algae

Eukaryotic microorganisms are excellent candidates for the study of specific histone and nonhistone chromosomal proteins, and can provide valuable information as to the structure, function, and evolution of the eukaryotic chromatin fiber. Recent studies suggest that the chromatin of slime molds, ciliates, and fungi is organized into nucleosomes by an interaction of four types of histones with the DNA. In contrast, little is known about the structure and composition of chromatin in eukaryotic algae. Studies on the chromatin of these organisms are of interest because heterotrophic forms often cannot be distinguished from protozoa, and are thought to be progenitors of protozoa and fungi.

The four unicellular algae studied in the present report are the two uninucleate dinoflagellates *Crypthecodinium cohnii* and *Peridinium trochoideum*, the binucleate dinoflagellate *Peridinium balticum*, and the Chloromonad *Olisthodiscus luteus*. *C. cohnii* and *P. trochoideum* are devoid of histones. Both organisms contain a single basic histone-like protein that migrates to a position similar to that of H4 in urea-acrylamide gels. *P. balticum* contains a typical dinoflagellate nucleus ("dinokaryotic") and a second nucleus that is more eukaryotic in ultrastructure ("eukaryotic"). The "eukaryotic" nucleus has four histones that co-migrate with four histones from calf thymus in urea-acrylamide gels, as does *O. luteus*. In the case of *O. luteus* the histones have been examined by three additional gel systems: SDS slab gels; two-dimensional Triton gels, and two-dimensional SDS gels. Due to differential migration patterns in the four gel systems, it is not yet possible to determine which of the five vertebrate histones is missing.

Some of the above nuclei were examined for the presence of nucleosomes using electron microscopy. The chromatin of *O. luteus* and the "eukaryotic" nucleus of *P. balticum* contain nucleosomes, while the chromatin of *C. cohnii* is not organized into beaded subunits.

Texas A & M University, Biology Department, College Station, Texas

R. KRUMLAUF AND G. A. MARZLUF

Genome Organization of *Neurospora Crassa*

Considerable genetic evidence is available which indicates that many sets of unlinked structural genes are coordinately controlled in *Neurospora*. We have examined the organization of repetitive DNA in the genome since it may play a role in gene regulation. We have demonstrated that the genome of *Neurospora crassa* contains 2.68×10^7 NBP and is comprised of 2% foldback, 8% repetitive, and 90% single-copy sequences. The repetitive component has a kinetic complexity of only 15,400 NBP and a repeat frequency of 140. The results of a variety of experiments all consistently indicate that the repetitive DNA occurs in long stretches of 10^4 NBP or longer. At most, only about 1% of the single copy sequences lie adjacent to repetitive DNA in fragments of 10,000 NBP length. Sensitive experiments using pure *Neurospora* rDNA isolated from cloned recombinant DNA indicates that rDNA comprises about 7% of the total genomic DNA and represents most (about 90%) of the repetitive component. It is repeated about 190 times. Since the rDNA units are believed to be tandemly repeated in long stretches and may all be located on a single chromosome, their clustering can largely account for the lack of interspersion of repetitive DNA with single copy sequences. These results argue against any regulatory role for repetitive DNA in *Neurospora*.

Ohio State University, Department of Biochemistry, Columbus, Ohio 43210

M. A. SCHAFER AND R. L. WHITE

Sequence Organization in the Nontranscribed Spacer within the rDNA of the Blowfly *Calliphora Erythrocephala*

Analysis of DNA from *Calliphora* by the method of Southern has revealed two size classes of nontranscribed spacer (NTS) associated with the rRNA genes. The molecular basis for the length difference has been studied by comparing cloned representatives of the different lengths.

Electron microscopic analysis reveals that heteroduplex molecules formed between the two size classes of NTS produce a single insertion/deletion loop of about 1 kb. In various molecules the loop is not found in a single position, but is distributed over a 1 kb DNA stretch. This indicates that the additional sequences in the longer NTS are a repetition of this 1 kb sequence in the shorter NTS.

Furthermore, the restriction site map determined by partial digestion with 4-base restriction enzymes demonstrates the 1.2 kb area in the cloned DNA containing the small NTS to be composed of four 320 to 350 bp segments. These 4 "repeats" are increased to 7 in the larger NTS. Furthermore, one additional cloned segment has been found to contain 10 repeats. These units are defined by the restriction enzymes *Alu*I, *Sau*3A, and *Hha*I. The last enzyme also detects sequence divergence among these repeat units, since units 1, 2, and 3 have a *Hha* restriction site, but unit 4 does not. This pattern is maintained in the larger clones as well such that the larger NTS segments have the sequence 1234–234 and 1234–234–234.

We feel these observations may have bearing on the role of recombination in maintaining families of identical genes. The question arises, however, of why recombination as suggested by the size classes found apparently only occurs between 1 kb units and not among the 350 bp units. Limits on the length of the repetitive region also seem to exist, since out of 12 *Calliphora* rDNA clones 8 contained a sequence of four, 3 a sequence of seven, and only one clone was found with 10 repeats.

University of Massachusetts Medical Center, Worcester, Massachusetts 01605

H. J. EDENBERG

Enzymes Associated with
Simian Virus 40 Chromosomes

We are studying the enzymology of Simian Virus 40 DNA replication, an excellent model of DNA replication in mammalian cells. Our current approach is to analyze the enzymes associated with SV40 chromosomes extracted under conditions that allow continuation of DNA synthesis *in vitro*.

We have shown in earlier studies that DNA polymerase α is associated specifically with replicating SV40 chromosomes. Further, DNA polymerase α is the only polymerase resistant to the nucleotide analog 2′, 3′-dideoxythmidine-5′-triphosphate; SV40 DNA replication *in vitro* is also resistant to this analog. Our evidence suggests that DNA polymerase α is responsible for all detectable SV40 DNA replication *in vitro*.

Since RNA primers are utilized in SV40 DNA replication, we have looked for the presence of an RNA polymerase activity associated with SV40 chromosomes. We have found such an activity, but believe that it is involved in transcription rather than replication, since replication *in vitro* is not sensitive to α-amanitin. Our transcription complex sediments just slightly faster than mature SV40 chromosomes, is stimulated by added salt, and is sensitive to low levels of α-amanatin (suggesting that it is RNA polymerase II). The yield of RNA synthesized *in vitro* by our complex is severalfold higher than in previously reported complexes.

We have also detected both an endonucleolytic activity capable of cutting the endogenous viral chromatin, and a superhelix-relaxing activity, in the region of the sucrose gradient that contains SV40 chromosomes. Work on these activities is continuing.

This work was supported by grants from NIH (GM 25681), a Biomedical Research Support Grant, an American Cancer Society Institutional Grant, and the Grace M. Showalter Trust.

Indiana University School of Medicine, Department of Biochemistry, Indianapolis, Indiana 46223

D. S. PARRIS, R. A. F. DIXON
AND P. A. SCHAFFER

Marker Rescue of Herpes Simplex Virus Type 1 *ts* Mutants: Correlation of the Genetic Map with One or Two Arrangements of HSV-1 DNA

The herpes simplex virus (HSV) genome is composed of a long (L) and a short (S) region of unique DNA each bounded by inverted repeated sequences. Due to inversions of the L and S regions by intra- or intermolecular recombination, four structural forms of HSV exist in approximately equimolar quantities in all virus populations. However, it is not known whether all four genome arrangements are equally infectious and in equilibrium with one another during replication and recombination.

The effects of the unique structural features of HSV DNA on recombination analysis was studied by comparing the relative positions of five temperature-sensitive (*ts*) mutants on the physical and genetic maps of HSV-1, strain KOS, constructed by marker rescue and two-factor crosses, respectively. The results of these studies demonstrate a good correlation between the order of mutants on the physical and genetic maps in the L region of the genome. The data also confirm that markers in L are only loosely linked to markers in S on the genetic map. Furthermore, marker rescue data demonstrate that the published KOS linkage map is consistent with only one or two arrangements of HSV DNA.

D. S. Parris, Harvard Medical School, Division of Viral Oncology, Sidney Farber Cancer Institute, Boston, Massachusetts.

Genome Structure of Avian Acute Leukemia Viruses

Avian acute leukemia viruses are a group of retroviruses that cause different forms of leukemia, including myeloblastosis, myelocytomatosis, endotheliomas, and erythroblastosis in fowls. They are usually defective and require helper virus for growth. They transform bone marrow cells and, sometimes, fibroblasts in tissue culture. Avian erythroblastosis virus (AEV) is one of these viruses. AEV has a 28S (2×10^6 daltons) RNA genome while its helper virus has a 35S (3×10^6 daltons) RNA. AEV genome has deletion in *gag*, *pol*, and *env* genes. However, complementary DNA (cDNA) made from helper virus RNA protects only 50% of 28s RNA, suggesting the presence of AEV-specific "transforming" gene sequences. By mapping heteroduplex between helper viral cDNA and 28S AEV RNA, we characterized these transformation-specific sequences as a 3.25 kb contiguous segment localized in the middle of the 6 kb AEV RNA genome. By studying the gene products of AEV, we reached the following conclusions. (1) The transformation-specific sequences code for 2 proteins, p75 and p40, which were detected by *in vitro* translation of AEV RNA in a reticulocyte-lysate system. AEV contains a new class of transforming gene unrelated to *src*. (2) The *gag* gene starts at 1 kb from the 5'-end of the RNA genome in sarcoma and leukemia viruses.

Similar genome structure has been detected in another leukemia virus avian reticuloendotheliosis virus (REV). REV contains yet another new class of transforming gene.

University of Southern California School of Medicine, Department of Microbiology, Los Angeles, California 90033

N. J. ALEXANDER, D. K. HANSON
H. R. MAHLER, and P. S. PERLMAN

Further Studies of the Mosaic Structure of the Mitochondrial *Cob-Box* Gene in *Saccharomyces Cerevisiae*

Recent studies of the region of the mitochondrial genome coding for apocytochrome *b* (p30) have shown it to contain a single mosaic gene consisting of at least four coding regions (exons) separated by at least three noncoding regions (introns). A major class of exon mutants lacks p30, retains cyt ox subunit I (cox I), and has one extra protein smaller than p30. Based on partial protease digestion patterns, these extra proteins appear to be premature termination fragments of p30, and their size correlates with the map position of the responsible mutation. Intron mutants lack p30 and cox I but have one or several extra proteins larger or smaller than p30 that appear to be largely unrelated to p30 by fingerprint analysis. Based on such genetic and phenotypic studies, it has been proposed that the gene is transcribed from its *oli2*- proximal end toward the *oli*-proximal end. As a further test of that model, we have constructed and analyzed the profile of mitochondrially synthesized polypeptides for more than 28 double mutants. Double mutants containing two exon mutations exhibit the protein phenotype of the mutant closest to *oli2*. Phenotypes of strains containing an exon and an intron mutation differ depending on which mutation is *oli2*- proximal. When the exon mutation is *oli2*- proximal, a recombinant phenotype is seen: the p30 fragment typical of the exon mutant is present, but all extra proteins characteristic of the intron mutant are extinguished and cox I is absent, a characteristic of the intron mutant. This demonstrates that the cox I deficiency of intron mutants is not caused by the intron-specific extra polypeptide. When the intron mutation is *oli2*-proximal, the double mutant exhibits the intron phenotype exclusively. These results indicate that the extra proteins made by intron mutants are the result of translation of sequences that would normally be excised in wild-type cells by RNA processing events. Over-all these studies clearly demonstrate a polar effect on one marker on another and support the model for the structure and direction of transcription of the *cob-box* gene.

Supported by NIH Grants GM-19607 and GM-12224.

N. J. Alexander, Ohio State University, Columbus, Ohio 43210

R. D. VINCENT, P. S. PERLMAN
R. L. STRAUSBERG, AND R. A. BUTOW

Physical Mapping of Determinants
Affecting the Size of the *var1* Protein

Over the last several years, we have used a combination of petite deletion mapping, physical mapping, and the zygotic gene rescue procedure of Strausberg and Butow to show that determinants affecting the size of the *var1* protein map within a restriction fragment, *Hinc*II-10, of 2,800 base pairs (bp), located on the yeast mitochondrial genome. Tzagoloff has recently found that base sequences sufficient to encode the *var1* protein appear to be absent from that region; thus, the location of the *var1* gene and the physical basis of the alleles mapped in *Hinc*II-10 remain to be determined. It now appears that gene rescue is a trans complementation phenomenon that assays for determinants affecting the size of the *var1* protein rather than the presence of the intact structural gene. Our mapping studies have defined the location of alleles of the *var1* locus responsible for at least six different forms of that protein. To better understand the molecular basis of the *var1* alleles, we now have prepared fine structure restriction site maps of this region for six different strains using 15 enzymes. The six maps are very similar, but differ chiefly in terms of the presence or absence of six short insertions. Insertion *z* does not appear to be associated with any *var1* allele since it is inherited independently of the size of *var1*; it maps *eryl*-proximal to the *var1* alleles and contains no extra restriction sites. Insertions x and y are located between the seryl-tRNA gene and the *var1* alleles and contain an extra HinfI site. Insertion *a* appears to be an inverted duplication of a 46 bp GC-rich segment found in HincII-10 from all strains. Insertions b1 and b2 have not yet been separated from each other genetically and contain no additional sites. We have studied the inheritance of insertions *a*, *bl*, and *b2* and can understand four allelic forms of *var1* in terms of combinations of them. However, neither a clear correspondence between the insertions and *var1* alleles nor the mechanisms by which the size of the *var1* protein is affected by them has been established.

Supported by grants GM-26546 from NIGMS and HD-00431 from NICHHD.

R. D. Vincent, University of Texas Health Sciences Center, Department of Biochemistry, Dallas, Texas

M. J. W. CHANG AND A. KOESTNER

Promutagenic 0^6-Ethylguanine in Rat Tissues Following Transplacental Inoculation of Ethylnitrosourea

N-eth-yl-N-nitrosourea (ENU) is a potent single-dose transplacental neurocarcinogen in rats. Under optimal conditions 100% of offspring develop neoplasms of the nervous system. The molecular mechanism of carcinogenesis of nitrosoureas is not clearly understood. It has been demonstrated that the persistence of one of the DNA adducts, O^6-alkylguanine (0^6-alkylBua), correlates well with tumor incidence in target organs. 0^6-alkylGua, if not removed enzymatically, will affect the helical structure of nucleic acids and cause base mispairing during DNA replication and RNA transcription.

Two groups of three pregnant CD Fisher rats were inoculated intravenously with freshly dissolved (1-^{14}C) ENU at the dose of 75 mg/kg b.wt. (S.A. $= 15.4$ mCi/mMole) in the 21st day of gestation. DNA-adducts of different maternal organs and organs of progeny were analyzed from separately pooled samples at 3 hours and 7 days after ENU inoculation.

Results of the experiment revealed that the amount of ethylation of DNA guanine at the N^7 and/or O^6 position was about the same in the maternal brain, kidney, and liver and the corresponding organs of the fetus at the 3 hours time point. At the 7-day time point, however, the promutagenic O^6-EtGua was retained only in maternal brain and kidney and in the brain of the offspring. Considering that fast DNA replication of many neuroectodermal precursor cells in the developing neonatal brain as compared to the mature maternal brain, this study further substantiates the hypothesis that a carcinogenic lesion must be fixed by at least one round of DNA replication before its removal in order to manifest its eventual "neoplastic phenotype."

Ohio State University, Department of Veterinary Pathobiology, Columbus, Ohio 43210

P. D. LIPETZ, C. S. LOWNEY
D. MHASKAR, AND R. W. HART

DNA Superhelicity: A Possible Mechanism for Polyamine Modulation of Prokaryotic Nucleic Acid Synthesis

Nucleic acid synthesis is quantitatively and qualitatively modulated by both DNA superhelicity and intracellular polyamine accumulation. The DNA superhelicity of prokaryotes is modulated by two opposing enzyme systems: DNA gyrase, which enhances negative DNA superhelicity, and nicking-closing (N-C) enzyme (also called omega protein), which relaxes negatively supercoiled DNA. We hypothesize that polyamines act as specific organic cations to exert coordinate control over these opposing enzyme processes.

We demonstrate that, of the two most common prokaryotic polyamines, spermidine, but not putrescine, inhibits the $MgCl_2$ requiring N-C activity contained in a protein preparation isolated from *Micrococcus luteus*. Neither spermidine nor putrescine alone can stimulate N-C activity. The addition of physiological concentrations of K^+ only slightly altered this modulation. *M. luteus* gyrase is stimulated by comparable concentrations of spermidine. These results indicate that spermidine accumulation may exert coordinate control over opposing enzyme activities to maximize negative DNA superhelicity.

When spermidine and putrescine are combined the *in vitro* N-C activity is determined by the molar ratio of putrescine and spermidine. At maximal $MgCl_2$ activation, molar ratios of [putrescine]/[spermidine] above approximately 6.5 completely prevent spermidine from inhibiting N-C activity.

Since DNA superhelicity modulates DNA replication and the relative binding of RNA polymerase to different promoter sites, this may represent a mechanism by which nucleic acid synthesis is quantitatively and qualitatively modulated.

Ohio State University, Department of Radiology, Columbus, Ohio 43210

C. DAVID ZARLEY AND THOMAS J. BYERS

Restriction Enzyme Analysis of Mitochondrial DNA from Three Species of *Acanthamoeba*

Techniques were developed for isolating large amounts of mitochondrial (mt) DNA, relatively free of nuclear DNA, from *Acanthamoeba astronyxis, A. castellanii*, and *A. palestinensis*. The mt DNA of each species had a unique fragment pattern when digested with any restriction enzyme used in this study. The sizes of the mt DNA genomes, calculated by summing the sizes of restriction fragments, were 43.7 ± 0.95 kilobases (kb) *A. astronyxis*, 40.8 ± 1.03 kb for *A. castellanii*, and 35.8 ± 2.80 kb for *A. palestinensis*.

Two cell lines of *A. castellanii* (OS2 and OS4) produced identical results when digested with any of the restriction enzymes used. There also was good agreement in fragment number and size of fragments (as % total genome) with a third line analyzed in another laboratory.

A. castellanii and *A. palestinensis* had similar numbers of fragments for several enzymes, whereas *A. astronyxis* had typically fewer fragments. Digestions of *A. castellanii* and *A. palestinensis* with *Hae* III (recognizing 5′-GGCC-3′) yield about the number of fragments expected for genomes of their size and G+C content. *A. astronyxis* yields only about one-third of the fragments expected from digestion with *Hae* III or enzymes that recognize other combinations of 2 Gs and 2 Cs. Reasons for this result are not clear.

The following maps, using double digestion or partial digestion techniques, were constructed for mt DNA; *Hha* I, *Hpa* II and *Eco* RI, for *A. astronyxis*; *Bam* HI, *Sal* I, and *Eco* RI, for *A. castellanii*; and *Kpn* I for *A. palestinesis*.

Distinction of species in the genus *Acanthamoeba* is a serious problem. It is proposed that restriction enzyme analysis of mt DNA could be useful as a tool in identification of these species.

C. David Zarley, Department of Biology, Jordan Hall 138, Indiana University, Bloomington, Indiana 47401

M. WESOLOWSKI and M. MONNEROT

tRNA Genes of Yeast Mitochondria: Genetic and Physical Maps

In all eukaryotic organisms including yeast, the mitochondria have a specific protein synthesizing apparatus of their own, distinct from that of the cytoplasm.

Most, or all, of the protein components of the machinery are nuclear coded, while mitDNA has the information for mitochondrial rRNAs and tRNAs, and probably all mRNA translated in the mitochondria. The previous genetic and physical maps of mitRNA genes have been precised and extended. A coretention/codeletion analysis of 15 genetic markers and 23 tRNA genes on mtDNA from 95 rho$^-$ allowed us to determine the position of some yet unordered tRNA genes and show that yeast mitDNA possesses tRNA genes for all twenty amino acids and at least 3 isoacceptor genes.

Since very few restriction sites were known in the C-P region (15 Kbp), which carries 18 of the 23 tRNA genes, we established a more detailed physical map of this region by using multi-sites enzymes (*Mbo* I, *Tac* I, *Hin*f I); then we could correlate defined restriction fragments with known sets of tRNA genes by comparing the genetic and physical maps of several rho$^-$.

It appears that 15 among these 18 tRNA genes are found in 3 distinct areas. Each covers no more than 2 Kbp, and 2 of them are clearly spaced by at least 2.5 Kbp.

It will be interesting to know whether each area is transcribed as a unit.

M. Wesolowski, Institut Curie, Section de Biologie, Orsay, France

LAWRENCE E. ALLRED

A 26s Ribonucleoprotein Particle
Similar to Histone Messenger

A 26s RNA-protein particle has been found that rapidly incorporates labeled uridine in the presence of actinomycin D concentrations that inhibit rRNA synthesis. Uridine uptake is immediate and linear. The labeled RNA found first with the 26s particle becomes associated with the 40s ribosomal subunits after approximately 30 min. The purified RNAs obtained from the 26s and 40s bands after 1 hr label were found to cosediment at 10s on 5–25% linear sucrose gradients. The 40s association of the RNA may be the translational initiation complex. The transfer of the labeled material from the 26s to the 40s position was prevented by sodium fluoride, which inhibits initiation.

The RNA of the 26s RNP was found to be similar to histone messenger in sedimentation character, labeling kinetics in the cytoplasm, association with small polyribosomes, and in having a low quantity of polyadenylic acid.

The appearance of the 26s RNPs in the cell cycle was determined. Synchronous cells were labeled with 3H uridine for 2 hr prior to mitotic selection in the presence of colcemid. These mitotic cells, exposed to label during G_2, were found to have high quantities of 26s RNP material. When these cells were allowed to progress into G_1, the 26s RNP material disappeared from the gradients.

A possible interpretation of these findings is that the cells package newly synthesized histone mRNA as 26s RNPs in the cytoplasm. Pretranslational regulation may involve a 30 minute delay following appearance in the cytoplasm before the message is translated. Finally, cytoplasmic histone message may be selectively destroyed at the beginning of each cell cycle forcing the cell to accumulate fresh histone message in the cytoplasm before a new DNA synthesis phase can begin.

This work was performed in the laboratories of Dr. Ronald M. Humphrey and Dr. Ralph B. Arlinghaus, both of the University of Texas System Cancer Center, M. D. Anderson Hospital and Tumor Institute.

Ohio State University, Department of Pharmacology, Columbus, Ohio 43210

J. C. BAGSHAW

Histone Gene Expression in Developing Brine Shrimp

We have initiated a study of the structure and expression of histone genes in the brine shrimp, *Artemia salina*, a biological system that has practical advantages for studies of specific gene expression. Histones are synthesized in developing larvae in concert with a synchronous wave of DNA replication, but neither DNA replication nor histone synthesis occurs during development of post-gastrula embryos prior to hatching. Whole cell RNA was extracted and hybridized in excess to nick-translated DNA of the recombinant plasmid pCh22, which carries sea urchin histone genes. The results indicated that larvae contained detectable histone mRNA, but embryos did not. Total polysomal RNA from larvae was fractionated by sucrose gradient centrifugation and the recovered RNA was translated in a wheat germ system. Core histones (H2A, H2B, H3, and H4) were the predominant products of translation of 7-16 S RNA. When larvae at the peak of histone synthesis were labeled with ^{32}P for 4 hr, several prominent RNA bands in the 7-16 S range were detected by gel electrophoresis and autoradiography. At least four of these labeled bands were distinct from RNAs similarly labeled in newly hatched larvae, before the period of peak histone synthesis. *Artemia* DNA was digested with several restriction endonucleases, and the fragments were separated by agarose gel electrophoresis, transferred to nitrocellulose, and hybridized to ^{32}P-labeled pCH22. Preliminary results suggest that the histone gene repeat unit in *Artemia* is about 7,400 base pairs in length, and contains one cleavage site each for *Sal* I and *Bam* I and two for *Eco* RI.

Wayne State University School of Medicine, Department of Biochemistry, Detroit, Michigan 48201

JEN-FU CHIU, C. SCHWARTZ
AND P. COMMER

Regulation of Alpha-Fetoprotein Gene Expression in Developing Liver and Hepatoma

Alpha-fetoprotein (AFP) is an oncofetal protein that is normally present in the serum of fetal and hepatoma-bearing animals of many species. It is present in trace amounts in the serum of adult rodents and humans. The control of AFP expression in normal liver and hepatoma provides a powerful model for study on gene activation and repression during development and neoplastic transformation.

AFP mRNA is isolated either from rat yolk sac or Morris hepatoma 7777 by the method of indirect immunoprecipitation of AFP specific polysomes. Upon examination of the translational activity of AFP mRNA in the cell-free micrococcal nuclease-treated reticulocyte lysate system, of the profile of purified AFP mRNA on polyacrylamide gels, and of the hybridization kinetics of AFP mRNA and cDNA, it was found that this AFP mRNA is not extensively contaminated with other species of mRNA. In order to examine the sequence homology between AFP mRNA isolated from rat yolk sac and hepatoma 7777, heterologous RNA-cDNA hybridization reactions were performed to allow a determination of the degree to which the AFP mRNA sequence from normal rat yolk sac and hepatoma 7777 were shared. The kinetics are essentially similar to homologous hybridization. These findings suggest that there has been little nucleotide sequence change in the hepatoma AFP mRNA.

The glucocorticoid hormones strongly inhibit AFP synthesis in newborn mouse and rat livers, and in Morris hepatoma cells. We have measured the level of AFP mRNA in liver cytoplasm. This would allow us to determine whether the decrease in the rate of AFP synthesis in newborn animals following hormonal administration is due to a fall in the tissue levels of its functional mRNA or to a hormonally mediated decrease in translational efficiency of this mRNA. Hybridization of cytoplasmic RNAs with AFP cDNA indicated that the dexamethasone-treated livers contained AFP mRNA only to 1% of that in the control animals. Nuclear RNAs isolated

University of Vermont College of Medicine, Department of Biochemistry, Burlington, Vermont 05405 from control and hormone-treated livers were also hybridized to AFP cDNA. Again the level of AFP mRNA sequences was considerably lower in the dexamethasone-treated livers. These experimental results demonstrate that glucocorticoids appear to exert a negative effect on gene expression at the level of transcription or post-transcription.

DEBORAH K. HANSON, HENRY R. MAHLER
JEANETTE A. JOHNSON, AND MAURICE CLAISSE

The Nature of Novel Polypeptides Accumulating in Intron Mutants in the *Cob-Box* Region of the Mitochondrial Genome

In *Saccharomyces cerevisiae* the *cob-box* region of mitochondrial DNA is responsible for the specification of cytochrome *b* of complex III of the respiratory chain. It is now known that this gene is characterized by a mosaic organization, in which segments containing structural information for the cytochrome *b* apoprotein (exons) are interspersed with noncoding regulatory sequences (introns).

Many exon mutants display new polypeptides of $M_r < 30,000$ whose fingerprints exhibit extensive homology with that of cytochrome *b*. Intron mutants accumulate one or more novel polypeptide products with molecular weights that may be greater than or less than 30,000 daltons. Mutants in *box* 7, an intron, and *boxes 2* and *9*, exons flanking the *box* 7 intron, synthesize multiple new polypeptides of $M_r = 23,000$ (p23), 26,000 (p26), and 35,000 (p35). Fingerprints of the p23 species of selected mutants from *boxes 2, 7,* and *9* indicate that they are closely related to each other and show extensive homology to fingerprint patterns of cytochrome *b*. Such is the case for fingerprints of p35s from *box 2* and *7* mutants.

Although several mutants in *boxes 9, 7, 2,* and *6* synthesize a polypeptide of similar mobility ($M_r = 26,000$), fingerprints reveal distinct differences in their sequences. Fingerprints of p26s from selected *box 9* and *box 2* mutants display a pattern of fragments that resemble each other, but show very little, if any, homology to fingerprints of cytochrome *b*. Fingerprints of another *box 2* mutant show more overlap with those of cytochrome *b* and the p26 from a representative *box 6* exon mutant. Since similar products are produced by an intron and its flanking exons, it is likely that these products accumulate as a result of processing aberrations and that these processing events involve sequences in exon portions of the gene as well as those of introns.

Deborah K. Hanson, Indiana University, Department of Chemistry and the Molecular, Cellular, and Developmental Biology Program, Bloomington, Indiana, 47405

I. M. LEFFAK

Effects of Cytosine Arabinoside on Chromatin Synthesis

Asynchronous cultures of exponentially growing cells incubated with the anti-tumor drug cytosine-1-b-D-arabinofuranoside (Ara C) show 95%-98% inhibition of the rate of DNA synthesis and 50-60% inhibition of the rate of histone synthesis, although total cellular protein synthesis is largely unaffected. Consistent with earlier experiments suggesting a dramatic decrease in the half-life of total histone mRNA induced by Ara C, our data indicate that the inhibition of histone protein synthesis is uniform across all five histone classes. The histone that, in the presence of Ara C, is synthesized in excess of the normal histone/DNA stoichiometry appears to accumulate in the cytoplasm, and in the nucleoplasm either free or loosely bound to chromatin.

By several criteria, normal nucleosomes are synthesized during Ara C incubation, although this chromatin shows enhanced susceptibility to nuclease digestion. The nucleosome structure and nuclease sensitivity of the preexisting chromatin in Ara C-treated cells is unaffected by the drug treatment, however.

Wright State University School of Medicine, Department of Biological Chemistry, Dayton, Ohio 45435

SHU-LEN H. LIU, LEE-JUN C. WONG
AND GEORGE A. MARZLUF

Characterization of Abundant Messenger RNA Species in *Neurospora Crassa*

We have found that polysomal poly(A)$^+$ RNA of *Neurospora crassa* is distributed into three abundance classes by RNA-cDNA hybridization analysis. From the observed kinetics of its hybridization with cDNA, the high abundance class was estimated to contain about 10 different poly(A)$^+$ mRNAs. *In vitro* translation of the polysomal poly(A)$^+$ RNA in a reticulocyte system yielded about 10 to 15 major proteins (separated by SDS-gel electrophoresis). These same major proteins are also found *in vivo* in soluble cellular protein.

We have compared the proteins synthesized *in vitro* from three different fractions of polysomal RNA, i.e., total RNA, poly(A)$^+$ RNA, and poly(A)$^-$ RNA. The results showed that each RNA fraction produced an identical pattern of major proteins although their translational efficiency was different. These results indicate that *Neurospora* possesses abundant messenger RNAs that encode 10 to 15 major proteins and that the pattern revealed by *in vitro* translation accurately reflects in *in vivo* situation. Furthermore, it appears that all of the abundant mRNAs are represented in the poly(A)$^+$ RNA population. We are now examining the RNAs produced under particular growth conditions to determine whether any new abundant mRNAs are synthesized under such specific conditions.

Ohio State University, Department of Biochemistry, Columbus, Ohio 43210

RICHARD K. MEISTER AND LEE F. JOHNSON

Rapid Changes in mRNA Content Following Perturbations in Protein Synthesis

A number of laboratories have shown that when mammalian cells are exposed to inhibitors of protein synthesis, the content of cytoplasmic mRNA increases significantly. We have previously shown that this increase is particularly striking in resting (G_o) mouse 3T6 fibroblasts that have been serum-stimulated to reenter the cell cycle. We found that stimulation in the presence of puromycin or cycloheximide led to a doubling of poly(A) (+) mRNA content within 1–2 hr. Messenger RNA content then remained constant for at least 10 hr. In this report we show that continuous exposure to both serum and cycloheximide are required to maintain this elevated mRNA level. Removal of either leads to an equally rapid decrease in poly(A) (+) mRNA content. If cycloheximide is withdrawn at either 2 or 10 hr following serum stimulation in the presence of the drug, allowing the rapid (< 30 min) restoration of the rate of protein synthesis, we observe that poly(A) (+) mRNA content decreases within 2 hr to a level nearly equal to that found in resting cells prior to stimulation. If the drug is withdrawn but the serum stimulus is not, the rapid decrease in poly(A) (+) mRNA content is followed by an increase parallel to that which occurs in cultures stimulated in the absence of drug, but displaced from the latter by an interval approximately equal to the length of exposure of the drug. These results show that the mammalian cell is able to decrease as well as increase its content of poly(A) (+) mRNA in response to drug-induced perturbations in the rate of protein synthesis. The changes in poly(A) (+) mRNA content occur extremely rapidly and may represent an attempt by the cell to correct the perturbation.

Ohio State University, Department of Biochemistry, Columbus, Ohio 43210

J. M. RICE, J. B. PERKINS
P. C. KIMBALL, AND D. II. DEAN

Infectivity of Restriction Endonuclease-Treated Integrated Rauscher Leukemia Virus Proviral DNA

We are examining the integration of Rauscher leukemia virus proviral DNA into mouse and rat cellular DNA by restriction endonuclease analyses and Southern elution gel-filter hybridization. An essential element in these studies has been the use of restriction endonucleases that do not cut the provirus so that direct enumeration of provirus integration sites can be determined. In order to assess whether RLV proviral DNA contains various restriction sites, we have tested the ability of different restriction endonucleases to inactivate the infectivity of this DNA. In this preliminary study we report the use of a DNA transfection assay to test the infectivity of integrated RLV proviral DNA after treatment with a series of restriction endonucleases. A line of NIH-3T3 cells chronically infected with RLV was used as the source of integrated RLV DNA. Normal NIH-3T3 cells were used as recipient cells for transfection.

The ability of each of the various restriction endonucleases to cut within the integrated RLV DNA provirus was tested by measuring the infectivity of the DNA following enzyme treatment. Infectivity of each DNA preparation was determined by assay of each transfected NIH-3T3 culture for sedimentable viral reverse transcriptase and XC plaque assay. Restriction endonucleases *Eco* RI, *Sma* I, and *Xba* abolished the infectivity of RLV proviral DNA, which suggests that the RLV DNA contain at least one site sensitive to each of these enzymes. Treatment with restriction endonucleases *Sal* GI, *Bam* HI, *Hind* III, *Sac* I, and *Kpn* I did not abrogate the infectivity of the RLV proviral DNA. Therefore, the RLV proviral DNA does not contain specific sites for these enzymes. Control transfection using untreated RLV proviral DNA were uniformly positive while DNAase treatment of the RLV DNA abolished its infectivity. Uninfected NIH-3T3 cell DNA was not infectious.

J. M. Rice, Battelle's Columbus Laboratories, Columbus, Ohio 43201

Isolation and Characterization of Actin Genes in *Drosophila Melanogaster*

A preselected portion of the *Drosophila* recombinant DNA library, cloned into bacteriophage lambda Charon 4, was screened for actin genes. The well-known high evolutionary conservation of actin from different species provided the rationale for this selection. It was hoped that this conservation is reflected at the level of the genome and that a heterologous approach would, among other things, generate a highly purified probe with respect to actin. Hence, polysomal poly A^+ mRNA highly enriched for actin mRNA was isolated from skeletal muscle of 12–14 day chick embryos. In the cell-free rabbit reticulocyte *in vitro* translation system well over 50% of the proteins synthesized from this RNA was actin, indicating that the mRNA preparation was indeed enriched for actin sequences.

A recombinant, B7, gave a positive signal to the ^{32}P-cDNA probe made to actin-enriched chicken mRNA. It was shown to contain an actin gene by additional independent criteria. Restriction enzyme mapping, using either the chicken mRNA or one of the actin coding fragments from the B7, indicates that the 17.5 kb *Drosophila* DNA insert contains a single actin gene located near the center of the insert. The actin coding as well as the immediately adjacent sequences appear to be rich in GC as indicated by an unusually high number of Sal I sites. Hybridization kinetics of the actin coding sequence indicates that the gene is repeated 1–6 times per genome. Since there are at least three different types of actin in *Drosophila*, it appears that each type is coded for by no more than 1 or 2 copies of the gene. Thermal stability studies show that the actin gene present in B7 cross-hybridizes with other actin genes, all of which display a higher Tm than the over-all *Drosophila* DNA. The actin coding sequence appears to label five disperse loci on the polytene chromosome.

Wayne State University, Department of Biological Sciences, Detroit, Michigan 48202

SCOTT A. WHALEN AND STANLEY G. SAWICKI

Analysis of the Synthesis of Late Adenovirus Proteins in Permissive and Restricted Host Cells

During the replication cycle of adenovirus type 2 (Ad2), it is only after the onset of viral DNA synthesis that a class of Ad2-specified messenger RNAs, termed the late-mRNAs, begins to be transcribed and translated. By 18–20 hours p.i., these late-mRNAs are responsible for more than 80% of the proteins synthesized in productively infected cells. We are examining different types of cultured cells that are either permissive or restricted in their ability to replicate Ad2 in order to find cells that vary in their expression of late genes and in the extent to which host protein synthesis is shut off. Cells are infected with ad2 at 10-200 pfu/cell and labeled for 1 hour with ^{35}S-methionine or ^{14}C-amino acids at 18–30 hr p.i. Whole cell extracts are run on SDS-PAGE and the incorporation of label into late Ad2 proteins determined. Permissive cells such as HeLa, human kidney, and Vero cells synthesized almost exclusively late Ad2 proteins beginning 18–20 hr p.i. In CV-1 cells, which are restricted in their ability to produce Ad2 virions, large amounts of hexon and IV_{a2} are synthesized, but relatively much smaller amounts of other late Ad2 proteins are synthesized; nevertheless, viral protein synthesis predominates over host protein synthesis. In rhesus monkey cells, which are highly restricted in their ability to produce Ad2 virions, very small amounts of viral proteins, mainly hexon, are synthesized, although there is a marked depression of protein synthesis. This depression is proportionately reduced with increasing multiplicity of infection. In human embryonic lung cells, which are permissive for Ad2 production, late Ad2 protein synthesis does not appear to greatly diminish host protein synthesis, and late Ad2 protein synthesis is apparent only with high multiplicity of infection. We are currently attempting to determine whether the transcription, processing, turnover, or translation efficiency of late Ad2 mRNAs varies in cells of different origin.

Supported by ACS, Ohio Branch.

Medical College of Ohio, Department of Microbiology, Toledo, Ohio.

Index